Biochemistry

Oxford

0865- 275268

Progress in Protein-Lipid Interactions

PROGRESS IN PROTEIN–LIPID INTERACTIONS

VOLUME 1

An international review series designed to critically
evaluate actively developing areas of research in protein-lipid interactions

Series Editors

A. Watts

Biochemistry Department
South Parks Road
Oxford OX1 3QU
England

and

J.J.H.H.M. De Pont

Department of Biochemistry
University of Nijmegen
Postbus 9101
6500 HB Nijmegen
The Netherlands

ELSEVIER
AMSTERDAM · NEW YORK · OXFORD

Progress in Protein–Lipid Interactions

Editors

A. Watts
Biochemistry Department
South Parks Road
Oxford OX1 3QU
England

and

J.J.H.H.M. De Pont
Department of Biochemistry
University of Nijmegen
Postbus 9101
6500 HB Nijmegen
The Netherlands

1985
ELSEVIER
AMSTERDAM · NEW YORK · OXFORD

ISBN 0-444-80630-X (volume)
ISSN 0168-9614 (series)

Published by:
Elsevier Science Publishers B.V. (Biomedical Division)
P.O. Box 211
1000 AE Amsterdam
The Netherlands

Sole distributors for the USA and Canada:
Elsevier Science Publishing Company, Inc.
52 Vanderbilt Avenue
New York, NY 10017
USA

Printed in The Netherlands

Preface

'Protein-lipid Interactions' has been initiated with the aim of allowing authors to present not a comprehensive review but a highly critical assessment of specific aspects of research into protein-lipid interactions in membranes. Active specialists have been asked to *freely* reappraise past work in the light of our present understanding, give new, possibly unpublished data, and then identify those areas which they believe show greatest potential for our future understanding of membrane biology.

The series will be aimed at established workers wishing to gain a fresh insight into their own interest in the field of protein-lipid interactions, or new researchers wanting to learn about other active areas of the topic. The editors have made an attempt to maintain a balance between the physical and biochemical approaches to the study of protein-lipid interactions.

In this first volume, one view of the many available theories used to describe protein-lipid interactions has been presented by Jim Abney and Jack Owicki. Magnetic resonance methods, probably still the most controversial, yet direct, approach to describing the dynamic interactions between proteins and lipids, are described by Myer Bloom and Ian Smith for deuterium nuclear magnetic resonance and Derek Marsh for spin-label electron spin resonance. Ben de Kruijff and his colleagues describe the way in which phosphorus nuclear magnetic resonance can identify lipid polymorphism in membranes and also monitor the modulation of these effects by proteins.

Bob Clegg and Winchil Vaz have described fluorescent methods, backed by some theoretical treatment, used to determine the relative movements of lipids and proteins in membranes. David Schachter has given his view of the function of intestinal plasma membranes in relation to lipid fluidity. Roberto Bisson and Cesare Montecucco conclude the first volume with a survey of the use of photoreactive phospholipids to study lipid-protein interactions. Although most work so far has approached protein-lipid interactions from the lipid viewpoint, as reflected in the composition of this volume, some attempt has been made to balance function and structure which loosely implies biochemical and biophysical studies. We hope to reach a broad readership and promote interdisciplinary discourse on this fascinating topic.

List of contributors

J.R. Abney
Department of Biophysics and Medical Physics, University of California, Berkeley, CA 94720, U.S.A.

R. Bisson
C.N.R. Center for the Study of the Physiology of Mitochondria and Laboratory of Molecular Biology and Pathology, Institute of General Pathology, University of Padova, Via Loredan 16, 35100 Padova, Italy

M. Bloom
Department of Physics, University of British Columbia, Vancouver, BC, Canada V6T 2A6

R.M. Clegg
Max Planck Institut für Biophysikalische Chemie, D-3400, Göttingen-Nikolausberg, F.R.G.

P.R. Cullis
Department of Biochemistry, University of British Columbia, Vancouver, BC, Canada V6T IW5

C.J.A. van Echteld
Department of Cardiology, University Hospital, State University of Utrecht, Catharijnesingel 101, Utrecht, The Netherlands

P. van Hoogevest
CIBA-Geigy Ltd., Agricultural Division, R & D Seeds/Biotechnology, Schwarzwaldallee 211, CH-4002 Basel, Switzerland

M.J. Hope
Department of Biochemistry, University of British Columbia, Vancouver, BC, Canada V6T IW5

J.A. Killian
Department of Biochemistry, State University of Utrecht, Padualaan 8, 3584 CH Utrecht, The Netherlands

B. de Kruijff
Institute of Molecular Biology, State University of Utrecht, Padualaan 8, 3584 CH Utrecht, The Netherlands
D. Marsh
Max Planck Institut für Biophysikalische Chemie, Abt. Spektroskopie, D-3400 Göttingen, F.R.G.
C. Montecucco
C.N.R. Center for the Study of the Physiology of Mitochondria and Laboratory of Molecular Biology and Pathology, Institute of General Pathology, University of Padova, Via Loredan 16, 35100 Padova, Italy
J.C. Owicki
Department of Biophysics and Medical Physics, University of California, Berkeley, CA 94720, U.S.A.
A. Rietveld
Department of Biochemistry, State University of Utrecht, Padualaan 8, 3584 CH Utrecht, The Netherlands
D. Schachter
Department of Physiology, Columbia University, College of Physicians and Surgeons, New York, NY 10032, U.S.A.
I.C.P. Smith
Division of Biological Sciences, National Research Council, Ottawa, Canada K1A OR6
A.T.M. van der Steen
Organon Teknika, Veedijk 58, B-2300 Turnhout, Belgium
T.F. Taraschi
Hahnemann Medical College, Department of Pathology, 230 N. Broad Street, Philadelphia, PA 19102, U.S.A.
W.L.C. Vaz
Max Planck Institut für Biophysikalische Chemie, D-3400 Göttingen-Nikolausberg, F.R.G.
A.J. Verkleij
Institute of Molecular Biology, State University of Utrecht, Padualaan 8, 3584 CH Utrecht, The Netherlands

Contents

Chapter 2

by

M. Bloom and I.C.P. Smith

Chapter 4

Chapter 5

Watts/De Pont (Eds.)
Progress in Protein-Lipid Interactions

CHAPTER 1

Theories of protein-lipid and protein-protein interactions in membranes

JAMES R. ABNEY and JOHN C. OWICKI

Department of Biophysics and Medical Physics, University of California, and Division of Biology and Medicine, Lawrence Berkeley Laboratory, Berkeley, CA, 94720 U.S.A.

I. Introduction

About 15 years ago it became apparent to a number of biochemists and biophysicists that biological membranes and model membranes behave in many ways like two-dimensional fluids. It also became apparent that a substantial fraction of membrane-associated proteins are integrally embedded in the bilayer membrane, and that some completely traverse it. These realizations were synthesized into the fluid-mosaic model of biological membranes (Singer and Nicolson, 1972), a paradigm that holds sway today with some modifications. One such modification, for example, is to include the immobilizing effects of cytoskeletal interactions.

The intervening years have brought substantial increases in our knowledge about membranes, due in large part to the application of new or refined experimental methods. There have been concomitant efforts on the part of theorists to construct detailed molecular or thermodynamic models that integrate and interpret the experimental work. The largest number of theoretical papers have dealt with pure lipid bilayers, especially the lipid bilayer phase transition. These have been reviewed well elsewhere (e.g., Nagle, 1980), and we do not consider them here except as they relate to our main topics: theoretical models of protein-lipid and protein-protein interactions in membranes.

Abbrevations: DMPC, dimyristoylphosphatidylcholine; DMR, deuterium magnetic resonance; DPPC, dipalmitoylphosphatidylcholine; DSC, differential scanning calorimetry; EPR, electron paramagnetic resonance; SPT, scaled particle theory; T_c, temperature of lipid bilayer phase transition.

We have focused on the influences that intrinsic membrane proteins have on membrane lipids, the effects of the lipids on the proteins, and the phase behavior and lateral distribution of both components. Unfortunately, spatial constraints have required the omission of a number of major topics. We do not discuss the mechanisms of lateral diffusion or the relationship of the diffusion coefficient to molecular geometry and environment. Nor do we discuss the behavior of cholesterol in membranes, in spite of the fact that cholesterol is much like a very small intrinsic membrane protein and has been the subject of a large amount of research.

We have concentrated on the most detailed statistical-mechanical and thermodynamic treatments. Our goal has been to give a reasonably comprehensive picture of the insights that theory has provided into a number of important membrane phenomena. We have also tried, in the spirit of this series, to assess deficiencies in the theoretical treatments and to recommend areas where more theoretical work seems most needed.

Finally, it is worth stating our opinion of the rôle of theory in this field. Clearly, predictive power and agreement with prior experiments are important for any scientific theory. Nevertheless, biological membranes – and even model membranes – are so complicated that any tractable theory must become only a caricature of the real system. If the caricature leads to recognizable consequences, then we hope that its few remaining features are indeed the most important ones for explaining the observations. Given the rampant approximations, conclusions must be rather tentative. It is then useful to examine several differing treatments of a phenomenon, to see whether some consensus emerges.

II. Effects of proteins on membrane lipids

1. SUMMARY OF EXPERIMENTAL RESULTS

A substantial body of experiments, beginning with the electron paramagnetic resonance (EPR) results of Jost et al. (1973), has shown that intrinsic membrane proteins alter the properties of nearby phospholipids in the membrane. We will take the liberty of summarizing these findings below without detailed attribution, directing the interested reader to review articles for further information (Chapman et al., 1979; Jost and Griffith, 1980; Marsh and Watts, 1982; Seelig et al., 1982; Silvius, 1982; Marsh, 1983).

Magnetic-resonance experiments have shown remarkably similar results for different proteins. Typically, EPR experiments on spin-labeled lipids show distinguishable signals from bulk lipid and lipid associated with the protein as discussed in Chapter 4 of this Volume. The latter, usually called boundary lipid or annular lipid, shows little motional averaging on the EPR time scale (ca. 10–100 ns). In this it is more like solid than fluid lipid, probably reflecting a motional restriction more than a conformational ordering. Estimates of the amount of motionally perturbed lipid range from one to at most a few layers of lipid around the protein. By 'solid' and 'fluid' we mean lipid with motional properties typical of temperatures below and above the lipid phase transition, respectively.

More recent experiments using deuterium magnetic resonance (DMR) (see Chapter 2) to study deuterated lipids have demonstrated that, on the DMR time scale of about 10–100 μs, boundary lipid shows *more* motional averaging than does fluid lipid in the absence of proteins. Exchange between boundary and bulk lipid is fast on this time scale, so separate signals are not resolved.

Additional studies have been performed with optical techniques, some of which are discussed in Chapters 5 and 6. Protein perturbations of *trans/gauche* conformational ratios have been deduced from Raman spectra of C-H and C-C stretches in lipid chains. Fluorescence-polarization studies have also given information about both chain dynamics and conformational order.

One of the few safe generalizations that can be made from the spectroscopic studies field is that the influence of proteins on the conformation and dynamics of lipids is complicated. Any comprehensive explanation must involve the different mechanical degrees of freedom of the lipids and also the resulting wide range of molecular dynamics. Both spectroscopic methods and theoretical models involve averaging over some degrees of freedom on some time scale. To compare experimental and theoretical results, it is important for these averagings to be similar.

The above methods, plus differential scanning calorimetry (DSC), have been used to determine the effect of proteins on the phase behavior of the membrane. Proteins always weaken the intensity of the first-order lipid phase transition. They may cause it to broaden, and may induce other continuous transitions above or below it in temperature, depending on the system. The apparent temperature of the main transition may be increased, decreased or left unchanged within experimental error.

Several points are important for connecting such results with theoretical treatments. First, the results are best interpreted in terms of a temperature-composition phase diagram (composition being protein lateral density). If the pure lipid has a phase transition, the Gibbs phase rule guarantees the existence of lateral phase separation in some part of the temperature-composition plane near the transition for pure lipid. Second, the existence of first-order (discontinuous) transitions and sharp phase boundaries is an important question theoretically. However, it is difficult to distinguish experimentally between truly discontinuous and merely rather sharp continuous behavior in lipid systems. Finally, relationships between transitions observed for protein-lipid mixtures and transitions in pure lipids are often obscure unless a series of protein concentrations is studied so that the behavior of the system can be extrapolated back to the pure lipid.

Electron microscopy often reveals lateral phase separation into protein-rich and protein-free domains at low temperatures. This is reminiscent of the familiar freezing out of impurities from a crystalline phase. The importance of determining the lateral distribution of the proteins in any experiment must be stressed, since most interpretations of the data will depend strongly on the presence or absence of aggregation.

Theoretical work in this field has been much influenced by the early dominance of experiments using magnetic resonance. Accordingly, the effects of proteins on conformational order in lipids have been well studied theoretically. The influences of proteins on lipid phase transitions have also been extensively modeled. In the rest of this section we will summarize most of the published theories in chronological order.

2. MARČELJA (1976)

a. Description of theory

The first detailed theory of the effects of proteins on membrane lipids was published by Marčelja (1976) as an extension of a theory of order and phase transitions in lipid bilayers (Marčelja, 1974). The model is constructed on a triangular (hexagonal) lattice, with one lipid alkyl chain per site (a lattice constant of 0.5–0.7 nm). Each protein occupies a hexagonal region of the lattice with a side length of three to five sites. Interactions between chains and between chains and proteins are determined by a conformational order parameter, S, analogous to that observed in EPR and DMR. For each chain i of length n segments, S_i is the average of the segmental order parameters, each of which in turn depends on the angle, θ_j, between the normal to the jth segmental H-C-H plane and the director (usually, the normal to the bilayer).

$$S_i = (1/n) \sum_{j=1}^{n} \frac{[3\cos^2(\theta_j) - 1]}{2} \tag{1}$$

$S_i = 1$ corresponds to an all-*trans* ordered chain, and $S_i = 0$ statistically denotes complete disorder.

The energy of a particular conformation of chain i is the sum of three terms:

$$E = -\Phi_i(n_{tr,i}/n)S_i + E_{int,i} + \pi_{int}A \tag{2}$$

The first term on the right represents interactions between chains and between chains and proteins. Φ_i is the molecular field at i due to its six neighboring sites, with

$$\Phi_i = \sum_{k=1}^{6} \begin{cases} (V_o/6)\langle (n_{tr,k}/n) S_k \rangle & \text{if k = lipid} \\ V_{1p} & \text{if k = protein} \end{cases} \qquad (3)$$

V_o and V_{lp} are the lipid-lipid and lipid-protein coupling parameters, and $n_{tr,k}$ is the number of *trans* segments in chain k. The angle brackets denote the ensemble average. The molecular field (or mean field) is thus an approximate way to represent the interaction of a molecule with its surroundings, as described by the average properties of the system. It notably does not include the details of local fluctuations about the average.

In this treatment, interactions between chains thus are favorable when V_o is large and positive, when the chains are predominantly *trans*, and when the order parameters are high. Relating the energy to the product of order parameters reveals the roots of the theory in the Maier-Saupe treatment of isotropic-nematic transitions in liquid crystals (Maier and Saupe, 1959). When dispersion attractions between rod-like molecules are expressed as a function of molecular orientation, the product of order parameters is the leading term in a series expansion. It is likely, however, that inter-chain repulsions are more important than attractions in determining conformational order (Nagle, 1980). For this reason, and because of the ad hoc inclusion of the fraction of *trans* states (n_{tr}/n), it is difficult to attach more than phenomenological significance to the value of V_o.

The protein interacts most favorably with an ordered lipid. Marčelja explores the range $V_{lp} = 0.4V_o - 0.8V_o$; the protein thus acts like a rigid lipid that has moderately low to moderately high order.

The second term in Eqn. 2 is the internal energy of the chain, determined from the sequence of *trans*, *gauche*$^+$ and *gauche*$^-$ states; only these three states are allowed each segment. Since the model only explicitly considers the nonpolar core of the membrane, it is necessary to introduce a third term with an effective lateral pressure, π_{int}, to simulate the hydrophobic interface and other area-dependent effects. The subscript int here stands for 'internal', since this is not the true thermodynamic (external) lateral pressure on the bilayer.

Computationally, the model is worked out by generating a partition function for each chain, based on the average conformations of its neighbors and the 3^n conformations available to it. Since the partition functions of adjacent chains are coupled through the molecular field, conformational self-consistency is obtained iteratively.

This seminal theory has undoubted historical importance, and it defined the field to a considerable extent. Its analysis of lipid conformational states is still the most thorough in a theory of protein-lipid interactions. Its disadvantages include high computational cost, the unphysical lattice representation of the lipid, and the mean-field Maier-Saupe interaction, which emphasizes attractions when, in our opinion, repulsions dominate.

b. Results

Results are reported for a lipid with $n = 12$ segments and V_o chosen to give a phase transition at 25.2°C. The perturbation of lipids by protein is found to be sensitive to V_{lp} and rather insensitive to the size of the protein. The correlation length for the decay of the perturbation radially outward from an isolated protein in a fluid bilayer is two to three lattice sites, the larger value occurring at temperatures close to the phase transition.

A protein-rich membrane is modeled by a replicated 'unit cell' consisting of a protein (five lattice sites per side) and one surrounding layer of lipids, so that about 55% of the lattice sites are covered by protein. The sharp lipid phase transition is abolished, though considerable broad temperature dependence of lipid order remains. This effect, plus the analogous result for lipids adjacent to an isolated protein, is shown in Fig. 1.

3. SCHROEDER (1977B)

a. Description of theory

Building on work involving the effects of external fields on molecular order in liquid crystals, Schroeder (1977a, b) developed a formalism for investigating the effects of proteins on the ordering of lipids. The theory applies when the order in a pure lipid bilayer is described by any of a large class of mean-field theories.

The order parameter is taken to be S, defined as above, though the theory

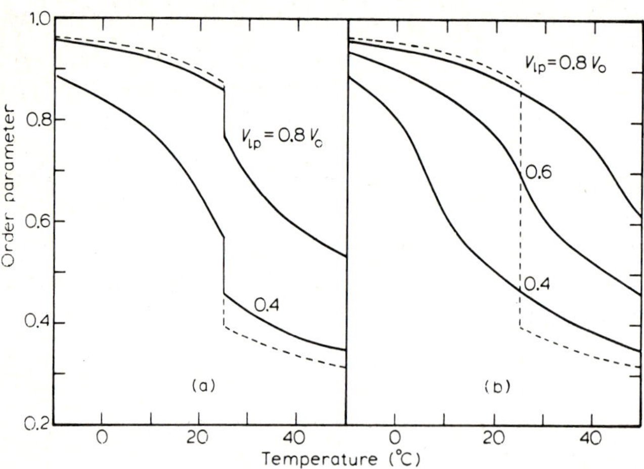

Fig. 1. Order parameter for layer of lipids immediately adjacent to protein, as a function of temperature. (a) Isolated protein; (b) concentrated protein; --- is the order parameter for unperturbed lipids. From Marčelja (1976).

holds for other definitions. The model is two-dimensional: proteins are circles of radius r_o, and lipids are treated as a continuum. The protein acts as a short-ranged external field, in the sense that its only effect is to fix the value of the order parameter to be S_o at the protein-lipid interface. In the unperturbed bilayer, $S = S_\infty$.

We will not repeat the development of the theory here. The mean-field treatment of lipid order leads to the definition of a correlation length for the decay of spatial fluctuations in the order parameter. All lengths can be expressed in reduced units (denoted by a tilde, ~) when divided by this correlation length.

b. Results

If the order-parameter perturbations are sufficiently small, the equations can be linearized to give the following simple result for the dependence of the order parameter on radial distance from an isolated protein:

$$S(\tilde{r}) = S_\infty + \frac{(S_o - S_\infty)}{K_0(\tilde{r}_o)} K_0(\tilde{r}) \tag{4}$$

K_0 is the zero-order modified Bessel function of the second kind, and $K_0(\tilde{r}) \approx (\pi/2\tilde{r})^{1/2}\exp(-\tilde{r})$ for $\tilde{r} \gg 1$.

When more than one protein is present, the order parameter at any point in the membrane is taken as the strict superposition of the perturbations from the individual proteins according to Eqn. 4. This is a good approximation only when proteins are distantly spaced (compared to the correlation length), since otherwise nonlinearitics become important and the boundary conditions at the protein surfaces are seriously violated. For two proteins at $\tilde{\mathbf{r}}_1$ and $\tilde{\mathbf{r}}_2$, superposition gives:

$$S(\tilde{\mathbf{r}}) = S_\infty + \frac{(S_o - S_\infty)}{K_0(\tilde{r}_o)} [K_0(|\tilde{\mathbf{r}} - \tilde{\mathbf{r}}_1|) + K_0(|\tilde{\mathbf{r}} - \tilde{\mathbf{r}}_2|)] \tag{5}$$

Fig. 2 shows contours of constant S about two proteins with lipids perturbed according to this equation.

Since the theory makes no estimate of the correlation length, it cannot be compared directly with experiment. The main point of the paper is to discuss lipid-mediated interactions between proteins, and we will return to that in Section IV.

4. LANDAU THEORIES: OWICKI ET AL. (1978), OWICKI AND McCONNELL (1979) and JÄHNIG (1981a, b)

a. Description of theories

Owicki and co-workers (Owicki et al., 1978; Owicki and McConnell, 1979) have

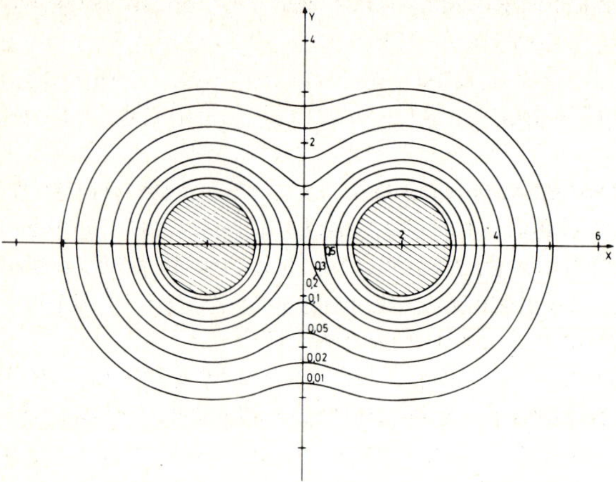

Fig. 2. Contours of constant order-parameter perturbation about two proteins with radius $\tilde{r}_o = 1$ and center-to-center separation $4\tilde{r}_o$, from Eqn. 5. The perturbation is presented as a fraction of that at the protein surface, i.e., $(S(\tilde{\mathbf{r}}) - S_\infty)/(S_o - S_\infty)$. From Schroeder (1977b).

elaborated a theory of lipid-protein interactions with two main features. First, as in Schroeder's model, the protein is a rigid geometric constraint on adjacent lipids; the lipids are treated as a continuum. Second, lipid-lipid interactions are described with a mean-field theory devised by Landau (see Landau and Lifshitz, 1969) for ferromagnetism and later extended by de Gennes (1974) for phase transitions in liquid crystals.

The protein constraint may reflect a necessity to match the thickness of the bilayer to the thickness of the hydrophobic part of the protein perimeter, to avoid exposure of nonpolar moieties to the aqueous environment. Since bilayers have low compressibility (in a three-dimensional sense), thickness and surface area per lipid are tightly coupled. The constraint is modeled in terms of area, specifically, in terms of an order parameter, u, defined by the area per lipid molecule in the plane of the bilayer, A:

$$u = (A - A_f)/(A_s - A_f) \tag{6}$$

The subscripts f and s refer to the areas in the fluid and solid phases of a lipid bilayer at the phase transition temperature, T_c, in the absence of proteins. For pure phosphatidylcholines, the area change is typically 20–30% (Janiak et al., 1976). There, u, changes from 1. to 0. with increasing temperature at the phase transition. Further decreases in u at higher temperatures are small relative to this change.

At the boundary between protein and lipid the order parameter is fixed at some value, u_0. This is the principal adjustable parameter in the theory.

In the Landau theory, the free-energy density is expressed as a truncated power series in the order parameter:

$$G(u) = a_1u + a_2u^2 + a_3u^3 + a_4u^4 \tag{7}$$

The observed value of u is that which minimizes $G(u)$. The expansion coefficients depend on temperature and lateral pressure. They can be chosen phenomenologically to reproduce the main thermodynamic characteristics of the lipid phase transition. Alternatively, the coefficients can in principle be derived through fundamental statistical-mechanical arguments (Jähnig, 1979).

The equation is easier to work with when cast in reduced (dimensionless) energy, lateral pressure, and temperature units, again indicated by tildes.

$$\tilde{G}(u) = \tilde{\pi}u + \left(\frac{\tilde{T}}{2}\right)u^2 - u^3 + \left(\frac{1}{2}\right)u^4 \tag{8}$$

Under most conditions the thermodynamic (external) lateral pressure, $\tilde{\pi}$, on the membrane is close to 0, and is henceforth neglected. The temperature, $\tilde{T}$, is both scaled and shifted: its value is 1 at the phase transition, with 0 estimated to lie 20°C lower.

Figure 3 shows $\tilde{G}(u)$ at different temperatures. There are in general two local minima, corresponding to fluid and solid membranes ($u = 0$ and $u \approx 1$, respectively). Above $\tilde{T} = 1$ the fluid state is more stable; below that temperature the

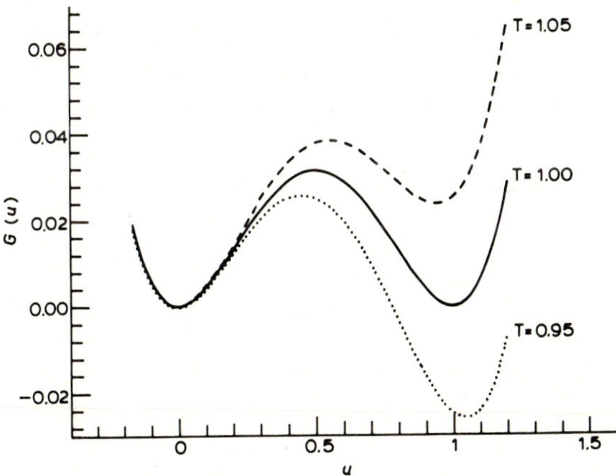

Fig. 3. Landau free-energy density of a spatially homogeneous lipid bilayer as a function of order parameter. This is Eqn. 9 in the text, without the gradient term. Results are shown for three temperatures: above ($\tilde{T} = 1.05$), at ($\tilde{T} = 1.0$), and below ($\tilde{T} = 0.95$) the bilayer phase transition. The value of u corresponding to the global minimum of G gives the observed phase at each temperature. From Owicki and McConnell (1979).

solid is found. At the transition temperature, $\tilde{T} = 1$, the two phases have equal free energy.

The treatment so far has neglected the effects of spatial inhomogeneities in u, which are important to include in treating lipid-protein interactions. These can be simply included by adding an elastic energy term to $\tilde{G}(u)$, i.e., one proportional to the square of the spatial gradient of u. The coefficient for this term can be subsumed into a reduced length unit equal to the correlation length for spontaneous order-parameter fluctuations in a bilayer. This is likely to be about 1.0–1.5 nm.

We can finally calculate the free-energy density at any point in a membrane, given the reduced temperature, order parameter and order-parameter gradient:

$$\tilde{G}(u, \nabla u) = \left(\frac{\tilde{T}}{2}\right)u^2 - u^3 + \left(\frac{1}{2}\right)u^4 + \left(\frac{1}{2}\right)|\nabla u|^2 \tag{9}$$

In principle, the values of u in a membrane can be obtained for any static lateral distribution of membrane proteins by minimizing the free-energy, subject to the condition that $u = u_o$ at lipid-protein boundaries:

$$\int\int G[u(\tilde{\mathbf{r}}), \nabla u(\tilde{\mathbf{r}})] d^2\tilde{\mathbf{r}} = \text{minimum} \tag{10}$$

A difficult computational problem in general, this is simplified by placing cylindrical proteins of radius $\tilde{r}_o$ (circular in projection) on a triangular lattice (see fig. 4). If the hexagonal regions are approximated as circles of radius $\tilde{r}_1$, the problem

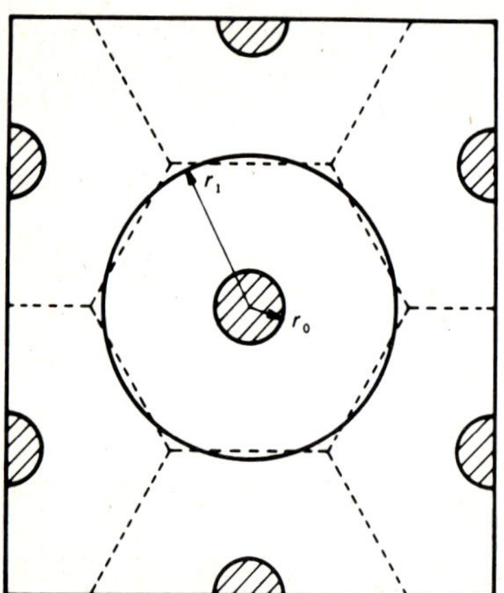

Fig. 4. Lattice of proteins used in Landau theories to analyze effects of proteins at concentrations above limiting dilution. The circle of radius $\tilde{r}_1$ has the same area as the hexagonal 'unit cell' of lipids surrounding the central protein. From Owicki and McConnell (1979).

is reduced to one (radial) spatial variable. The integral equation then can be replaced conveniently by a second-order nonlinear boundary-value differential equation by using the Euler-LaGrange equation (Sokolnikoff and Redheffer, 1966).

$$u'' + u'/\tilde{r} - \tilde{T}u + 3u^2 - 2u^3 = 0 \tag{11}$$

$$u(\tilde{r}_o) = u_o;\ u'\ (\tilde{r}_1) = 0$$

Above, $'$ denotes differentiation with respect to $\tilde{r}$. The $u(\tilde{r})$ that satisfies this equation is also the solution to Eqn. 10.

Jähnig (1981a, b) published an alternate Landau development of the problem, based principally on the conformational order parameter, S. In the limit $S = 0$ for the unperturbed fluid bilayer, and for relatively dilute proteins, he was able to derive analytical expressions that complement the numerical work of Owicki et al. He also used the Landau theory to analyze the influence of proteins on thermal fluctuations in bilayers.

In all the Landau work the system was constrained to be in a single phase. Comparison with experiments that are sensitive to lateral phase separation is therefore not necessarily straightforward. We (Abney and Owicki) are currently computing full phase diagrams for this theory.

There are three main disadvantages to the Landau treatment. First, as in Schroeder's theory, the lipids are treated as a continuum. This is questionable at the molecular level, but at least the characteristic spatial variations in order parameters occur over distances greater than the diameter of a lipid chain (see below). Second, as the theory is usually applied it is phenomenological; it gives only limited insight into the molecular physics of the lipid phase. Third, the proteins are not allowed translational freedom (a limitation shared with some of the other theories discussed here).

The theory does have several strong points. There is experimental evidence that it gives a good description of the bilayer phase transition (Mitaku et al., 1983). It is computationally and conceptually simple, permitting a thorough analysis of its implications. It is not over-parametrized, having only one important adjustable parameter for protein-lipid interactions, u_o.

b. Results

Many of the Landau results are similar to those obtained by Marčelja with a quite different theory. The order parameter near an isolated protein decays outward to the prevailing unperturbed value in an approximately exponential manner. This $1/e$ decay length is close to the coherence length for spontaneous fluctuations (ca. 1.0–1.5 nm) and is maximal at T_c.

Fig. 5 shows the order-parameter profiles about a representative small protein

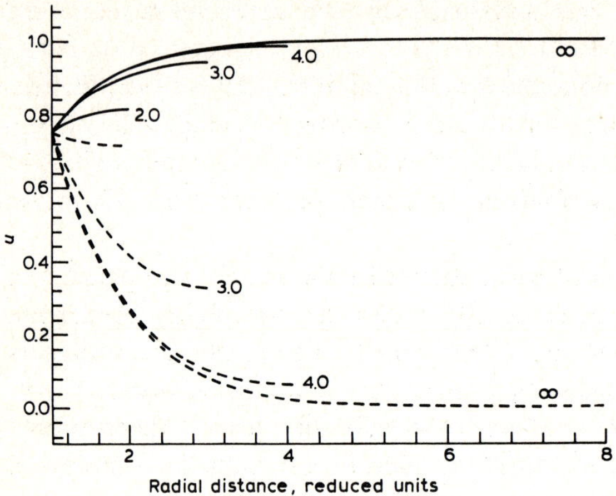

Fig. 5. Profiles of Landau order parameter going radially out from a protein ($\tilde{r}_o = 1.0$; $u_o = 0.75$), for various annulus radii ($\tilde{r}_1 = 2.0, 3.0, 4.0$ and ∞), for temperatures above and below the pure lipid phase transition. Broken curves, $\tilde{T} = 1.2$; solid curves, $\tilde{T} = 0.999$. From Owicki and McConnell (1979).

($\tilde{r}_o = 1.0$, $u_o = 0.75$) at four different concentrations and at temperatures above and below T_c. It can be seen that separations between protein centers as low as eight correlation lengths ($\tilde{r}_1 = 4$) yield profiles close to those for infinite dilution. By six correlation lengths, deviation from infinite-dilution behavior is marked.

Above infinite dilution, the proteins affect T_c and decrease the transition enthalpy. Values of u_o above 0.5 increase T_c, while values below 0.5 lower it. The special case $u_o = 0.5$ has no effect on T_c. Jähnig's analytical result, in reduced units, is

$$\Delta \tilde{T}_c = \frac{1}{\tilde{r}_1^2 - \tilde{r}_o^2} (2\tilde{r}_o + 1) \left(2 \frac{S_o}{\Delta S^t} - 1\right) \tag{12}$$

here $\tilde{r}_1$ and $\tilde{r}_o$ are as in Fig. 4, S_o is the value of S fixed at the protein boundary, and ΔS^t is the discontinuity in S at T_c in the unperturbed lipid bilayer. Both the numerical and analytical treatments predict shifts in T_c of no more than a few degrees.

Increasing the protein concentration weakens the lipid phase transition, causing it to disappear as a first-order transition at some critical concentration. For the protein in Fig. 5, this occurs when 16% of the membrane surface area is protein ($\tilde{r}_1 = 4.8$). The analytical approximation for the critical lipid-to-protein molar ratio is

$$(L/P)^c = 2(2\tilde{r}_o + 1)/\tilde{r}_L^2 \tag{13}$$

where $\tilde{r}_L$ is the radius of a lipid. This agrees well with the numerical result quoted above.

Owicki and McConnell find that the protein induces a broad endothermic order-disorder transition superimposed on the first-order transition. The discontinuous and broad transitions do not, however, represent the separate melting of bulk and boundary lipids. All the lipids in the system participate in both transitions.

Jähnig has shown that the critical point should be associated with peaks in a variety of membrane properties: heat capacity, lateral compressibility, membrane permeability and lateral diffusion. The latter is true to the extent that the process is enhanced by density fluctuations. We believe that it would be worthwhile to attempt to observe critical fluctuations experimentally for a protein reconstituted with a lipid that exhibits a phase transition. Such observations in a biological membrane, where lipid heterogeneity would probably eliminate critical phenomena, are more problematic.

In other work based on the Landau treatment, Jähnig et al. (1982) have explicitly treated the simultaneous effects of protein perturbation of lipid conformational order and lipid director (i.e., tilt). Both perturbations are taken to decay exponentially away from some fixed value at the protein boundary. These perturbations should be statically and dynamically distinct. The correlation length for director fluctuations should be at least as long as that for conformational order. In liquid crystals, director fluctuations often decay slowly and may have nearly macroscopic range (de Gennes, 1974). One would expect director fluctuations also to be much slower than conformational fluctuations, which take place on the nanosecond time scale.

The dynamical differences imply that experimental methods with different time scales can be used to differentiate between the effects. This has been done for melittin, Ca^{2+}, Mg^{2+}-ATPase, and cytochrome oxidase (Jähnig et al., 1982). The authors conclude that all three are similar: $S_o = 0.51$ ($S = 0.36$ in the unperturbed fluid bilayer) and the lipid tilt angle at the protein boundary is 20–30° from the bilayer normal. It was not possible to estimate the correlation length for director fluctuations. However, we note that the analysis of positional correlations of proteins in electron micrographs (see Section IV) might allow this to be done.

5. PINK AND CHAPMAN (1979); LOOKMAN ET AL. (1982)

a. Overview

The complexity of intermolecular interactions is the greatest difficulty that any statistical-mechanical theory of a condensed phase (such as a membrane) must face. Historically, one of the most important simplifications is the approximation in which molecules are confined to the nodes of a lattice and interactions take place only pairwise between adjacent lattice sites. Further, each molecule exists only in a small number of discrete states, so that intermolecular interaction energies can assume only a small number of discrete values. The familiar Ising model of magnetism in one example of the genre, and some of these features appear in Marčelja's (1976) model as well.

Pink and co-workers have used this approach to derive a model of the lipid-bilayer phase transition and have extended the model to account for behavior of a variety of other membrane-bound molecules (for a review, see Caille et al., (1980)). How proteins influence the state of membrane lipids was studied by Pink and Chapman (1979) using a version of the model similar to one that had been used to investigate the effects of cholesterol on membrane lipids (Pink and Carroll, 1978). That work is the subject of the rest of this section.

b. Description of theory

In the simplest version of the theory, each site of a triangular planar lattice is occupied by one of three types of molecules: a lipid chain in its ground state, G; a lipid chain in its excited state, E; or a protein, P. The six possible types of nearest-neighbor interactions are accounted for by six coupling-constant parameters, J_0 through J_4 and J_p, as shown in Table 1.

Differences between intra-chain conformational energies are accounted for by assigning an energy, U_e, to the excited-state chain, with the zero of energy representing the ground-state chain. Since the excited state is really a conglomerate of many conformations, it is assigned a degeneracy, D. As in Marčelja's model, area-dependent effects are treated by introducing an internal lateral pressure, π_{int}, together with estimates of the projections of the areas of the three components in the plane of the membrane: A_g, A_e, and A_p.

The full Hamiltonian for the system is thus:

$$\begin{aligned} H = -(J_0/2) \sum_{\langle ij \rangle} (G_iG_j + 2J_2G_iE_j + J_3E_iE_j \\ + 2J_1G_iP_j + 2J_4E_iP_j + J_pP_iP_j) \\ + \sum_i [U_eE_i + \pi_{int}(A_gG_i + A_eE_i + A_pP_i)] \end{aligned} \tag{14}$$

In the sums, lattice sites are labeled i and j; $\langle ij \rangle$ is the sum over all nearest neighbors. The subscripted G's, E's and P's are projection operators for the appropriate states. Thus, e.g., when lattice site i is occupied by a G chain, $G_i = 1$ and $E_i = P_i = 0$. The excited-state degeneracy appears in the counting of states, not in the Hamiltonian.

An obvious drawback of this formulation is that a protein is represented as being the same size as a lipid chain. Pink and Chapman do some calculations with a Hamiltonian modified to ameliorate this. Rather than expand a protein to

TABLE 1
Parameters for interactions between ground-state lipids G, excited-state lipid E, and proteins P (Pink and Chapman, 1979)

	G	E	P
G	$-J_0$	$-J_0J_2$	$-J_0J_1$
E		$-J_0J_3$	$-J_0J_4$
P			$-J_0J_p$

multiple lattice sites (as does Marčelja), they collapse two or three lipids onto one lattice site. Both intra- and inter-site terms in the Hamiltonian become correspondingly more complex than in Eqn. 14.

Another variation of the model allows the proteins to bind lipids in one fixed state, either G or E. In interactions involving three of the six neighboring sites, the protein is taken to behave like the putative bound lipid.

The behavior of any of these systems is in principle calculated by generating a partition function, which requires evaluating the Hamiltonian for all possible configurations of the system. The protein concentration can be controlled by including a term in the protein chemical potential in the Hamiltonian.

Even the Hamiltonian in Eqn. 14 is too complicated to allow an exact description of the system it represents. Two levels of approximation are used by Pink and Chapman: a first-order perturbation of a linearized mean-field Hamiltonian, and a cluster approximation. In the latter, interactions are calculated in detail for a cluster of, e.g., four adjacent lattice sites. Interactions involving sites outside the cluster are handled in a mean-field approximation.

This theory is attractive in its robustness. One can calculate a wide variety of ensemble-average properties of the system. Within the cluster approximation there is some ability to calculate correlations between properties at nearby sites. Finally, the lateral distribution of proteins is not fixed (as in the Marčelja and Landau treatments), so information on protein interactions and phase separations results in a very natural way. Against these advantages must be set the large number of adjustable parameters, the unphysical constraint of the lattice, the restriction of the lipid to only two states, and the obscure connection between the interaction parameters and intermolecular interactions.

c. Results

Results are obtained with lipid-lipid parameters set to reproduce the behavior of dimyristoyl- or dipalmitoylphosphatidylcholine (DMPC or DPPC).

Many of the general results are similar to and in agreement with the predictions of the theories discussed in previous sections: with increasing protein concentration, lipid phase-transition enthalpies decrease, disappearing (as sharp first-order transitions) at sufficiently high protein content. Transition temperatures (or, more precisely, phase boundaries) can be above or below the T_c of pure lipid, depending on whether the choice of parameters for protein-lipid interactions favors ground- or excited-state lipids adjacent to the proteins. The correlation length for the decay of conformational perturbation around a protein corresponds to about one layer of lipid molecules.

The most interesting novel results are phase diagrams for several sets of parameters, reproduced here in Fig. 6. Proteins with parameters similar to those determined for cholesterol by Pink and Carroll (1978) have phase diagrams similar to those in Fig. 6A. If interactions between the protein and excited-state lipids are made extremely unfavorable, eutectic behavior can occur (Fig. 6B). Both Figs. 6A and B are computed for small (lipid-chain sized) proteins. Figure 6C is the phase diagram calculated for a larger protein that binds three ground-state lipids and behaves like cholesterol in its other interactions. Note the critical mixing point (arrow) that results in critical behavior analogous to that predicted by the Landau treatment.

To date there is insufficient experimental data to make satisfying comparisons with these phase diagrams. One well established point was mentioned earlier, however: lateral separation into distinct protein-rich and protein-poor phases has been observed frequently.

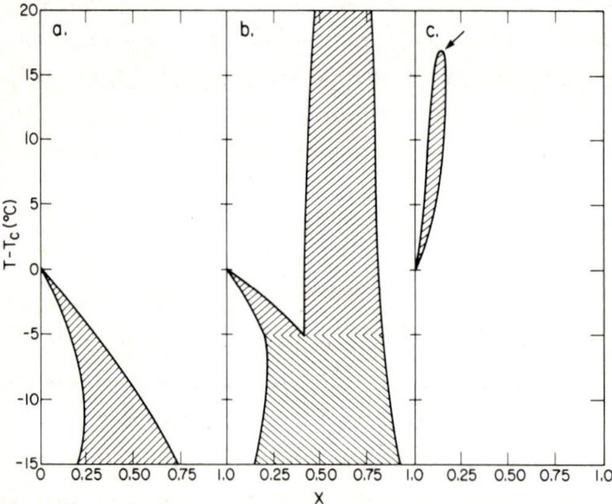

Fig. 6. Temperature-composition phase diagrams for protein-lipid mixtures, computed for different parameter choices in the lattice model of Pink and Chapman (1979). x is the mole fraction of protein, and two-phase regions are indicated by hatching. (a) A small cholesterol-like protein in DPPC (J_1 = 0.40, J_4 = 0.005, J_p = 0.607); (b) eutectic behavior in DMPC (J_1 = 0.35, J_4 = −0.05, J_p = 0.7); (c) phase diagram for a protein that presents a bound ground-state lipid to three of its lattice neighbors and behaves like cholesterol to the other three. After Pink and Chapman (1979).

d. Computer simulation of the lattice model

More recently, Lookman et al. (1982) have reported computer simulations based on a variation of the lattice model described above. The Monte-Carlo calculation should be a more accurate realization of the Hamiltonian than the methods used by Pink and Chapman. Lookman et al. presented some analysis of the influence of protein on lipid order and thermodynamics. Since their primary emphasis was on the lateral distribution of proteins, however, we will delay discussing this work until Section IV.

6. SCOTT AND COE (1982)

a. Description of theory

Scott and Coe (1982) have extended a statistical-mechanical treatment of lipid bilayers (Scott and Cheng, 1978; Scott, 1981) to the consideration of lipid-protein interactions. Their model represents lipids and proteins by their projections onto the plane of the bilayer. The interactions in the resultant two-dimensional system, treated with Scaled Particle Theory (SPT) (Helfand et al., 1961; Scott, 1980), are assumed to represent those in the actual membrane.

Proteins are simply represented as circles, lipids as rectangles of length a and width b, with circularly rounded ends. The width of each lipid projection is held constant and the length varied to represent different lipid conformations. A discrete number of possible lipid conformations and their orientation is then created, and each is assigned a statistical weight.

Having thus assigned parameters to the various protein and lipid states, it is possible to write down an expression for the Gibbs free energy of the system. This includes contributions from internal chain conformational free energy, van der Waals attractions, and hard-core repulsions between molecules. The semi-empirical van der Waals term is

$$E_{\mathrm{vdW}} = \frac{-C}{(A_{\mathrm{L}} - A_{\mathrm{o}})^{3/2}} \tag{15}$$

A_{L} is the average area of a lipid molecule, A_{o} is fixed at 0.38 nm^2, and C is an adjustable strength parameter. This equation effectively assumes that proteins act like lipids. Since this term is the only source of intermolecular attractions in the theory, all specific effects of the proteins must appear as repulsions.

It is the hard-core interactions that are treated with SPT. This theory, which has been widely used in the theory of liquids, assumes ideal mixing of the component molecules. The chemical potential of each component is calculated as the free-energy change for adding an additional molecule to the system.

This is done by considering the addition of a particle of 'scaled' dimensions αa and βb to a system of particles of regular size. It is possible to examine the limiting cases $\alpha, \beta \rightarrow 0$ and $\alpha, \beta \rightarrow \infty$ directly. The insertion term is then constructed by assuming it has some polynomial form in α and β and matching coefficients with the limiting forms found above. The final expression is evaluated for the case $\alpha = 1, \beta = 1$, thus giving the free energy of addition of a regular sized particle.

We will not reproduce the final equation for the SPT free energy here, but instead note that it is a computationally tractable expression involving the mole fractions of all components. The observed free energy of the system is obtained by minimizing the overall free-energy expression with respect to the mole fractions of components and the membrane area. As in the Landau treatment, phase transitions take place when there exist two minima of equal free energy, implying the existence of phase equilibrium.

Scott and Coe generated the temperature and enthalpy change of the phase transition, while constraining the system to remain in a single phase. The authors sought to reproduce experimental observations in which proteins caused large decreases in transition enthalpies without much effect on transition temperatures.

b. Results

Initially, nonspecific interactions between proteins and various lipid states were assumed. Large decreases in transition temperatures were seen: about 18°C for 0.5 mol% of a protein of radius 1.5 nm. In the presence of protein, the phase transition is between a low-temperature phase with high conformational order and nematic-like orientational order in the rodlike projections, and a high-temperature phase with lower conformational order and no orientational order. Since the protein interferes with orientational order, it relatively stabilizes the disordered phase and lowers the transition temperature.

The authors also analyzed a model in which ordered lipid states were favored by the protein. Favorable interactions were chosen by allowing the lipid to interact with the protein via a reduced area. This is equivalent to saying that closer contact is allowed, possibly through surface roughness, chain folding, or tilting into or around protein protrusions. This lowers the SPT free energy by reducing overlap repulsions.

It was found that favoring the all-*trans* configurations reduced the depression of the transition temperature. To achieve satisfactory agreement with experiment, however, it was necessary to allow the favorable interaction to extend to a rather unphysical number of lipids (135, while geometric considerations dictate that only about 25 can be in direct contact with the protein).

This is the only theory that directly addresses the behavior of the protein as an incommensurate impurity in an ordered phase. It would be helpful to see the full temperature-composition phase diagram for the system, a calculation that would not be difficult. Finally, given the assumption of homogeneous protein distribution, the model cannot easily examine protein-protein forces or aggregation.

7. O'LEARY (1982)

a. Description of theory

In an effort to assess the nature of the conformationally excited states that appear at temperatures above the lipid-bilayer phase transition, Jackson (1976) constructed a theory in which all excited states were β-coupled *gauche* kinks (sequences such as g^+tg^-). Recently, O'Leary (1982) extended the theory to include the effects of proteins and cholesterol on the transition. The basic idea was that proteins alter the probability of occurrence of kinks in adjacent lipid chains. This was achieved by adjusting the energy and entropy of kink formation.

In the original Jackson model, all kinked chains have an intra-chain conformational energy of $2E_g$ due to the presence of two *gauche* bonds. In addition, an isolated kink is assigned an energy penalty, dependent on the area, A, per chain, to account for unfavorable steric interactions with neighboring unkinked chains. This penalty has a strength parameter, w. Additional kinks that form next to a pre-existing kink are not assigned a penalty, since they can nestle spoonwise into the vacancy created by the initial kink. An entropy parameter, P, is also associated with kink formation. Other area-dependent interactions, such as van der Waals interactions between chains, repulsions between polar head groups, and hydrophobic interactions at the aqueous-nonpolar interface are accounted for with a semi-empirical expression.

The overall free energy of the bilayer can be computed simply, as a function of A, from the free energies discussed above. In the constant-pressure ensemble ($\pi = 0$) the observed area is that which minimizes the free energy. Multiple minima as a function of A can lead to phase transitions, as in the Landau and SPT treatments above.

O'Leary assumed that proteins have two effects on the lipids. First, they introduce free volume through their rough surfaces. This reduces the steric penalties for kink formation adjacent to proteins, leading to a reduction in w. Second, proteins prevent the lateral propagation of kinks and thus reduce the entropy of kinking. To model this, P was reduced for kinks adjacent to proteins. Decreasing w increases the stability of kinks; decreasing P decreases it.

The number of lipid-protein contacts was determined in the following mean-field way. Both molecules were represented as circular disks, with the proteins having twice the area of the lipids. The work of Dodds (1975) on packing arrangements of such disks was used to characterize the contacts, without reference to any specific protein-lipid interactions. The values of w and P in the theory were made to depend linearly on the number of protein-lipid contacts per lipid. In this theory w (and to a lesser extent P) is the principal adjustable parameter.

The kink model is not one of the more successful quantitative models of the lipid phase transition. It does treat steric interactions between chains reasonably realistically, but it severely restricts the number and type of excited states. The extension to include protein-lipid interactions does not remove these problems. Nevertheless, it does have the virtue of giving physical motivation for the chosen form of the protein-lipid interactions. The physical picture is fairly clear, and the model is computationally tractable.

It is not possible to analyze positional correlations among molecules in this model. The radial decay of lipid perturbation outward from the protein cannot be computed. In fact, there is no distinction between bulk and boundary lipid at the level of computing the properties of the system. Only one class of lipids exists, an average of the two.

b. Results

The principal result for protein-lipid mixtures is the graph of heat capacity vs. temperature shown in Fig. 7. This resembles a DSC result except that the system was constrained to be a single phase throughout. The protein has lowered the temperature of the first-order transition of the simulated DPPC to about 302°K from 305°K in the absence of protein (the experimental T_c is 314°K).

A second transition that is continuous but sharp has appeared at 304°K. It would be hard to differentiate from a first-order transition experimentally. The higher transition is associated with a lateral expansion of the bilayer without large

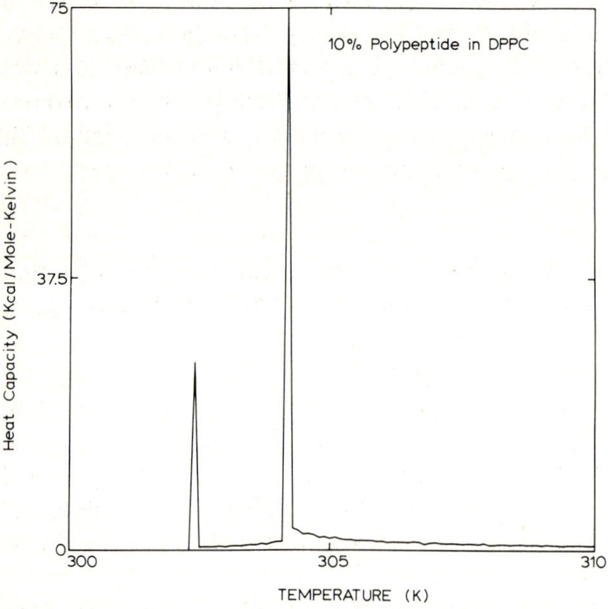

Fig. 7. Heat capacity vs. temperature in the O'Leary (1983) model for a 10 mol% protein impurity in DPPC (the model gives a bulk lipid transition temperature of 305°K). Lateral phase separation is suppressed for the calculation. The transition near 302°K is discontinuous; the protein-induced transition near 304°K is sharp but continuous. From O'Leary (1983).

increases in the fraction of kinked lipids (T. O'Leary, personal communication).

As the author pointed out, the theory has shown that the observation of a second transition need not be linked either to phase separation or to the separate melting of boundary lipid. Similar conclusions can be drawn from the Landau results (see Section II, 4).

The sharp continuous transition is not observed for all combinations of protein concentration and parameters. At high protein concentrations the first-order transition becomes very weak, but it does not disappear. In this, the theory differs from the others described in this section.

8. PEARSON ET AL. (1984)

a. Overview

Given the basic model of a lipid continuum and proteins that impose boundary conditions on the order parameter, there are several ways to obtain approximate analytical descriptions of the effects of proteins on lipid order. Schroeder (1977b) linearized the mean-field description of the lipid free energy. Jähnig (1981a) kept the nonlinear Landau free energy but assumed a simple form for the spatial behavior of the order parameter. Pearson et al. (1984) have used an approach rather like Schroeder's.

Starting with a simple empirical free-energy function, they obtain an analytical solution for the radial decay of the order-parameter perturbation about an isolated protein. As with Schroeder and Jähnig, a superposition of these solutions describes the effects of multiple proteins. However, the superposed perturbations are modulated by a term that depends on the lateral density of proteins. This should extend the validity of the approximation to higher protein densities.

b. Description of theory

The physical nature of the order parameter, ϕ, used in the theory is left open. The free-energy density of a bilayer at position $\mathbf{r}$, relative to that of an unperturbed membrane with order parameter ϕ_∞, is

$$G[\phi(\mathbf{r}), \nabla\phi(\mathbf{r})] = k_1|\nabla\phi(\mathbf{r})|^2 + k_2(\phi(\mathbf{r}) - \phi_\infty)^2 \tag{16}$$

The constants k_1 and k_2 determine the strengths of the two elastic forces that tend to keep the order parameter constant in space and at the bulk lipid value, ϕ_∞. This same function was used by Kleeman and McConnell (1976) in a brief discussion of protein-lipid interactions.

The quantity $(k_1/k_2)^{1/2}$ defines a correlation length for spatial variations of the order parameter. Its value could be determined experimentally or from additional theoretical considerations that fix k_1 and k_2. Reduced length units are defined as above.

As in the Schroeder and Landau treatments, proteins are represented as circles with radius $\tilde{r}_o$, and the order parameter is constrained to be ϕ_o at the protein-lipid boundary. The thermodynamics of the system are then calculated by minimizing the free energy, as in Eqn. 10.

c. *Results*

Some useful expressions describing the perturbation of lipid order parameters by proteins have been derived. For an isolated protein, the radial decay of order parameter is the same Bessel function expression obtained in Schroeder's treatment, Eqn. 4.

For finite protein dilutions, the convenient Bessel function behavior of the order parameter cannot hold exactly. Pearson et al. show, however, that it is possible to construct a superposition treatment that at least does not violate the boundary conditions at $\tilde{r}_o$ very seriously if protein densities are not too high and protein-protein interactions are not too strong. For N proteins, each at position $\tilde{\mathbf{r}}_i$, the order parameter at $\tilde{\mathbf{r}}$ is approximately

$$\phi(\tilde{\mathbf{r}}) = \phi_\infty + c_N \sum_i K_0(|\tilde{\mathbf{r}} - \tilde{\mathbf{r}}_i|) \tag{17}$$

$$c_N = \frac{\phi_o - \phi_\infty}{K_0(\tilde{r}_o) + 4\pi\tilde{\varrho}\tilde{r}_o K_1(2\tilde{r}_o)} \tag{18}$$

Here $\tilde{\varrho}$ is the (reduced) lateral density of proteins and K_1 is the first-order modified Bessel function of the second kind. It is apparent that high values of $\tilde{\varrho}$ tend to reduce the amount of perturbation obtained in the superposition approximation. This correction is in the right direction.

Finally, the average order parameter in the membrane, at low protein density, is given by

$$\langle \phi \rangle = \phi_\infty + c_N 2\pi\tilde{\varrho}\tilde{r}_o K_1(\tilde{r}_o) \tag{19}$$

The above development was not compared with experimental determinations of order parameters or heat capacities. This would, in any case, be difficult to do unless ϕ were specified more fully. Instead, the theory was used to analyze lipid-mediated interactions among proteins, as is described in Section IV of this review. Additional information can be found in Edelman's thesis (Edelman, 1978).

9. LIPID MIXTURES. OWICKI ET AL. (1978)

a. Description of theory

The theories discussed so far have focused on perturbations of lipid order. A complementary problem arises when the lipids are a mixture of chemically distinct forms and the protein binds one preferentially: how far out from the protein do compositional perturbations extend, and what other effects on the system can be expected? Owicki et al. (1978) have developed a simple qualitative approach motivated by the Landau treatment, which describes the dependence of lipid free energy on composition with regular solution theory (Lewis and Randall, 1961).

The protein is taken as a cylinder of radius r_0, projected to a circle. The binary mixture of lipids is treated as a continuum, with x the mole fraction of one component. The protein holds the lipid composition at its boundary fixed at some value x_0.

The regular solution theory expression for lipid-lipid interactions is:

$$G(x, \nabla x) = Wx\,(1 - x) + kT(x \ln(x) + (1 - x)\ln(1 - x)) + E|\nabla x|^2 \quad (20)$$

W is a parameter that measures the nonideality of the solution. In the absence of protein, the model gives an upper critical mixing point at $x_{crit} = 0.5$, $T_{crit} = W/2k$. In other words, there is complete lipid miscibility for $T > T_{crit}$; lateral phase separation occurs for some compositions at lower temperatures. No attempt is made to include lipid order in the treatment.

E is a parameter that measures the cost of spatial variations in composition. The quantity $(E/W)^{1/2}$ is a natural length unit for such fluctuations, estimated to be 0.3 nm. It can be used to define reduced length units, denoted with tildes as before. The variational Euler-Lagrange equation associated with Eqn. 20 is

$$x'' + \frac{1}{\tilde{r}}x' + 2(x - x_\infty) + (kT/W)\ln\frac{x(1 - x_\infty)}{x_\infty(1 - x)} = 0 \quad (21)$$

Above, x_∞ is the bulk lipid composition.

b. Results

Numerical results for an isolated protein are shown in Fig. 8. It can be seen that the compositional perturbation has a very limited spatial extent – about one lipid layer – unless the lipid is very near the critical mixing point. For small compositional perturbations, Eqn. 21 approximates a Bessel equation, and analytical solutions can be obtained along the lines of Eqn. 4. In the critical region the correlation length for compositional fluctuations radially outward from the protein is proportional to

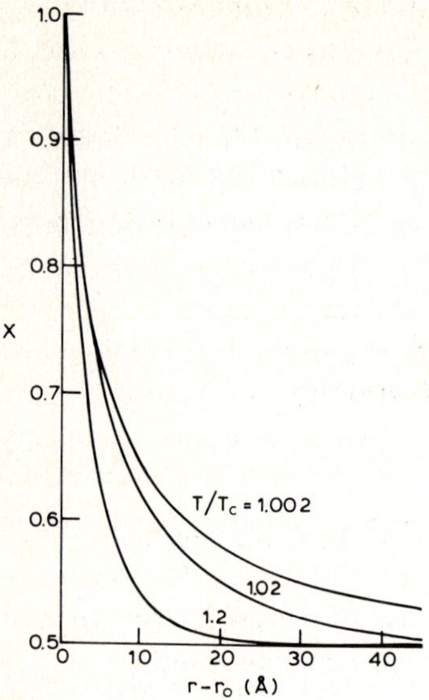

Fig. 8. Radial profile of lipid composition about protein in regular-solution theory of Owicki et al. (1978). Curves are solution to Eqn. 21 for $x_o = 0.99$, $x_\infty = 0.5$, $r_o = 1.5$ nm, with reduced length unit $(E/W)^{1/2} = 0.3$ nm. Results are plotted for three temperatures, $T/T_{crit} = 1.2$, 1.02 and 1.002. The increase of the correlation length for compositional fluctuations in the critical region is evident. From Owicki et al. (1978).

$$(T - T_{crit})^{-1/2}|x_\infty - x_{crit}|^{-1}$$

Although lipid phase diagrams with critical mixing points have been published (e.g., Wu and McConnell, 1975), we are aware of no direct experimental evidence of critical mixing phenomena in membranes. Finally, we note that a satisfyingly comprehensive treatment of the lipid mixtures problem would use one of the extant theories of order and phase separation in lipid mixtures. The computations would be fairly complex but would unify the treatments of lipid order, composition, and phase behavior of all the components.

10. SUMMARY

In an important sense, the theories presented above share a common point of view about protein-lipid interactions. Most start with a theory of the lipid bilayer phase transition, which is parametrized independently of the presence of proteins. The direct interaction between protein and lipid is taken to be very short ranged. It consists of a relative or absolute preference for some subset of the lipid states that are associated with the description of the pure lipid system. Because of this underlying similarity, there is more broad agreement than might be imagined from surveying the different approaches.

Specifically, proteins weaken the phase transition; there is ample experimental evidence to support this. Depending on the choices of parameters, transition temperatures can rise, fall, or remain constant; all three types of behavior have been observed experimentally. Those theories capable of estimating correlation lengths generally agree that one or two layers of lipid around a protein are significantly perturbed. This also is in satisfactory agreement with most magnetic-resonance and DSC experiments.

III. Effects of lipids on the activity of membrane proteins

1. OVERVIEW

Lipids can have significant effects on the properties of membrane proteins. Principal among these is the lipid-mediated alteration of protein activity. As in the previous section, we begin by briefly summarizing some pertinent experimental work. Further information and detailed references are available in review articles (e.g., Sandermann, 1978; Klein, 1982).

Protein activity can be modulated in two major lipid-dependent ways. The first is through the binding of specific lipids. These lipids appear to play a variety of regulatory roles, with details depending on the protein and lipids in question. Besides satisfying cofactor requirements, specific lipid binding could be involved in optimizing lipid packing and fitting hydrophobic and hydrophilic regions at the protein-lipid interface, and in stabilizing protein conformation through electrostatic interactions and hydrogen bonding. This can be a difficult experimental area; the results obtained have a complicated dependence on the experimental methods employed. To date, the binding requirements of a vast number of membrane proteins have been analyzed.

The second way results from the effects of global lipid state. Protein activity may display abrupt, temperature dependent changes. Similar behavior is often seen with membrane transport and unfacilitated permeation. Many of these temperature-dependent changes are presumed to result from concomitant changes in lipid phase and lateral distribution. Although unproven, this correlation is intuitively reasonable; a fluid membrane provides a smaller barrier to permeation and conformational changes and allows much greater lateral mobility than a solid one. The most important experimental systems have included sarcoplasmic reticulum $Ca^{+}K^{+}$-ATPase, $Na^{+}K^{+}$-ATPase, cytochrome oxidase and bacterial sugar transport.

There have been a variety of theoretical and quasi-theoretical treatments of these two phenomena. Unfortunately, an extensive overview is beyond the scope of this review. We propose, instead, to look primarily at work that makes explicit use of the protein-lipid interaction theories developed in Section II. We have divided the remainder of this section into two parts, dealing first with activation by the binding of specific lipids, and secondly, with the effects of global lipid state.

2. ACTIVATION OF PROTEINS BY BINDING OF SPECIFIC LIPIDS

The most obvious role of protein-lipid interaction theories in this area is the examination of specific lipid binding, relegating the concomitant effects on protein activity to more selective discussions. Preferably, such theories should predict the attributes of this binding (perhaps from the input of fundamental molecular information), rather than assuming them.

Proceeding along these lines, it should be possible to generalize many of the models in Section II to encompass lipid composition as well as lipid order. This might involve assuming different strengths for the direct interactions between proteins and annular lipids of various molecular and conformational kinds. The interplay of lipid-lipid and protein-lipid interactions would then determine the behavior of the annular lipid. To date, however, no such treatment has appeared.

In the absence of any analysis through protein-lipid theory, the primary theoretical treatments of this area have come through mass-action modeling. In general, such models assume a fixed number of lipid or alkyl chain binding sites at the protein-lipid interface. The distribution of lipid species between this annulus and the bulk phase, expressed in terms of relative binding constants, may be determined from fundamental statistical considerations (e.g., Lee, 1983) or by fitting the model to experimental results (e.g., Brotherus et al., 1981). These models make predictions on the composition of the first shell of boundary lipids, and should be compared with the treatment in Section II.9 (Lipid Mixtures), which fixed the first shell and looked at the composition in successive ones.

3. EFFECTS OF GLOBAL LIPID STATE

a. *Overview*

The phase behavior of lipid bilayer membranes has been exhaustively characterized (e.g., Lee, 1977a, b). In one-component systems, there is a well-defined crystalline to liquid-crystalline (solid to fluid) phase transition that occurs over a very small temperature range around T_c. The transition is arguably first order. Lipid mixtures show more complicated behavior, with the possibility of lateral phase separations into domains of different composition and fluidity. The addition of proteins into either system may lead to further lateral separations into protein rich and protein poor regions, particularly at low temperatures or around the phase transition.

By definition, we are interested in alterations in protein activity that arise due to changes in the lipid environment and not due to temperature-dependent conformational changes that are intrinsic to the protein. Experimentally this may be tested by studies in different lipid systems or by spectroscopic monitoring of protein conformation. Because an understanding of the simpler single-lipid case seems more accessible, and is surely relevant to lipid mixtures, we will focus on it. After introducing the assumptions and basic methodology (i.e., Arrhenius analysis), we will look at the Landau treatment of Jähnig and Bramhall (1982).

b. Arrhenius analysis

The customary method of analyzing reaction rates is via the Arrhenius plot. Arrhenius (Eisenberg and Crothers, 1979) postulated that the reaction rate constant may be expressed as

$$k = A\,\exp(-E_a/RT) \tag{22}$$

The frequency of properly oriented collisions between reacting molecules is given by A, a weakly temperature-dependent term called the frequency factor. The fraction of those colliding molecules possessing the necessary activation energy, E_a, for reaction is given by $\exp(-E_a/RT)$, a Boltzmann factor. The frequency of successful encounters is then given by the product, k.

The Arrhenius equation may be put into a form that allows a convenient, graphical analysis of experimental results.

$$\ln(k) - \ln(A) - E_a/RT \tag{23}$$

Plotting $\ln(k)$ versus $1/T$ – the Arrhenius plot – gives a straight line with slope $-E_a/R$ in regions where A and E_a are constant. Changes in A or E_a produce alterations in the line that are subject to various interpretations. Representative plots may be seen in Figs. 9 and 10.

Since kinetic results are usually interpreted from the graphed data, it is clearly of the utmost importance that curve-fitting be done as carefully as possible. The pitfalls of methods such as Scatchard and Michaelis-Menten analyses are well known. A number of the difficulties and dangers in Arrhenius analysis are described by Silvius and McElhaney (1981), Houslay and Stanley (1982) and Klein (1982).

Figure 9 is a representative Arrhenius plot for Na^+K^+-ATPase in four single-lipid systems. Each plot displays an apparently abrupt change in slope at some temperature, T_{break}. The rate itself changes continuously. This suggests an abrupt change in activation energy, E_A, balanced by a compensatory change in A. That is,

$$\ln\left(\frac{A_+}{A_-}\right) = \frac{E_{a,+} - E_{a,-}}{RT} \tag{24}$$

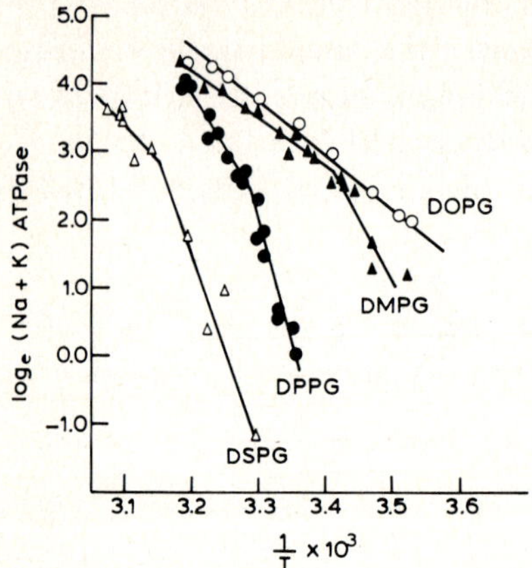

Fig. 9. Arrhenius plot of rabbit kidney Na^+K^+-ATPase activity in four pure phosphatidylglycerol systems. The various lipid systems are (together with acronym, transition temperature in pure lipid and break temperature, respectively): distearoylphosphatidylglycerol (DSPG, 54°C, 44.3°C), dipalmitoylphosphatidylglycerol (DPPG, 40°C, 31.4°C), dimyristoylphosphatidylglycerol (DMPG, 23°C, 20.2°C), and dioleoylphosphatidylglycerol (DOPG, no break observed, no transition on temperature range examined). The phase transition temperatures reported are for pure lipid; they are presumably modified in the presence of protein. The DMPG results were multiplied by 2.3 before plotting. From Kimelberg and Papahadjopoulos (1974).

where + and − refer to states immediately above and below the transition, respectively. For this simple system, the break temperature has been loosely correlated with the temperature of the phase transition (Kimelberg and Papahadjopoulos, 1974).

For purposes of contrast, Fig. 10 shows a representative Arrhenius plot for lipid mixtures typical of biological systems, in this case β-galactoside transport in elaidic acid-enriched membranes from *E. coli*. Much greater complexity is revealed. There are now two breaks and one actual discontinuity in the plot. In addition to the features seen here, deviations from linearity are not uncommon. Most workers assign the low and high temperature breaks to the beginning and ending of lateral phase separations.

c. Jähnig and Bramhall (1982)

Jähnig and Bramhall (1982) used the Landau theory to address two questions pertinent to protein activity and unfacilitated permeation in single-lipid systems.

(1) What is the physical significance of the very large activation energies seen at low temperatures?

(2) How does the activation energy change without a discontinuity in protein activity?

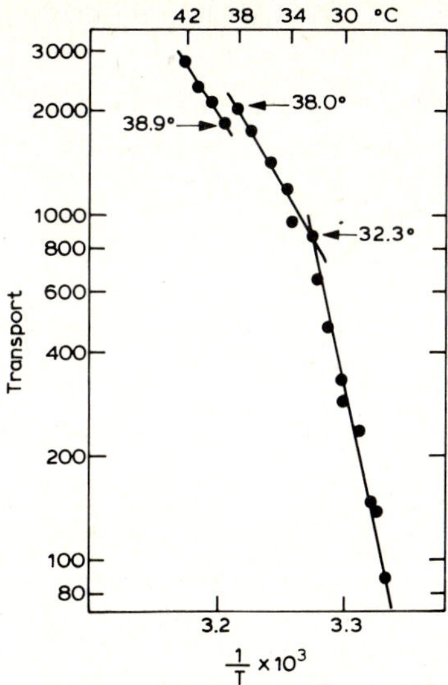

Fig. 10. Arrhenius plot of β-galactoside transport in *E. coli* grown at 37°C in elaidic acid-enriched medium. The break temperatures indicated in the plot should be compared with the temperature of the onset (37.7°C) and conclusion (32.1°C) of lateral phase separation as determined by 2,2,6,6-tetramethylpiperidine-1-oxyl (TEMPO) partitioning. From Linden et al. (1973).

We emphasize aspects of their development most pertinent to protein activity.

For this work, the rate equation was rewritten in a form familiar from the Eyring absolute rate theory (Eisenberg and Crothers, 1979). That is, for constant temperature and volume,

$$k = A \exp(-\Delta F/RT) \tag{25}$$

where $\Delta F = \Delta U - T\Delta S$, the Helmholtz free energy. This formulation is almost identical to that of Arrhenius, differing primarily in that it explicitly considers the reaction entropy, ΔS. It is then possible to model the system in terms of this equation and the Landau formulation and reproduce the observed behavior.

In essence, Jähnig and Bramhall attribute the break to cooperative effects between lipids. The break is assumed to occur at the phase transition temperature, T_c. It is also assumed that for activity or transport, 20–50 lipid molecules in the neighborhood of the protein must be disordered. Physically, an enzyme might only function in a locally fluid environment. Similarly, unfacilitated permeation might require or invoke a local disordering. By definition, no disordering is necessary above the phase transition. Below it, disordering requires energy which must be added to that needed for activation. Hence, question **(1)** is answered in terms of this assumption.

Above the phase transition, quantification is simple and ΔF is just the protein's activation energy (in the case of protein activity) or a residual, amphiphilic interaction energy (in the case of transport or permeation). This term is assumed to be independent of temperature and is denoted $\Delta F_{\text{amphiphilic}}$, ΔF_a for short. Below the transition, the energy also includes a temperature dependent term representing the lipid disordering energy, ΔF_{melt}. Thus,

$$\Delta F = \begin{cases} \Delta F_a + \Delta F_{\text{melt}} & \text{if } T < T_c \\ \Delta F_a & \text{if } T > T_c \end{cases} \tag{26}$$

The Landau theory is invoked to get an expression for ΔF_{melt}; it is not necessary to evaluate ΔF_a. The calculations give an equation for the free energy of melting *per lipid*, Δf_{melt},

$$\Delta f_{\text{melt}} = \begin{cases} \Delta q \dfrac{(T_c - T)}{T_c} & \text{if } T < T_c \\ 0 & \text{if } T > T_c \end{cases} \tag{27}$$

where Δq represents the latent heat of melting per lipid. Neglecting interfacial energies and assuming a sharp boundary to the perturbation, ΔF_{melt} and Δf_{melt} are simply related according to

$$\Delta F_{\text{melt}} = \nu \Delta f_{\text{melt}} \tag{28}$$

where ν is the number of perturbed lipids. So the expression for k may be written

$$k = \begin{cases} A'' \exp(-(\Delta F_a + \nu \Delta q)/RT) & \text{if } T < T_c \\ A' \exp(-\Delta F_a/RT) & \text{if } T > T_c \end{cases} \tag{29}$$

where $\exp(-\nu \Delta q/RT_c)$ has been incorporated into A' and renamed A''.

Then

$$\ln(k) = \begin{cases} \ln(A'') - (\Delta F_a + \nu \Delta q)/RT & \text{if } T < T_c \\ \ln(A') - \Delta F_a/RT & \text{if } T > T_c \end{cases} \tag{30}$$

which, when graphed, duplicates the observed plots with a break at T_c. In fact, choosing ΔF_{melt} to be any affine function of temperature that vanishes at and above T_c will reproduce this behavior. Of course, the advantage of the function derived here is its transparent physical significance. At the phase transition,

$\Delta F_{melt} = 0$, implying that $T_c \Delta S_{melt} = \Delta U_{melt}$, the so-called 'energy-entropy compensation' (Lumry and Rajender, 1970). It is the continuity of ΔF at all temperatures that insures the continuity of k, thus answering question **(2)**.

The above analysis assumes that the system behaves as a single component, so that no phase separation occurs. As mentioned earlier, however, the system is a binary mixture of lipid and protein (in the case of enzyme activity); lateral phase separation must occur in the vicinity of the lipid phase transition. Lateral phase separation automatically implies continuous Arrhenius plots with breaks at the onset and completion of phase separation. This would seem to diminish the importance of the Landau analysis aimed at explaining the continuity.

Nevertheless, it may often be the case that the two-phase temperature region is narrow or the protein concentrations differ greatly in the two phases. An analysis of protein activity based solely on phase separation would then predict features in Arrhenius plots that would be difficult to distinguish experimentally from discontinuities. Integrating the Landau treatment with the phase diagram under these conditions is an attractive explanation for observations of continuous Arrhenius plots with breaks.

IV. Protein-protein interactions

1. OVERVIEW

Inhomogeneous lateral distributions of intrinsic membrane proteins are frequently observed; striking examples include the densely packed plaques found in eukaryotic gap junctions (Loewenstein, 1981) and in the purple membranes of halophile bacteria (Stoeckenius et al., 1979). In general, lateral segregation of proteins and other membrane components contributes to the functional specialization of different regions of membrane. These may reflect tissue organization (e.g., apical vs. basolateral epithelial surfaces) or the spatial localization of multienzyme metabolic pathways (e.g., photosynthesis in thylakoid membranes).

All the many kinds of intermolecular interactions that are biologically important elsewhere presumably operate between membrane proteins, although little is known about the subject. The forces may range from rather nonspecific effects, such as electrostatics, to the interactions of specific complementary binding sites. In addition, the presence of the lipid bilayer creates the possibility of a unique type of interaction, as will be seen below.

We have divided this section of the review into two parts. The first describes how theory, mostly statistical mechanics, can be used to extract information on protein-protein interactions from electron micrographs. The second reviews the attempts to predict some aspects of the lateral distribution of proteins starting from fundamental considerations.

2. ANALYSIS OF ELECTRON MICROGRAPHS

a. Statistical versus statistical-mechanical analyses

Freeze-fracture electron microscopy provides a way of directly visualizing the positions of many larger membrane-bound proteins. Such positions are unavailable in the study of three-dimensional fluids*. These distributions are rich in information and may be used in studies at many levels. There is thus a strong interest in developing a formalism capable of quantifying and interpreting protein distributions. Two routes have developed, statistical and statistical-mechanical, that differ in their techniques and in their underlying intent.

Purely statistical treatments (e.g., Melhorn and Packer, 1976; Donnell and Finegold, 1981) employ quantitative criteria to assess the degree of clustering in membranes. The criteria are not based on intermolecular interactions, and there is typically a minimal attempt to correlate findings with any underlying physical basis. Statistical descriptions are most useful in studies of gross morphology. Since this review is biophysical, we will not consider these descriptions further.

The statistical-mechanical treatments assume that protein distributions are dictated by interprotein forces. They seek to quantify the relationship between the distribution and the forces. This is usually done by computing 'distribution functions' involving two or more proteins and relating these to the forces via a hierarchy of coupled integrodifferential equations. Under various conditions and assumptions these hierarchies may be closed and inverted to yield the forces. The following sections will focus on this development.

b. The radial distribution function and the Percus-Yevick equation

As noted, the statistical-mechanical treatments are based on distribution fuctions. A distribution function gives the probability of certain equilibrium arrangements of molecules in a fluid; it may be determined analytically from a knowledge of the interaction energy between molecules, or it may be computed empirically from a knowledge of the molecular positions. More detailed developments of the ideas in Sections (i) and (iii) below may be found in any book containing a discussion of fluid mechanics (e.g., Hill, 1956; McQuarrie, 1976).

i. Theoretical treatment. The probability density, $\varrho^{(n)}(\mathbf{r}_1,\ldots,\mathbf{r}_n)$, of finding a particle at $\mathbf{r}_1$, ..., and a particle at $\mathbf{r}_n$ in a fluid of N indistinguishable particles at equilibrium may be expressed in terms of the total interaction energy, U_N, and partition function, Z_N, according to

* Incidentally, this marks one of the few times that those of us who study membranes have a distinct experimental advantage over our colleagues who study more classical aqueous systems.

$$\varrho^{(n)}(\mathbf{r}_1, \ldots, \mathbf{r}_n) = \frac{N!}{(N-n)!} \frac{\int \ldots \int \exp(-U_N/kT) d\mathbf{r}_{n+1} \ldots d\mathbf{r}_N}{Z_N} \tag{31}$$

This is a trivial result of basic equilibrium statistical mechanics. In the absence of interaction (i.e., $U_N = 0$),

$$\varrho^{(n)}(\mathbf{r}_1, \ldots, \mathbf{r}_n) = N(N-1) \ldots (N-n+1) \frac{V^{N-n}}{V^N} \approx \frac{N^n}{V^n} \equiv \varrho^n \tag{32}$$

where ϱ is the number density. This expression approaches ϱ^n exactly for $N \gg n$, which is almost always the case. It is then possible to define a correlation function, $g^{(n)}$ $(\mathbf{r}_1, \ldots, \mathbf{r}_n)$, by

$$\varrho^{(n)}(\mathbf{r}_1, \ldots, \mathbf{r}_n) = \varrho^n g^{(n)}(\mathbf{r}_1, \ldots, \mathbf{r}_n) \tag{33}$$

which gives a simple measure of the deviation of $\varrho^{(n)}(\mathbf{r}_1, \ldots, \mathbf{r}_n)$ from ϱ^n.

With increasing n, correlation functions become more and more complex. For $n = 1$, $g^{(1)}(\mathbf{r}_1) = 1$, an expression of the fact that a particle is equally likely to be found at any point within a homogeneous fluid. For $n = 2$, we encounter the pair distribution function, $g^{(2)}(\mathbf{r}_1, \mathbf{r}_2)$, the most important member of the series. The pair distribution function may be determined experimentally, it carries a simple physical interpretation, and, by involving only minimal assumptions, it can be related to several thermodynamic functions. The pair distribution function is customarily rewritten in a conditional form, $g(\mathbf{r})$ – the radial distribution function – which gives the relative probability of finding a molecule at $\mathbf{r}$ *given* that there is a molecule at the origin. That is,

$$\begin{aligned} &\text{prob(mol at } \mathbf{r} \textit{ given} \text{ mol at } \mathbf{0}) \\ &= \frac{\text{prob(mol at } \mathbf{r} \textit{ and} \text{ mol at } \mathbf{0})}{\text{prob(mol at } \mathbf{0})} \\ &= \frac{\varrho^{(2)}(\mathbf{r}, \mathbf{0})}{\varrho^{(1)}(\mathbf{0})} = \varrho^2 \frac{g^{(2)}(\mathbf{r}, \mathbf{0})}{\varrho} \equiv \varrho g(\mathbf{r}) \end{aligned} \tag{34}$$

The angular aspects of $g(\mathbf{r})$ may be removed by integration over the angular variables in the system, yielding $g(r)$, the radial distribution function. Finally, higher order distribution functions ($n \geqslant 3$) are seldom used, although Braun et al. (1984) have found an important application of a specialized three-particle case; see Section d.

ii. Experimental treatment. We have seen that the distribution of particles in a fluid may be calculated theoretically from their total interaction energy, which, of course, depends in turn on the forces acting between them. At the same time, the simple physical interpretation of distribution functions allows them to be empirically determined from the knowledge of molecular positions afforded by freeze-fracture pictures. The pair distribution function can then be determined as the average ratio of particle density in an annulus of inner and outer radii, $r - dr/2$ and $r + dr/2$, respectively, to the bulk density, where r is centered around each protein. Averaging over annuli of nonzero thickness is necessary because of the small number of proteins, 0(1000), typically contained in an electron micrograph. Higher order distribution functions may also be determined, although the necessary computational effort is much greater.

A plot of an empirical $g(r)$ for connexons in a gap junction is shown in Fig. 11. It is typical of $g(r)$ seen for fluids. For small values of r, $g(r) = 0$, a reflection of overlap repulsions. At intermediate values of r, $g(r)$ oscillates, indicating regions where the surrounding particles are denser or sparser: coordination shells. These oscillations decay to unity for large values of r, where there are no longer any correlations with the central particle.

iii. Percus-Yevick analysis. The quantity $(g(r_{12}) - 1)$ can be regarded as a measure of the total correlation between particles 1 and 2 (separated by r_{12}), as reflected, for instance, in its decay to zero at large separations. Ornstein and Zernike resolved this total correlation into two contributions. The first contribution, called the direct correlation function $c(r_{12})$, is a measure of the direct correlations between particles 1 and 2. The second contribution comprises correlations

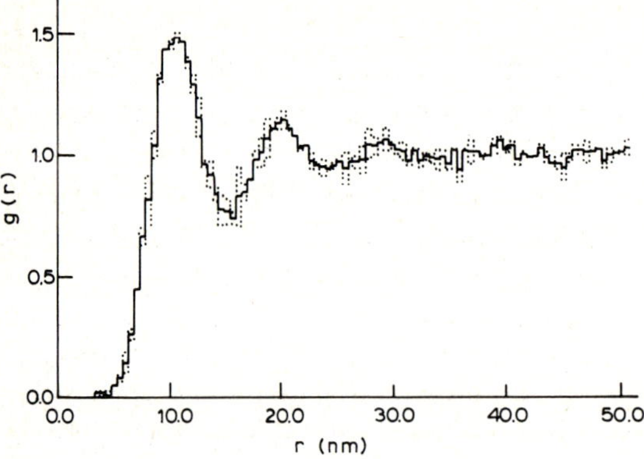

Fig. 11. Radial distribution function, $g(r)$, for connexons in a mouse liver gap junction that displayed liquid-like particle arrangements. The dotted error envelope represents the standard deviations obtained by performing the analysis separately on two parts of the plaque. The function was tabulated from a sample of 4319 particles at a lateral density of 0.0093 per nm^2. From Braun et al. (1984).

transmitted between particles 1 and 2 by all third particles. It is expressed as the product of the direct effect of 1 on 3, $c(r_{13})$, times the total effect (direct plus indirect) of 3 on 2, $(g(r_{32}) - 1)$, weighted by the density, ϱ, and integrated over all third particle positions. Combining the two contributions then gives the Ornstein-Zernike equation, which defines $c(r)$ in terms of $g(r)$:

$$g(r_{12}) - 1 = c(r_{12}) + \varrho \int c(r_{13})[g(r_{32}) - 1]d\mathbf{r}_3 \tag{35}$$

The benefit of this equation stems from the fact that $c(r)$ exhibits a much simpler behavior than $g(r)$ (with a shorter range and no oscillations) and that approximate expressions exist that relate $c(r)$ to particle interactions. One such link is to write

$$c(r) = g(r) - g_{\text{indirect}}(r) \tag{36}$$

Now, $g(r)$ can be related to a so-called potential of mean force, $w(r)$, via the Boltzmann relation, $g(r) \equiv \exp\{-w(r)/kT\}$. If U_N (from Eqn. 31) is the pairwise sum over molecules interacting by a pair potential, $U(r)$, then $w(r) - U(r)$ approximates the indirect interaction and so $g_{\text{indirect}}(r) \equiv \exp\{-(w(r) - U(r))/kT\}$. This relation is good for low ϱ. Then

$$c(r) = g(r)[1 - \exp(U(r)/kT)] \tag{37}$$

This is the Percus-Yevick approximation, and when substituted into the Ornstein-Zernike equation, it gives rise to the Percus-Yevick equation:

$$g(r_{12}) \exp(u(r_{12})/kT) = 1 + \varrho \int g(r_{13})[1 - \exp(u(r_{13})/kT)][g(r_{32}) - 1]d\mathbf{r}_3 \tag{38}$$

The Percus-Yevick equation relates the radial distribution function, $g(r)$, which can be determined experimentally, to the pair potential, $U(r)$, in a computationally tractable way. The potential is, of course, directly related to the pair force acting between particles according to $f(r) = -dU(r)/dr$. Incidentally, it is also possible to determine $g(r)$ directly from a known potential; this forms the basis of several papers to be discussed in Section c.

A few words are necessary on the significance of the interprotein forces determined from freeze-fracture pictures. The distribution of proteins in the membrane is determined not only by forces acting directly between proteins, but also by a variety of other factors. The effects of these unseen degrees of freedom, particularly protein rotation in the membrane and lipid-mediated forces, are included in the pair force in an averaged way. Hence, one should properly refer to $f(r)$ as an *effective* pair force. In addition, the extent to which the calculated force represents the true effective pair force is limited by the validity of the underlying assumptions. The Percus-Yevick equation assumes pairwise additivity of the potential and assumes Eqn. 37 for the relation between $c(r)$ and $U(r)$. It is known to break down for sufficiently concentrated protein solutions and other fluids with high ϱ.

c. *Historical development*

The applications of the radial distribution function and the Percus-Yevick equation to the study of protein distributions in membranes began as early as 1974 and have been developed in a series of papers (Markovics et al., 1974; Perelson, 1978; Gershon et al., 1979; R.P. Pearson et al., 1979; Pearson et al., 1983). Each of these studies is *briefly* summarized below with an emphasis on its contributions to the development of the technique.

Markovics et al. (1974) used $g(r)$ to analyze the distribution of pore patterns on nuclear membranes. The paper discussed the computation of $g(r)$, the sources and sizes of errors and, for comparative purposes, used a Percus-Yevick analysis to compute $g(r)$ for model hard-core and soft-core potentials. The authors noted that particular pore patterns give rise to characteristic functional forms of $g(r)$, relating this to the critical idea that movement in response to mutual interactions determines the distribution.

Subsequent work saw the refinement of the technique and its application to a variety of systems. Perelson (1978), studying the distribution of surface immunoglobulin on B lymphocytes, applied the (perturbing) technique of ferritin markers to enhance visualization, introduced automatic particle digitization, and discussed the difficulties to be encountered in employing curved specimens. He also made the important connection between lateral distribution and the movement and aggregation necessary for the function of the surface immunoglobulin. Gershon et al. (1979), working with mouse fibroblasts and simulated random particle distributions, demonstrated that the deviations of $g(r)$ from unity typical of proteins are significant above noise. R.P. Pearson et al. (1979), studying human erythrocytes, made the noteworthy introduction of a three particle angular distribution function to study angular correlations in molecular positions.

Most recently, Pearson et al. (1983) have applied $g(r)$ and the Percus-Yevick analysis of model potentials to the significance of bacteriorhodopsin and rhodopsin distribution functions in reconstituted systems. Such systems allow simple and better characterized manipulations*, thereby making them suitable for controlled experiments and a critical examination of the theories. The authors indirectly obtained information on the interprotein forces (i.e., repulsive vs. attractive) and related these to membrane structure and protein function.

* Although unfortunately, reconstitution does introduce large surface curvature and the possible loss of protein asymmetry.

d. Higher-order distribution functions and the BGY equation

The use of higher-order distribution functions does permit the determination of effective pair forces from electron micrographs without recourse to approximations such as the Percus-Yevick equation. Braun et al. (1984), in a study of connexon distribution in conducting gap junctions, employed the Born-Green-Yvon (BGY) hierarchy (e.g., Hill, 1956; McQuarrie, 1976) for this purpose. With the assumption of pairwise additivity, all of the information obtainable from the hierarchy about pair forces is contained in its second equation. This integrodifferential expression relates $g(r_{12})$, $f(r_{12})$, and a three-particle distribution function $\varrho(r_{13},\theta;r_{12})$, where θ is the angle between $\mathbf{r}_{12}$ and $\mathbf{r}_{13}$. The quantity $\varrho(r_{13},\theta;r_{12})$ represents the probability of finding a particle at polar coordinates r_{13} and θ, given that there is one particle at the origin and a second at a distance r_{12} away at the polar angle $\theta = 0$.

The BGY equation reads

$$kT\frac{d(\ln(g(r_{12})))}{dr_{12}} = f(r_{12}) + \int_0^{\infty}\int_0^{2\pi} f(r_{13})\cos(\theta)\varrho(r_{13},\theta;r_{12})r_{13}d\theta dr_{13} \tag{39}$$

The left hand side represents the statistical mean force, $-dw(r)/dr$, acting on the particle at the origin. The equation states that this force arises from two sources: the direct interaction from the particle at $\mathbf{r}_{12}$ ($f(r_{12})$) and the component along $\mathbf{r}_{12}$ of the interaction from all other particles (the integral). In this sense the equation resembles the Ornstein-Zernike equation for total correlations. The difference is that it is written in terms of forces rather than measures of correlation, and that it takes into account the true distribution of third particles around the central pair, rather than assuming spherical symmetry. The force, $f(r)$, is the true effective pair force (in the sense discussed above).

This technique and the precise pair and triplet distribution functions obtained with the gap junction system allowed the first direct computation of an effective pair force between membrane proteins. This is shown in Fig. 12.

e. Critique of EM analysis

Freeze-fracture electron micrographs provide a convenient and unique method of analyzing protein-protein distributions. However, a few words of warning are necessary on two points, (1) the assumptions and usefulness of the theory, and (2) the quality and faithfulness of the pictures themselves.

The theory itself is applicable largely because of the paucity of detailed mechanistic information that enters into it. For this same reason, it does little to explain the physical origin of the observed forces without thoughtful analysis. In addition, the theory, in the form presented, assumes an equilibrium distribution of a single species of particle with no cytoskeletal attachments, a fact that limits its applicability to complex biological membranes.

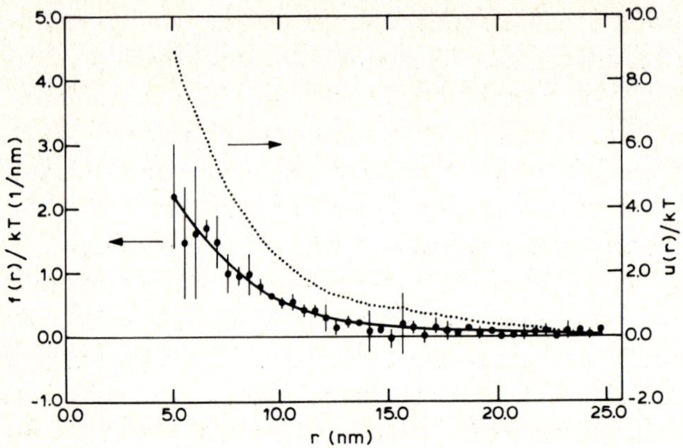

Fig. 12. Effective pair force, $f(r)$, between connexons in the gap junction analyzed for Fig. 11; computed using the BGY equation. The force was divided by the thermal Boltzmann energy, kT, where T was ≈310°K. The error bars were obtained as in Fig. 11. Below 5 nm not enough particle separations occurred to calculate the force in this region; the force must, however, be very large. In the range from 25 to 50 nm, not shown, the force is small and probably not significantly different from zero. The upper dotted line is the effective pair potential, $u(r)$, obtained by integrating the force over r. This is the pair energy required to compress two connexons from a reference separation of 25 nm in the plaque. From Braun et al. (1984).

The micrographs themselves must satisfy a variety of conditions. To insure that an instantaneous representation of the equilibrium protein distribution in the native membrane is being analyzed, the specimen must be frozen sufficiently rapidly to prevent rearrangement. The fracture pictures themselves should be free of plastic deformations and of sufficient quality to resolve each particle individually (e.g., using circular shadowing), and to indentify it as a single protein. A system containing a mixture of monomers, dimers, etc., would violate the single species assumption of this model.

3. THEORETICAL PREDICTIONS OF PROTEIN-PROTEIN INTERACTIONS

a. *Lipid-mediated interactions*

i. Overview. Four of the theories described in Section II explicitly predict an attraction between proteins that is mediated by the perturbation of the surrounding lipids (Marčelja, 1976; Schroeder, 1977b; Owicki and McConnell, 1979; Pearson et al., 1984). The cause is the same in all cases: the perturbed lipid is assigned an unfavorable free energy; protein aggregation decreases the total amount of boundary lipid and, hence, the free energy of the membrane. The argument is similar in style to those invoked for surface tension and for the tendency of hydrophobic solutes to aggregate.

If the overlap of regions of boundary lipid controls the interaction, at high protein densities it should show marked deviations from superposition of the interactions between pairs of isolated proteins. In other words, it should inherently be a multi-body effect that violates the usual pairwise-additivity approximations for potentials in statistical mechanics. This makes the phenomenon interesting from a physical standpoint, regardless of biological implications.

Until recently, there has been scant experimental evidence for the existence of this indirect interaction. We are excluding the segregation of proteins from solid lipid phases, which is well known and can technically be considered a lipid-mediated attraction. However, a recent paper (Pearson et al., 1983) has now demonstrated that the lateral distribution of membrane proteins can be sensitive to membrane thickness. This result is most obviously interpreted in terms of the lipid-mediated interaction, though one could also invoke thickness-dependent conformational changes in the proteins. Additional evidence has been obtained by Pearson et al. (1984) (see below).

ii. Marčelja (1976). Marčelja originated the idea of the lipid-mediated interaction and was first to demonstrate its existence theoretically. Figure 13 shows the free energy of two proteins as a function of the number of layers of lipid chains between them, relative to infinite separation. The strength of the effect, which is on the order of kT, depends on the size of the lipid perturbation (i.e., V_{lp}). The range is temperature dependent, being longest (several chains) close to the phase transition, which is at 25.2°C for this model.

iii. Schroeder (1977b). The principal result of Schroeder's work is an analytical expression for an interaction potential, $U(\tilde{r})$, due to lipid-mediated interactions between two proteins. It can be put into the following form:

$$U(\tilde{r}) = \frac{-U_o(S_o - S_\infty)^2}{K_0^2(\tilde{r}_o)} K_0(\tilde{r}) \tag{40}$$

U_o is a positive strength parameter that can be derived from the underlying mean-field theory. For large separations, the following asymptotic expression holds:

$$U(\tilde{r}) \approx -U_o(S_o - S_\infty)^2 \tilde{r}_o \left(\frac{2}{\pi\tilde{r}}\right)^{1/2} e^{-(\tilde{r} - \tilde{r}_o)} \tag{41}$$

Thus, Schroeder's work predicts an attraction with a range similar to the correlation length. Quantitative analysis of the range and strength require specification of the underlying mean-field model.

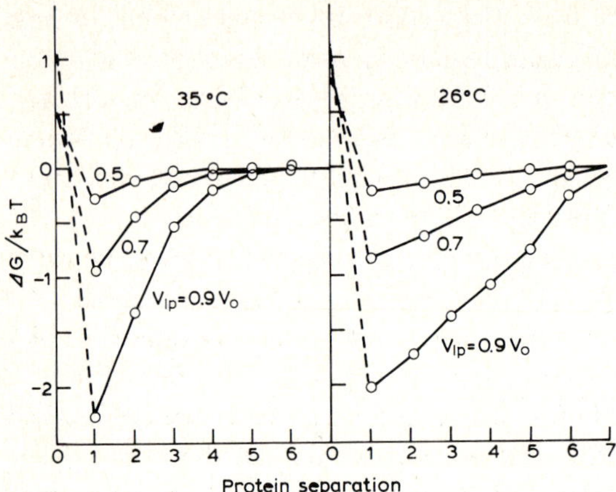

Fig. 13. Free energy (relative to infinite separation) of two proteins in Marčelja's (1976) theory. The abcissa is the number of layers of intervening lipid chains. The apparent repulsion for proteins in direct contact (dashed lines) is an artifact due to the absence of any direct protein-protein interaction in the theory. The protein side length is three lattice sites, and the transition temperature of the pure lipid system is 25.2°C. From Marčelja (1976).

iv. Landau theory of Owicki and McConnell (1979). The theories discussed above stress the idea of a pair potential. Multi-body aspects of the phenomenon can be investigated more directly by examining the dependence of the free energy of the system on protein density. Owicki and McConnell have approximated that calculation by examining the dependence of the Landau free energy on the separation of proteins that are fixed on a triangular lattice in a bilayer. Figure 14 shows the results for proteins of unit (reduced) radius at a temperature above T_c for the unperturbed bilayer. Since $u_o = 0.75$ in this example, the protein tends to stabilize the ordered phase.

For sufficiently large separations, the excess free energy becomes asymptotically equal to the total perturbation energy for an isolated protein. The order parameter decays toward the disordered state away from the protein. Near $\tilde{r}_1 = 4.8$ a first-order phase transition occurs between ordered and disordered lipid. For lower protein separations the lipid order increases toward the ordered state as r increases. As the proteins approach each other closely, the free energy decreases substantially due to a reduction of the amount of perturbed boundary lipid. When the proteins are in contact, all the boundary lipid is assumed to have been transferred to a pure lipid phase, and the excess free energy is zero*.

The depth of the attractive well in Fig. 14 is estimated to be on the order of a few kT. In this case its range is considerable: due to the phase transition, measur-

* Of course, close-packed disks cannot fill the plane; the calculation neglects the small amount of lipid that remains trapped in the voids between the proteins.

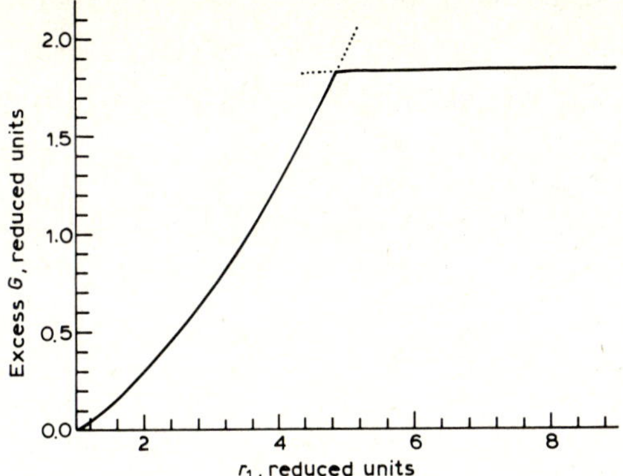

Fig. 14. Free energy per protein as a function of protein separation on a triangular lattice (Fig. 4) in the Landau theory of Owicki and McConnell (1979). This is a multi-body potential, not a pair potential as shown in the previous figure. The free energy is taken relative to that of unperturbed lipids; conditions are $\tilde{r}_o = 1.0$, $u_o = 0.75$, $\tilde{T} = 1.05$ (slightly above the phase transition). One reduced free-energy unit is on the order of one kcal/mol of protein. Note the concentration-driven phase transition that occurs near $\tilde{r}_1 = 4.8$. Dotted lines show metastable extensions of phases near the transition. From Owicki and McConnell (1979).

able effects begin at protein separations up to 9.6 correlation lengths, about 10–15 nm. The curvature of the free-energy plot at low protein separation is due to the saturation of the interaction, as is discussed more fully by Pearson et al. (1984).

According to the theory, interactions between proteins with similar values of u_o are attractive. With sufficiently different values of this parameter, however, proteins repel each other. This introduces a certain amount of specificity into the interaction, which could be exploited biologically.

v. Pearson et al. (1984). The pair potential for lipid-mediated interactions between two isolated proteins can be obtained by evaluating the free energy of the order-parameter profiles as a function of protein separation.

$$U(\tilde{r}) = 2E_1 \left[\frac{1 + M(\tilde{r})/M(0)}{(1 + K_0(\tilde{r})/K_0(\tilde{r}_o))^2} - 1 \right] \tag{42}$$

E_1 is the excess free energy of a membrane containing one protein. $M(\tilde{r})$ is a function defined in terms of Bessel functions, including $I_0(\tilde{r})$, the zero-order modified Bessel function of the first kind.

$$M(\tilde{r}) = 2\pi K_0(\tilde{r})[2\tilde{r}_o I_0(\tilde{r}_o) - 1] \qquad \text{if } \tilde{r} > \tilde{r}_o \tag{43}$$

$$\approx 2\pi\tilde{r}_o K_0(\tilde{r}_o) K_1(\tilde{r}_o) \qquad \text{if } \tilde{r} = 0$$

Figure 15 shows $U(\tilde{r})$ for several choices of $\tilde{r}_0$.

This expression is not appropriate at higher protein densities, due to multi-body effects. The authors have proposed a way to retain the useful feature of pair-wise additive potentials, while still taking multi-body aspects into account in an averaged way. If the effect of finite protein dilution is fundamentally to change the bulk order parameter from ϕ_∞ to $\langle\phi\rangle$, then the basic functional form for the pair potential in Eqn. 42 can be retained. The strength of the potential is changed by a multiplicative factor $(1 - Y)^2$, where Y is a density-dependent factor between 0 and 1 that can be, in principle, calculated from the theory.

The weakening or saturation of the pair potential with increasing protein density reflects the decrease in the driving force for aggregation, the difference between ϕ_0 and the effective bulk order parameter. This is important conceptually. Quantitatively, it is not clear at what protein densities the approximations involved begin to break down seriously. Nevertheless, this is the best analytical treatment of the problem to date.

The authors use $U(\tilde{r})$ (plus a hard core of radius $\tilde{r}_0$) and the Percus-Yevick approximation (see above) to generate the pair distribution functions $g(r)$. The computed functions are compared with $g(r)$ determined experimentally from freeze-fracture electron micrographs of *Acholeplasma laidlawii* membranes (James and Branton, 1973). The main goal is to find values of the correlation length and E_1 that provide a good fit; see Fig. 16 for representative results. The derived values of E_1 are 2–4 kT, and the correlation lengths are quite long, 6–10 nm. We cannot, however, consider this more than a very preliminary application of the theory. The particles in the micrographs probably represent more than one kind of pro-

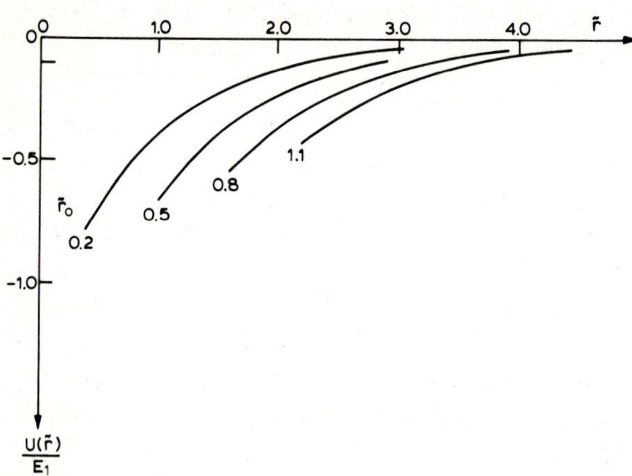

Fig. 15. Free energy (relative to infinite separation) of two proteins in the theory of Pearson et al. (1984) for several different protein radii $\tilde{r}_0$. The curves are solutions to Eqn. 38. The free-energy units are relative to E_1, the free energy due to lipid perturbations by an isolated protein. From Pearson et al. (1984).

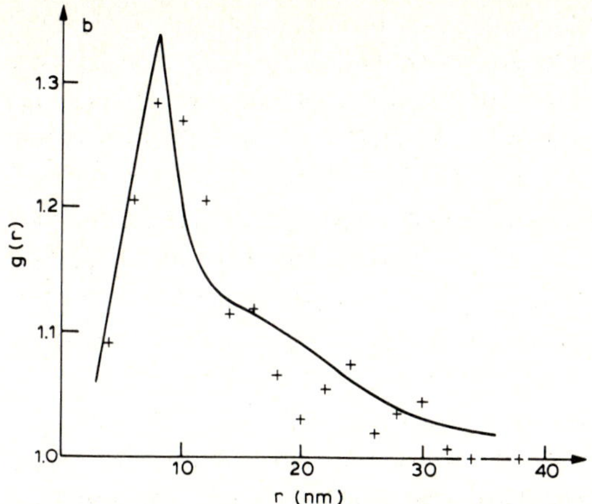

Fig. 16. Radial distribution function for particles in *A. laidlawii* membrane; from Fig. 1b in James and Branton (1973). Measured points are indicated by +'s. The theoretical fit (solid line) was generated by applying the Percus-Yevick approximation to the pair potential in Eqn. 38 with an added hard core of radius r_0. The relevant parameters are: particle density = 0.0025 per nm^2, r_o = 4.0 nm, E_1 = 2.3 kT, correlation length = 6 nm. From Pearson et al. (1984).

tein, and many kinds of interactions (e.g., electrostatic) probably influence the protein distribution.

Additional work from the same laboratory (S. Chan, personal communication) has shown long-range protein-protein repulsions in reconstituted cytochrome oxidase-phosphatidylcholine membranes. The range is too long for electrostatic repulsions and is thought to arise from lipid director perturbations. The symmetry of some director perturbations is such that similar proteins should repel each other; behavior quite different from that predicted by the theories described in this review.

b. Computer simulations

i. Overview. Discrepancies between theoretical results and accurate experimental results can have two important causes: inaccuracies in the fundamental theoretical model (e.g., the wrong choice of interactions in the Hamiltonian), and inaccuracies in working out the predictions of the model (e.g., mean-field approximations). This has also been noted by Nagle (1980). Theories of intermolecular interactions in membranes are generally vulnerable on both accounts. One is therefore often unsure whether disagreement with experiment reflects a basic inappropriateness of the model, or merely the effects of some bad approximations that were made to get concrete results.

Two classes of computer simulations – molecular dynamics and Monte Carlo – hold a special place in statistical physics. They enable one to get essentially exact numerical results for a model Hamiltonian. Remaining inaccuracies must therefore be an intrinsic property of the Hamiltonian. Moreover, the Hamiltonians that are susceptible to these treatments can explicitly reflect the many degrees of freedom involving the interactions of individual molecules by realistic potentials.

Molecular dynamics is generally synonymous with solving Newton's equations numerically for a system of 0(100) molecules (Wood and Erpenbeck, 1976). It is computationally expensive but produces dynamical as well as equilibrium results, the latter as a time average. To our knowledge, it has not been applied to protein-lipid systems.

Monte Carlo methods are more diverse (Valleau and Whittington, 1977), but typically use the Metropolis algorithm (Metropolis et al., 1953). This is essentially an efficient stochastic method for doing the multi-dimensional integrals (over molecular coordinates) that appear in ensemble averages. Monte Carlo calculations are often more economical than molecular dynamics, and two applications to membranes are discussed in Sections ii and iii.

Not dynamical, but only equilibrium quantities can be obtained by methods akin to the Metropolis algorithm. There is a temptation to compare successive Monte Carlo steps to the progression of time, a temptation reinforced by the connection between random walks and diffusion (a dynamical quantity). However, even a modest hope such as calculating ratios of diffusion coefficients at two temperatures involves questionable assumptions about the temperature invariance of the relationships between diffusion mechanisms and the Monte Carlo transition algorithm. Such simulations must be interpreted with extreme care.

There is a hybrid kind of calculation, usually called stochastic molecular dynamics, in which timescale differences are exploited. Slow processes (e.g., isomerizations) are simulated by molecular dynamics, while fast processes (e.g., collisions with solvent molecules) are simulated by stochastic (Brownian) force functions applied to the slow coordinates (Helfand, 1983). We have never heard of an application of this method to membranes, but we believe that this is the only practical way to implement molecular dynamics on a usefully realistic membrane model.

Other classes of simulations exist that are less transparently related to the intermolecular interactions in the system. One example is Finegold's (1976) dynamic model of protein aggregation in membranes, where diffusing proteins stick together irreversibly on contact, generating ever larger and more slowly diffusing aggregates. As usual, we choose to limit our discussion to the more statistical-mechanical approaches.

ii. Freire and Snyder (1982). (1) *Description of theory.* Freire and Snyder (1982) have extended a computer-based model of lipid mixtures (Snyder and Freire, 1980; Freire and Snyder, 1980a, b) to the study of protein-protein distributions in membranes. The system is represented on a triangular lattice. Proteins are regular hexagons that cover multiple sites, with one lipid molecule (both chains) implicitly occupying each remaining site. This is similar to Marčelja's arrangement, but in the present treatment only interactions between proteins are treated explicitly. In this respect the theory is similar to the analysis of electron micrographs discussed above, where only protein coordinates are available and lipid effects must appear in an averaged way as lipid-mediated protein-protein interactions.

The interactions between proteins are analyzed pairwise, by evaluating a Boltzmann factor or statistical weight, P, for each interaction. Three cases are distinguished. First, if the proteins overlap, $P = 0$ (i.e., infinite energy). Second, if the proteins are neighbors (their edges occupy some adjacent lattice sites), then P is assigned some fixed non-zero value. To model proteins that repel each other, $P < 1$; for attractions, $P > 1$; when $P = 1$, there are no interactions. Third, if the proteins are too far apart to count as neighbors, $P = 1$ (no interaction). The statistical weight of an entire configuration is the product of the P's determined from pairwise interactions.

(2) *Simulation procedure.* The Monte Carlo computation algorithm that the authors use is similar to that of Alexandrowicz (1971). Proteins are added successively to the system at positions chosen randomly. Though random, the addition process is biased by the statistical weights (P's) for proteins to occupy currently open sites, based on the proteins already present. In principle, this gives results identical to those obtained by the standard Metropolis (1953) Monte Carlo algorithm.

(3) *Results.* The model is sensitive to changes in protein size, protein/lipid ratio, and the assumed interaction potential between protein molecules in the lattice. The authors have determined three quantities of interest: (1) the protein radial distribution function, $g(r)$, (2) an aggregation state for protein molecules, and (3) the extent of protein domains, protein and annular lipid domains, and bulk lipid domains. Annular lipid is defined as lattice points in the first layer around a protein that are not occupied by another protein. The radial distribution function is the quantity of most fundamental physical importance; we shall concentrate on it here.

Freire and Snyder examine the behavior of $g(r)$ at constant protein size (protein side length, three sites) under changes in protein/lipid ratio and the sign and magnitude of the interaction potential. Within the approximations of the model, only the relative values of $g(r)$ have physical significance. Increasing the protein concentration increases the short-ranged order in the system, as measured by deviations of $g(r)$ from 1. In Fig. 17 it is shown that both attractive and repulsive neighbor interactions increase the overall order in the system over that seen in their absence under otherwise similar conditions. Attractions lead to greater coordination at close (nearest neighbor) separations while repulsions shift the coordi-

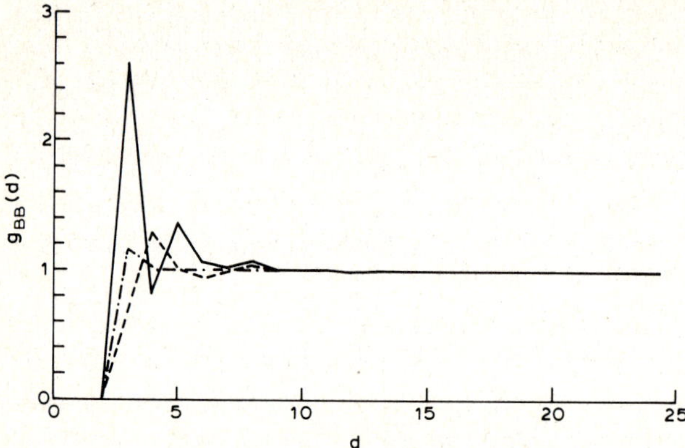

Fig. 17. Radial distribution functions computed by Freire and Snyder (1982) for three values of the statistical weight for protein neighbor interactions: $P = 5$ (— attractive), $P = 1$ (—·— no interaction), and $P = 0.2$ (--- repulsive). The units on the abcissa are lattice sites, with lipids taking up one site and proteins represented as hexagons with side length three. From Freire and Snyder (1982).

nation shell to greater radial separations. Sufficiently strong attractions can lead to the formation of large protein clusters (probably phase separation).

These results are all straightforward consequences of the model. The analysis of clustering leads to some more interesting results that are relevant to lateral diffusion, a subject we are not reviewing.

The strength of the work is that it implements a Hamiltonian exactly. The major weakness is that the model is ultra-simplified, with a parameter (P, for neighbor interactions) that is very hard to connect to experiment.

iii. Lookman et al. (1982) (1) *Description of theory.* This computer simulation analyzes protein-protein interactions and lateral phase separation for a version of Pink's lattice model with a Hamiltonian that differs from the one discussed in Section II.5. in three ways. First, the proteins are represented as hexagons with side lengths of three sites, and they are restricted to positions on a triangular lattice that is superimposed on the lattice of lipid chains (see Fig. 18). Second, the description of the ground and excited states of the lipids is somewhat different. Third, the lateral distribution of proteins is governed not by the lattice Hamiltonian, but instead by a heuristic algorithm that is intended to simulate protein diffusion.

(2) *Simulation procedure.* Transitions between lipid states are done according to the standard Metropolis Monte-Carlo procedure (Metropolis et al., 1953). Proteins are moved to realize a hypothesis made originally by Brûlet and McConnell (1976): a protein diffuses only through regions of fluid lipid; if the bulk of the lipid is solid, the protein may still diffuse by inducing a fluid state in the lipids near its boundary (the icebreaker hypothesis).

One of the six superlattice sites about a given protein is randomly chosen. If that site is occupied by another protein or by lipids that are not sufficiently fluid, no move is made. If the lipids are sufficiently fluid, the protein is moved to the new site and the displaced lipids are exchanged into the vacated protein site. The fluidity criterion for movement is that half the lipids are in the excited state, averaged over the recent history of the region. Not only lipids in the new protein site, but also the interstitial lipids between the two sites (Fig. 18) are considered.

The parameters for interactions between neighboring protein and lipid are adjusted, on the basis of DMR results, so that proteins tend to stabilize the excited lipid state. No parameter for protein-protein interactions is used, since protein motion is not governed directly by the lattice Hamiltonian. The lipid parameters are chosen to simulate DPPC (T_c = 41°C).

The cost of heuristically simulating a dynamic process in this paper is to muddy the definition of the underlying Hamiltonian for the system. Interpretation of the results is made correspondingly more difficult. These statements apply primarily to the results on protein distribution in this paper and also, to a lesser extent, to the results on lipid order.

(3) *Results.* Representative configurations are shown in Fig. 18, and an approx-

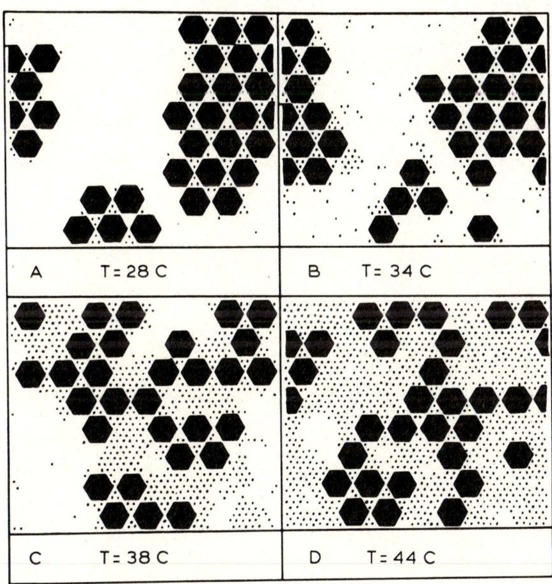

Fig. 18. Typical distributions of proteins and lipids in computer simulation by Lookman et al. (1982). The hexagons are proteins, the dots are excited-state lipids, and the blank areas ground-state lipids. At low temperatures the proteins are highly clustered, while at high temperatures they are more uniformly distributed. The lipid in this system simulates DPPC (T_c = 41°C). The mole fraction of protein is 0.028. The protein-lipid coupling parameters are chosen so that the proteins interact more favorably with excited-state than ground-state lipids. From Lookman et al. (1982).

imate phase diagram is shown in Fig. 19. For $T > T_c$, the system appears to consist of a single phase for all protein concentrations. The lipid is fluid (mostly excited chains); the protein lateral distribution shows no large-scale inhomogeneity.

Below a temperature designated T_K, which is ≈32°C in this case, the system separates into two phases: pure solid lipid and protein at its maximum lateral density, with associated interstitial lipid. For $T_K < T < T_c$ there is a two-phase region for sufficiently low protein concentrations (solid lipid plus a protein-rich fluid phase of variable composition). In the same temperature range, at sufficiently high protein concentrations, only the protein-rich fluid phase exists. Since these conclusions are obtained for a small system (at most 64 proteins), the authors note that extrapolation to the phase behavior of an infinite system must be approximate.

What causes the protein aggregation and phase separation? Lipids near the proteins tend to be fluid, and the proteins are constrained to move into fluid but not solid regions. The algorithm for moving proteins can very loosely be viewed as defining rate constants for the motion of proteins into fluid and solid regions, with the ratio being a partition coefficient. For this model, at low temperatures and appropriate protein concentrations, there should be lateral phase separation into solid lipid and fluid protein-rich phases. This is what the authors observe. A rigorous Monte Carlo simulation of the entire system presumably would have yielded similar results, unless a very unfavorable parameter for protein-protein interactions had been used.

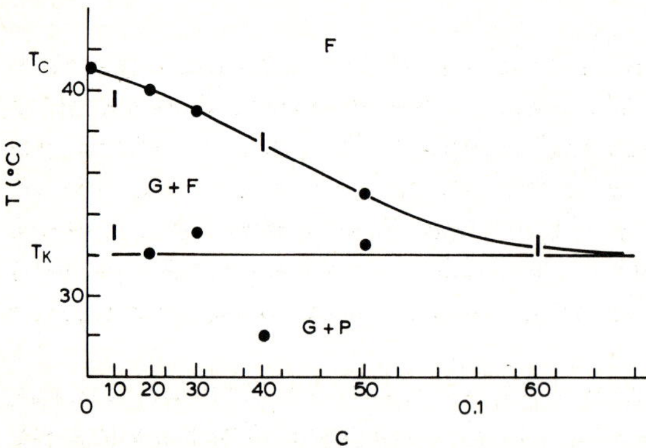

Fig. 19. Approximate phase diagram for a DPPC-protein bilayer, obtained by Lookman et al. (1982). The horizontal axis is protein concentration, expressed either as mole fraction of protein (lower numbers) or as number of protein molecules per lattice of 1600 sites (upper numbers). Parameters are the same as in the previous figure. From Lookman et al. (1982).

V. Thermodynamics of partition between membrane and water

1. OVERVIEW

Up until now, this review has dealt exclusively with interactions within the membrane. The partitioning of protein between membrane and water is also a problem worthy of note; we consider it here. Experimental work in this area has been comprehensively reviewed (e.g., Rothman and Lenard, 1977; Wickner, 1980); we will summarize a few of the important results.

Since protein synthesis occurs in the cytoplasm, membrane proteins must find their way into the membrane. For many proteins, insertion is accomplished co-translationally. For others, insertion is post-translational. Depending on the protein, this can occur without specific catalysis and may require transmembrane potentials.

The insertion itself is invariably asymmetric, irrespective of detailed mechanism. That is, for a particular type of protein, a constant orientation is established and maintained over long periods. It may be possible, however, to induce a change in the orientation of certain proteins by a reversal in membrane polarity (Kempf et al., 1982).

The detailed processes are both complicated and varied. Therefore, the goal of theory in this area is primarily to determine and quantify the important forces involved, particularly in the establishment and maintenance of an asymmetric membrane bound state. We consider two papers: Jähnig (1983) explores the forces involved in partitioning; Weinstein et al. (1982) consider protein asymmetry.

2. PARTITIONING. JÄHNIG (1983)

a. Description of theory

Jähnig (1983) identifies four major factors that affect the thermodynamics of partitioning of protein between aqueous and membrane phases. They are (i) the hydrophobic effect (Tanford, 1980), and changes in (ii) hydrogen bonding, (iii) protein entropy, and (iv) lipid order*. The final two effects are novel. We will consider each of these factors in turn, using Jähnig's numerical values to assess their importance. These values were computed for a small peptide, assuming either α-helical or totally hydrated random coil configurations, as summarized in Table 2.

The exposure of nonpolar amino acid sidechains to solvent provides a strong, mostly entropic, force favoring partition into the membrane. This is a manifestation of the hydrophobic effect. Jähnig chooses, after Richards (1977), to compute

* An additional factor which may be important is 'buried charge'. There is an energetic cost for transferring charge from the high-dielectric aqueous medium to the low-dielectric membrane.

TABLE 2

The free energy contributions of various forces involved in the water to membrane partitioning of a 20 residue hydrophobic peptide (Jähnig, 1983). The peptide conformation in the membrane is assumed to be α-helical; the aqueous conformation is given when important.

Force	ΔG *(kcal/mol)*	*peptide conformation (in solution)*
Hydrophic effect	−35	α-helix
	−55	random coil
Hydrogen bonding	0	
Internal protein entropy	0	α-helix
	+24	random coil
External protein entropy	+16	
Lipophobic effect	+2	
Total	−17	α-helix
	−13	random coil

the free energy change for water → membrane transfer on the basis of water/protein interfacial energies. The results are large and negative and are assumed to be entirely entropic. Unfortunately, estimating exposed areas (and thus total hydrophobicities) is subject to a great deal of ambiguity (Richards, 1977), leaving the analysis only semiquantitative. It can be said, however, that this force is felt most strongly by proteins that are denatured and thus have many nonpolar moieties exposed to water.

The large energy involved in hydrogen bonding is assumed to require that no net hydrogen bonds are broken upon transfer from the aqueous to the membrane phase. This requires that hydrogen bonds between protein and water be replaced by intra-protein bonds. Backbone (amide) hydrogen bonding can be assumed to be complete both in the water and for helical transmembrane segments. If only hydrophobic side chains are present (a reasonable assumption for the buried part of a membrane protein), hydrogen bonds between water and sidechains may be ignored. Thus, hydrogen bonding itself is not assumed to play an important role in protein partitioning.

Hydrogen bonding may, however, cause changes in the internal state of the protein that do give rise to more notable energetic effects. For molecules that are helical in both water and hydrocarbon, no hydrogen bonds need change and there is no net effect. Denatured proteins must, however, undergo a major structural change to adopt a conformation rich in hydrogen bonded secondary structure (e.g., α-helix) upon entering the membrane. This results in a considerable decrease in entropy and hence a large positive free energy change. Apparently fortuitously, this change is of such a sign and magnitude that it approximately cancels the additional *hydrophobic* free energy lost for coil to helix transitions. Thus, the net effects of hydrophobicity, hydrogen bond conservation, and internal protein state are largely independent of initial and final protein conformation.

The calculations to this point are an important part of the total analysis, and so deserve careful consideration. Jähnig has considered molecules that are either completely α-helical or totally hydrated random coils in solution. With this assumption, the effects of hydrophobicity and conformational restriction cancel to within 0.2 kcal/mol per residue (by Jähnig's calculations), an amount that would be significant for any real-sized peptide. More importantly, if the true conformation in solution is globular, the hydration of hydrophobic sidechains may be less than in the helix; the two effects may then be additive rather than compensatory.

While protein partitioning out of the aqueous phase results in an increase in water entropy, it is also accompanied by a considerable loss in protein mobility, and hence entropy. The size of this effect may be properly determined only via a statistical-mechanical treatment. In analogy with an ideal gas, the loss of translational freedom may be likened to transfer from an unbound volume, V_{free} (corresponding to the aqueous phase), to a bound volume, V_{bound} (corresponding to the membrane phase); hence for N molecules.

$$\Delta G_{\text{t}} = NkT \ln(V_{\text{free}}/V_{\text{bound}}) \tag{44}$$

The loss of rotational freedom may be similarly expressed. These calculations yield a positive free energy change with about one-half the magnitude of that determined for the hydrophobic effect. The concomitant enthalpy change is quite small.

As has been noted previously, the effects of protein molecules on lipid structure may be far reaching. This interaction, dubbed the 'lipophobic effect,' must surely be considered. In principle this should be possible using any of the lipid-protein interaction theories. However, the Landau theory proves to be a particularly easy approach. Jähnig finds that at low protein concentrations (i.e., in the analytical range of his linear approximations (Jähnig, 1981)), large entropic and enthalpic effects approximately cancel to give a very small free energy change, 1–2 kcal/mol. Similar estimates of the free-energy changes have been obtained by Owicki and McConnell (Fig. 14) and by Pearson et al. (1984). Thus, lipid effects do not seem to be as important a driving force as might have been thought a priori.

b. Results

These effects and any others combine to yield the total free energy for partitioning from the aqueous phase into the membrane. For membrane proteins, the membrane phase is favored and so we expect (and see) a large negative free energy change for this process. Jähnig's α-helix of 20 residues is seen to undergo a free energy change of between −13 and −17 kcal/mol, in reasonable agreement with experiment. It should be stressed that, due to the semiquantitative treatment of these effects, this result is subject to a great deal of uncertainty. However, it probably does provide a reasonable estimation of the magnitude of the total process and of the relative importance of the various contributions.

From this analysis, the hydrophobic effect provides the main driving force for insertion. This result follows the conventional wisdom. However, Jähnig also demonstrates that this force is significantly reduced by the entropy lost upon protein immobilization; this result is new and so might require modification of older results formulated without it (e.g., Engelman and Steitz, 1981).

3. ASYMMETRY. WEINSTEIN ET AL. (1982)

a. Description of theory

Weinstein et al. (1982) considered the related area of protein asymmetry. Noting that the (ubiquitous) presence of transmembrane potentials provides a strong force tending to orient an asymmetrically charged membrane protein, they developed a simple thermodynamic model based on electrostatics. Numerical results and experimental data were used to assess its validity.

In the model, it is necessary to assume or identify a charge asymmetry across the membrane. Only charges near the membrane are considered; multiple charged species are assumed to act as a cluster and exert a net effect. These interact with the 'microscopic' membrane potential, whose value gives the voltage difference between inner and outer membrane surfaces. It is then a trivial calculation to determine the free energy of orientation using the number of charges in each cluster and the microscopic membrane potential (Eisenberg and Crothers, 1979). The calculation is restricted to orientations that traverse the bilayer once, leaving the N and C protein termini on opposite sides; the generalization to multiple traversals is obvious. The pertinent equation expressing the free energy difference between correct and incorrect orientations is

$$\Delta G = zFV_{\mathrm{m}} \tag{45}$$

where z is the total charge, F is Faraday's constant, and V_{m} is the microscopic membrane potential. Because the membrane potential may be expressed in terms of a variety of parameters – ionic strength, pH, lipid composition and charge, among others – the model is easily adaptable to a variety of physical situations.

The free energy may be related to the probability of proper orientation via the Boltzmann expression

$$\frac{\text{correct}}{\text{incorrect}} = \exp(-\Delta G/kT) \tag{46}$$

Modest membrane potentials and clusters of as few as three elementary charges give rise to extremely polarized assemblies.

b. Results

These predictions were compared with known protein sequences; the results are displayed in Fig. 20. These five proteins, which all traverse the membrane one time, contain (predominantly) positive charge clusters on the cytoplasmic side and (predominantly) negative charge clusters on the extra-cytoplasmic side. As the cytoplasmic potential is negative relative to the extra-cytoplasmic surface, these findings are in agreement with expectation based on this model. Bacteriorhodopsin, which is thought to span the membrane seven times (Stoeckenius et al., 1979), does not fit.

4. CONCLUSIONS

These papers play an important defining role in the study of protein partitioning. They have elaborated many of the important thermodynamic forces at work in

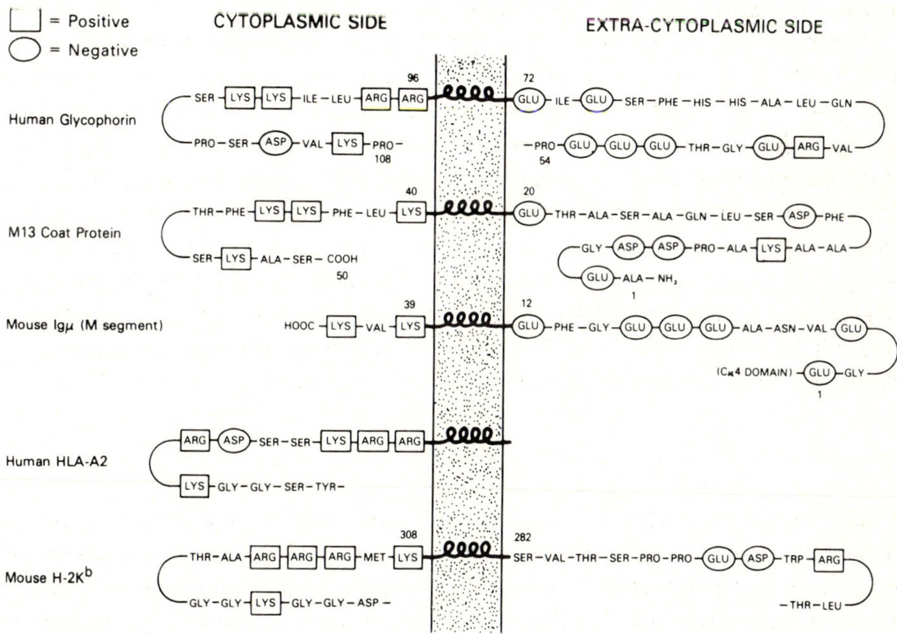

Fig. 20. Charge clusters in five simple transmembrane proteins. The precise placement of the proteins relative to the membrane is somewhat arbitrary. From Weinstein et al. (1982).

the establishment and maintenance of an asymmetric membrane bound protein state. In addition, they have made an assessment of their relative importance.

Before concluding this section, it is important to comment on the compatibility of thermodynamic and kinetic treatments of this or any area. The thermodynamic outcome of a process is the most favorable from a free energy standpoint and is necessarily attained in the limit of long times. However, processes can follow pathways with initial outcomes different from those predicted thermodynamically; these are kinetic results. Kinetic results are not stable indefinitely, although reversion may not happen on the time scale of biological interest.

Thermodynamic arguments are nevertheless interesting because they predict real forces at work in a system that must be overcome, irrespective of the mechanism. They may also be compatible with the specific mechanism; for example, many of the observations mentioned in the overview are in agreement with predictions made by the theories.

VI. Recommendations for future work

Writing this review has given us a chance to gain some perspective on the application of physical theory to problems involving proteins and lipids in membranes. We have developed some strong opinions about the kind of further research that will prove fruitful, as well as some opinions about approaches that will not be particularly useful.

We see little need for more mean-field treatments of protein-lipid interactions that are based on the molecular order that changes characteristically at the lipid bilayer phase transition. The theories in Section II cover this approach well and are in broad agreement. We do see at least three promising directions for related work to take.

First, it would be most useful to understand the direct interactions between a phospholipid and a membrane protein at the detailed molecular level. These interactions have been assumed (rather than derived) to date. Moreover, their molecular nature has generally been rather vaguely defined. Molecular mechanics, or computer modeling of molecular interactions with very accurate potential-energy functions (Weiner et al., 1984, and references therein), has been used to analyze enzyme-substrate interactions and may be helpful here. The method is by now advanced enough to make this feasible, at least for a transbilayer helical protein and a single phospholipid.

Second, existing theories can be expanded to include other types of perturbations. The paper by Jähnig et al. (1982) on lipid tilt and conformational order is a step in this direction. There is considerable literature on director fluctuations in liquid crystals (de Gennes, 1974) that probably can be applied here.

Third, the intermolecular interactions can be modeled more accurately by removing the constraints of lattices and mean-field analysis. Analytical analyses are always preferable to numerical ones, all else being equal, but we nevertheless believe that computer simulations offer the best hope for advancement in this regard.

At a more technical level, we would like to reiterate the importance of obtaining phase diagrams whenever possible, both in theoretical and experimental treatments. For experiments, it is also important to determine whether thermodynamic equilibrium is established.

The work treating lipid effects on protein activity would benefit from a thorough analysis of lateral phase separation. Questions of protein partition between lipid phases and protein activation by specific lipids might be addressed by generalizations of some of the theories in Section II, as we noted in Section III.

The analysis of interactions between membrane proteins is a biologically important field that is virtually virgin territory for both experimentalists and theorists. We think that freeze-fracture electron microscopy, as discussed in Section IV, is the experimental technique of choice for many such problems. Two technical improvements need to be made in reconstitution methodology to facilitate the theoretical analysis of the images. The proteins should be incorporated into larger (hence less curved) bilayers, and a systematic way should be developed to obtain highly asymmetric insertion of the reconstituted proteins. To date the latter problem has been dealt with on an empirical protein-by-protein basis.

So far, theoretical treatments of protein-protein interactions have involved the lipid-mediated interactions (Section IV.3) and the statistical-mechanical apparatus for obtaining information on intermolecular potentials or forces from electron micrographs (Section IV.2). There is substantial room for improvement in both these areas. Nor have other types of interactions (e.g., electrostatic) been characterized very well theoretically in protein-lipid systems.

Finally, we consider the prospects for understanding the partition of proteins between water and membranes, and the orientation of the proteins. Many of the important factors have been identified, but uncertainties in free-energy estimates will have to be reduced before decent experimental tests can be made.

Acknowledgements

We are pleased to acknowledge helpful conversations with B. Alder, S. Chan, E. Helfand and T. O'Leary. S. Chan kindly sent us a preprint of the paper by Pearson et al. (1984) We also wish to thank J. Braun and B. Scalettar for critical readings of the manuscript. J. Abney is an N.I.H. predoctoral trainee.

References

Alexandrowicz, Z. (1971) Stochastic models for the statistical description of lattice systems. J. Chem. Phys. 55, 2765–2779.

Braun, J., Abney, J.R. and Owicki, J.C. (1984) How a gap junction maintains its structure. Nature, 310, 316–318

Brotherus, J.R., Griffith, O.H., Brotherus, M.O., Jost, P.C., Silvius, J.R. and Hokin, L.E. (1981) Lipid-protein multiple binding equilibria in membranes. Biochemistry 20, 5261–5267.

Brûlet, P. and McConnell, H. (1976) Protein-lipid interactions: glycophorin and dipalmitoylphosphatidylcholine. Biochem. Biophys. Res. Commun. 68, 363–368.

Caille, A., Pink, D., de Verteuil, F. and Zuckermann, M.J. (1980) Theoretical models for quasi-two-dimensional mesomorphic monolayers and membrane bilayers. Can. J. Phys. 58, 581–611.

Chapman, D., J. Gomez-Fernandez and F. Goni, 1979 Intrinsic Protein-Lipid Interactions. Physical and Biochemical Evidence. FEBS Lett. 98, 211–223.

de Gennes, P.G. (1974) *The Physics of Liquid Crystals*. Oxford University Press, London.

Dodds, J.A. (1975) Simplest statistical geometric model of the simplest version of the multicomponent random packing problem. Nature 256, 187–189.

Donnel, J.T. and Finegold, L. (1981) Testing of aggregation measurement techniques for intramembranous particles. Biophys. J. 35, 783–798.

Edelman, J.B. (1978) *Statistical Mechanics of Biological Membranes: Protein Aggregation and Lipid Ordering*. Ph.D. Thesis, California Institute of Technology, Pasadena.

Eisenberg, D. and Crothers, D. (1979) *Physical Chemistry with Applications to the Life Sciences*. Benjamin/Cummings, Menlo Park, California.

Engelman, D.M. and Steitz, T.A. (1981) The spontaneous insertion of proteins into and across membranes: the helical hairpin hypothesis. Cell 23, 411–422.

Finegold, L. (1976) Cell membrane fluidity: molecular modeling of particle aggregations seen in electron microscopy. Biophys. Biochim. Acta 448, 393–398.

Freire. E. and Snyder, B. (1980a) Estimation of the lateral distribution of molecules in two-component lipid bilayers. Biochemistry 19, 88–94.

Freire, E. and Snyder, B. (1980b) Monte Carlo studies of the lateral organization of molecules in two-component lipid bilayers. Biochim. Biophys. Acta 600, 643–654.

Freire, E. and Snyder, B. (1982) Quantitative characterization of the lateral distribution of membrane proteins within the lipid bilayer. Biophys. J. 37, 617–624.

Gershon, N.D., Demset, A. and Stackpole, C.W. (1979) Analysis of local order in the spatial distribution of cell surface molecular assemblies. Exp. Cell Res. 122, 115–126.

Helfand, E. (1983) Brownian dynamics simulation of systems with transitions, particularly conformational transitions in chain molecules. Physica A 118, 123–135.

Helfand, E., Frish, H. and Lebowitz, J. (1961) Theory of the two- and one-dimensional rigid sphere fluids. J. Chem. Phys. 34, 1037–1042.

Hill, T.L. (1956) *Statistical Mechanics*. McGraw-Hill, New York, NY.

Houslay, M.D. and Stanley, K.K. (1982) *Dynamics of Biological Membranes*. John Wiley & Sons, New York, NY.

Jackson, M.B. (1976) A beta-coupled gauche kink description of the lipid bilayer phase transition. Biochemistry 15, 2555–2561.

Jähnig, F. (1979) Molecular theory of lipid membrane order. J. Chem. Phys. 70, 3279–3290.

Jähnig, F. (1981a) Critical effects from lipid-protein interaction in membranes. I. Theoretical description. Biophys. J. 36, 329–345.

Jähnig, F. (1981b) Critical effects from lipid-protein interaction in membranes. II. Interpretation of experimental results. Biophys. J. 36, 347–357.

Jähnig F. (1983) Thermodynamics and kinetics of protein incorporation into membranes. Proc. Natl. Acad. Sci. USA 80, 3691–3695.

Jähnig, F. and Bramhall, J. (1982) The origin of a break in Arrhenius plots of membrane processes. Biochim. Biophys. Acta 690, 310–313.

Jähnig, F., Vogel, H. and Best, L. (1982) Unifying description of the effect of membrane proteins on lipid order. Verification for the melittin/dimyristoylphosphatidylcholine system. Biochemistry 21, 6790–6798.

James, R. and Branton, D. (1973) Lipid- and temperature-dependent structural changes in *Acholeplasma laidlawii* cell membranes. Biochim. Biophys. Acta 323, 378–390.

Janiak, M., Small, D. and Shipley, G. (1976) Nature of the thermal pretransition of synthetic phospholipids: dimyristoyl- and dipal-mitoyllecithin. Biochemistry 15, 4575–4580.

Jost, P.C., Griffith, H., Capaldi, R.A. and Vanderkooi, G. (1973) Evidence for boundary lipid in membranes. Proc. Natl. Acad. Sci. USA 70, 480–484.

Jost, P.C. and Griffith, O.H. (1980) The lipid-protein interface in biological membranes. Ann. N.Y. Acad. Sci. 348, 391–407.

Kempf, C., Klausner, R.D., Weinstein, J.N., Van Renswoude, J., Pincus, M. and Blumenthal, R. (1982) Voltage-dependent transbilayer orientation of melittin. J. Biol. Chem. 257, 2469–2476.

Kimelberg, H.K. and Papahadjopoulos, D. (1974) Effects of phospholipid acyl chain fluidity, phase transitions, and cholesterol on sodium, potassium-stimulated adenosine triphosphatase. J. Biol. Chem. 249, 1071–1080.

Kleeman, W. and McConnell, H. (1976) Interactions of proteins and cholesterol with lipids in bilayer membranes. Biochim. Biophys. Acta 419, 206–222.

Klein, R.A. (1982) Thermodynamics and membrane processes. Q. Rev. Biophys. 15, 667–757.

Landau, L. and Lifshitz, E. (1969) *Statistical Physics,* 2nd. Edn., pp. 429–433. Addison-Wesley, Reading, MA.

Lee, A.G., (1977a) Lipid phase transitions and phase diagrams. I. Lipid phase transitions. Biochim. Biophys. Acta 472, 237–281.

Lee, A.G. (1977b) Lipid phase transitions and phase diagrams. II. Mixtures involving lipids. Biochim. Biophys. Acta 472, 285–344.

Lee, A.G. (1983) An overlapping site model for the lipid annulae of membrane proteins. FEBS Lett. 151, 297–302.

Lewis, G.N. and Randall, M. (1961) *Thermodynamics,* 2nd. Edn., pp. 282–290. McGraw-Hill, New York, NY. (Revised by K. Pitzer and L. Brewer).

Linden, C.D., Wright, K.L., McConnell, H.M. and Fox, C.F. (1973) Lateral phase separations in membrane lipids and the mechanism of sugar transport in *Escherichia coli.* Proc. Natl. Acad. Sci. USA 70, 2271–2275.

Loewenstein, W.R. (1981) Junctional intercellular communications: the cell-to-cell membrane channel. Physiol. Rev. 61, 829–913.

Lookman, T., Pink, D.A., Grundke, E.W., Zuckermann, M.J. and de Verteuil, F. (1982) Phase separation in lipid bilayers containing integral proteins. Computer simulation studies. Biochemistry 21, 5593–5601.

Lumry, R. and Rajender, S. (1970) Enthalpy-entropy compensation phenomena in water solutions of proteins and small molecules: a ubiquitous property of water. Biopolymers 9, 1125–1227.

Maier, W. and Saupe, A. (1959) Eine einfache molekular-statistische Theorie der nematischen kristallinflussigen Phase. Teil I. Z. Naturforsch. A 14, 882–889.

Marčelja, S. (1974) Chain ordering in liquid crystals. II. Structure of bilayer membranes. Biochim. Biophys. Acta 367, 165–176.

Marčelja, S. (1976) Lipid-mediated protein interaction in membranes. Biochim. Biophys. Acta 455, 1–7.

Markovics, J., Glass, L. and Maul, G.G. (1974) Core patterns on nuclear membranes. Exp. Cell Res. 85, 443–451.

Marsh, D. (1983) Spin-label answers to lipid-protein interactions. Trends Biochem. Sci. 8, 330–333.

Marsh, D. and Watts, A. (1982) Spin labeling and lipid-protein interactions in membranes. In: *Lipid-Protein Interactions*, Vol. 2, pp. 53–126. Editors: P. Jost and O. Griffith. Wiley-Interscience, New York, NY.

McQuarrie, D.A. (1976) *Statistical Mechanics.* Harper and Row, New York, NY.

Melhorn, R.J. and Packer, L. (1976) Analysis of freeze-fracture electron micrographs by a computer-based technique. Biophys. J. 16, 613–625.

Metropolis, N., Rosenbluth, A., Rosenbluth, M., Teller, A. and Teller, E. (1953) Equations of state calculations by fast computing machines. J. Chem. Phys. 27, 720–733.

Mitaku, S., Jipo, T. and Kataoka, R. (1983) Thermodynamic properties of the lipid bilayer transition. Pseudocritical phenomena. Biophys. J. 42, 137–144.

Nagle, J. (1980) Theory of the main lipid bilayer phase transition. Ann. Rev. Phys. Chem. 31, 157–195.

O'Leary, T.J. (1983) A simple theoretical model for the effects of cholesterol and polypeptides on lipid membranes. Biochim. Biophys. Acta 731, 47–53.

Owicki, J.C. and McConnell, H.M. (1979) Theory of protein-lipid and protein-protein interactions in bilayer membranes. Proc. Natl. Acad. Sci. USA 76, 4750–4754.

Owicki, J.C., Springgate, M.W. and McConnell, H.M. (1978) Theoretical study of protein-lipid interactions in bilayer membranes. Proc. Natl. Acad. Sci. USA 75, 1616–1619.

Pearson, L.T., Chan, S.I., Lewis, B. and Engelman, D.M. (1983) Pair distribution functions of bacteriorhodopsin and rhodopsin in model bilayers. Biophys. J. 43, 167–174.

Pearson, L.T., Edelman, J. and Chan, S.I. (1984) Statistical mechanics of lipid membranes: protein correlation functions and lipid ordering. Biophys. J., 45, 863–871.

Pearson, R.P., Hui, S.W. and Stewart, T.P. (1979) Correlative statistical analysis and computer modelling of intramembraneous particle distributions in human erythrocyte membranes. Biochim. Biophys. Acta 557, 265–282.

Perelson, A.S. (1978) Spatial distribution of surface immunoglobulin on B lymphocytes: local ordering. Exp. Cell Res. 112, 309–321.

Pink, D.A. and Carroll, C.E. (1978) A model of cholesterol in lipid bilayers. Phys. Lett. 66A, 157–160.

Pink, D.A. and Chapman, D. (1979) Protein-lipid interactions in bilayer membranes: A lattice model. Proc. Natl. Acad. Sci. USA 76, 1542–1546.

Richards, F.M. (1977) Areas, volumes, packing, and protein structure. Ann. Rev. Biophys. Bioeng. 6, 151–176.

Rothman, J.E. and Lenard, J. (1977) Membrane asymmetry. Science 195, 743–753.

Sandermann, H. Jr. (1978) Regulation of membrane enzymes by lipids. Biochim. Biophys. Acta 515, 209–237.

Schroeder, H. (1977a) Molecular statistical theory for inhomogeneous nematic liquid crystals with boundary conditions. J. Chem. Phys. 67, 16–25.

Schroeder, H. (1977b) Aggregation of proteins in membranes. An example of fluctuation-induced interactions in liquid crystals. J. Chem. Phys. 67, 1617–1619.

Scott, H.L. Jr. (1980) Scaled particle theory of hard rods in two dimensions: effect of variable rod size and attractive interactions. Phys. Rev. A 21, 2082–2086.

Scott, H.L. Jr. (1981) Phase transitions in lipid bilayers. A theoretical model for phosphatidylethanolamine and phosphatidic acid bilayers. Biochim. Biophys. Acta 648, 129–136.

Scott, H.L. and Cheng, W. (1979) A theoretical model for lipid mixtures, phase transitions, and phase diagrams. Biophys. J. 28, 117–132.

Scott, H.L. and Coe, T.J. (1982) A theoretical study of lipid-protein interactions in membranes. Biophys. J. 42, 219–224.

Seelig, J., Seelig, A. and Tamm, L. (1982) Nuclear magnetic resonance and lipid-protein interactions. In: *Lipid-Protein Interactions*, Vol. 2, pp. 127–148. Editors: P. Jost and O. Griffith. Wiley-Interscience, New York, NY.

Silvius, J. (1982) Thermotropic phase transitions of pure lipids in model membranes and their modification by membrane proteins. In: *Lipid-Protein Interactions*, Vol. 2, pp. 239–281. Edited by P. Jost and O. Griffith. Wiley-Interscience, New York, NY.

Silvius, J.R. and McElhaney, R.N. (1981) Non-linear Arrhenius plots and the analysis of reaction and motional rates in biological membranes. J. Theor. Biol. 88, 135–152.

Singer, S. and Nicolson, G. (1972) The fluid mosaic model of the structure of cell membranes. Science 175, 720–731.

Snyder, B. and Freire, E. (1980) Compositional domain structure in phosphatidylcholine-cholesterol and sphingomyelin-cholesterol bilayers. Proc. Natl. Acad. Sci. USA 77, 4055–4059.

Sokolnikoff, I. and Redheffer, R. (1966) *Mathematics of Physics and Modern Engineering*, p. 352. McGraw-Hill, New York, NY.

Stoeckenius, W., Lozier, R.H. and Bogomolni, R.A. (1979) Bacteriorhodopsin and the purple membrane of halobacteria. Biochim. Biophys. Acta 505, 215–278.

Tanford, C. (1980) *The Hydrophobic Effect: Formation of Micelles and Biological Membranes*, p. 233. Wiley-Interscience, New York, NY.

Valleau, J. and Whittington S. (1977) A guide to Monte Carlo for statistical mechanics: 1. Highways. In: *Statistical Mechanics, Part A: Equilibrium Techniques.* Editors: B. Berne Plenum Press, New York, NY.

Weiner, S., Kollman, P., Case, D., Singh, U., Ghio, C., Alagona, G., Profeta, S. Jr. and Weiner, P. (1984) A new force field for molecular mechanical simulation of nucleic acids and proteins. J. Am. Chem. Soc. 106, 756–784.

Weinstein, J.N., Blumenthal, R., van Renswoude, J., Kempf, C. and Klausner, R.D. (1982) Charge clusters and the orientation of membrane proteins. J. Membr. Biol. 66, 203–212.

Wickner, W. (1980) Assembly of proteins into membranes. Science 210, 861–868.

Wood, W. and Erpenbeck, J. (1976) Molecular dynamics and Monte Carlo calculations in statistical mechanics. Ann. Rev. Phys. Chem. 27, 319–348.

Wu, S. and McConnell, H. (1975) Phase separations in phospholipid membranes. Biochemistry 14, 847–854.

Note in proof

As is inevitable in writing a review of this magnitude, existing papers are inadvertently overlooked while new papers are constantly appearing. We would like to mention three such papers here. In the first, a lattice model based on that presented in Section II.5 was extended to the examination of lipid-mediated protein interactions (Tessier-Lavigne, M., Boothroyd, A., Zuckermann, M.J. and Pink, D.A. (1982) Lipid-mediated interactions between intrinsic molecules in bilayer membranes. J. Chem. Phys. 76, 4587–4599). It was reported that this interaction was of relatively short range and displayed different characteristics above and below the main lipid phase transition. Secondly, a more recent paper by the same group (Jan, N., Lookman, T. and Pink, D.A. (1984) On computer simulation methods used to study models of two-component lipid bilayers. Biochemistry 23, 3227–3231) corrected an error in the simulation algorithm employed by Freire and Snyder (1982), Section IV.3.b.ii of this article, to generate equilibrium properties of a protein-containing bilayer. The error manifests itself most strongly in Freire and Snyder's analysis of protein clustering, and should not significantly alter the qualitative results on the protein distribution function, $g(\mathbf{r})$, that we have presented here. Finally, Mouritsen and Bloom (Mouritsen, O.G. and Bloom, M. (1984) Mattress model of lipid-protein interactions in membranes. Biophys. J. 46, 141–153) have developed a thermodynamic model of the bulk phase behavior of lipid-protein systems. It is formulated around the idea that hydrophobic and hydrophilic regions of proteins and lipids must be matched in the membrane (a fundamental tenet of the Landau analysis, Section II.4), an effect that is measured in terms of an elastic energy cost to the bilayer. In the dilute protein limit, direct interactions between proteins may be ignored, so that with the inclusion of other lipid-protein interaction terms, real solution theory may be employed to describe the phase behavior of the system. Such behavior is expressed in terms of temperature-composition phase diagrams.

Watts/De Pont (Eds.)
Progress in Protein-Lipid Interactions

CHAPTER 2

Manifestations of lipid-protein interactions in deuterium NMR

MYER BLOOM[a] and IAN C.P. SMITH[b]

[a]*Department of Physics, University of British Columbia, Vancouver, BC V6T 2A6 and*
[b]*Division of Biological Sciences, National Research Council, Ottawa, K1A 0R6, Canada*

I. Introduction

1. DIFFRACTION AND NUCLEAR MAGNETIC RESONANCE ARE COMPLEMENTARY METHODS OF DETERMINING PROTEIN STRUCTURE

One of the most active areas of molecular biology is the effort to establish relationships between molecular structure and biological action and function. Usually, such questions regarding biological action and function are formulated in terms of protein structure. A majority of proteins operate in water and are very strongly influenced by 'protein-water' interactions. In the same way, lipid-protein interactions are expected to play an important role in the relationship between structure and action of integral membrane proteins.

The technique of X-ray diffraction has been responsible for almost all our knowledge on the geometrical structures of proteins up to now, e.g., detailed information on approximately 60 proteins, almost all of which are water soluble, was listed in a recent text on protein structure (Schultz and Schirmer, 1979, see their Table 5-2). By contrast, only a few integral membrane structures are known, and these have been determined with low resolution (Henderson, 1981; Unwin and Henderson, 1984). The reason for the relative paucity of information on integral membrane protein structures is that high resolution diffraction measurements of structure require protein crystals, i.e., the use of spatially ordered three-dimensional arrays of protein molecules. Systematic procedures for producing good three-dimensional crystals of integral membrane proteins have not yet been perfected; the successful elucidation of the structure of bacteriorhodopsin involved the imaginative use of image reconstruction methods on two-dimensional ordered arrays of bacteriorhodopsin molecules in their native membrane (Henderson and Unwin, 1975).

In recent years, high resolution proton nuclear magnetic resonance (^{1}H-NMR) has come into its own as a technique for determining the structure of water-soluble proteins. With the advent and refinement of two-dimensional Fourier Transform NMR (2D-FT-NMR), it has become possible to determine the average geometrical structure of small proteins in solution independently of diffraction methods. In the case of more complex proteins, high resolution ^{1}H-NMR has been used to check whether local structural features of proteins in solution are consistent with structures determined in protein crystals by X-ray diffraction. Structural information is obtained from the influence of the dipolar interaction between pairs of spins which varies as $Y_{2m}(\theta, \phi)/r^3$ for a given pair of spins, where Y_{2m} is a spherical harmonic and (r, θ, ϕ) are the spherical polar coordinates of the vector joining the pair. For proteins in water, the secular part ($m = 0$) of the dipolar interaction, which is proportional to $(3 \cos^2\theta - 1)$ is actually averaged to zero by the rapid isotropic tumbling of the protein molecules. As a result, the ^{1}H-NMR spectrum of proteins in water is primarily due to chemical shift and indirect spin-spin interactions rather than the much larger dipolar interactions. Thus, the spectral features are directly related to specific amino acids, i.e., to specific positions in the primary amino acid sequence of the proteins. Geometrical information is determined, using 2D-FT-NMR, through the influence of the dipolar interaction in producing correlated changes, as a function of the time between preparation and measurement pulses, in the intensities of ^{1}H-NMR lines associated with different amino acids. Thus, while diffraction methods exploit the translational symmetry of *ordered* arrays of protein molecules to extract structural information on individual proteins, 2D-FT high resolution NMR makes use of the rotational *disorder* associated with protein reorientation to obtain comparable information.

As illustrated in Fig. 1, the type of NMR techniques which are proving to be so powerful in the determination of the structure of proteins in solution cannot be applied to proteins in membranes at the present time. It is somewhat paradoxical that, while the successful use of diffraction methods is made difficult for proteins in membrane by their *low* translational symmetry, the high resolution methods are foiled by the *high* rotational order of individual proteins in bilayer structures. The reason for this is that the dipolar orientational order parameters,

$$S = \tfrac{1}{2}\langle 3 \cos^2\theta - 1\rangle \tag{1}$$

are not zero in non-isotropic environments such as phospholipid bilayer membranes. The resulting relatively large dipolar interactions between the myriad of

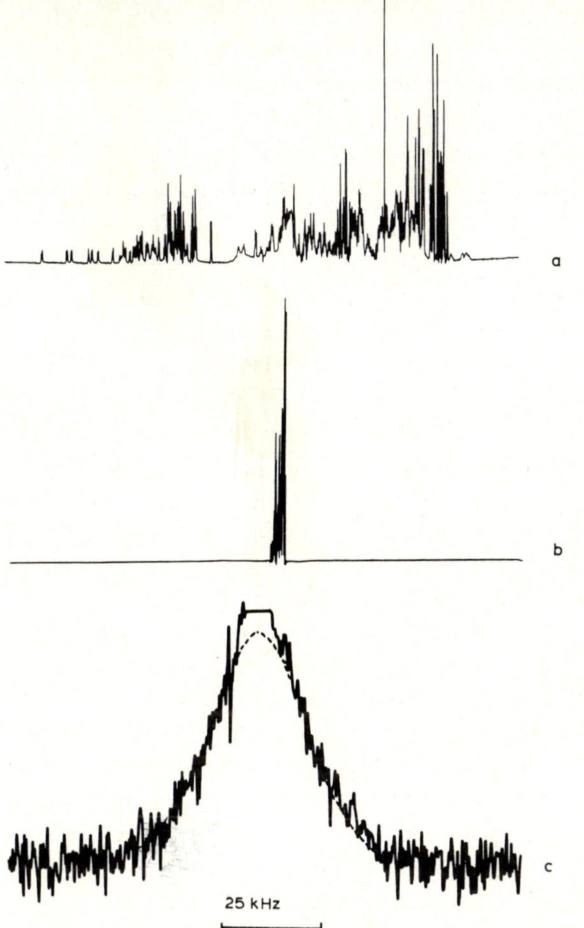

Fig. 1. Illustration of the different nature of the ^{1}H-NMR spectra of proteins in water and in membranes. (a) The 400 MHz ^{1}H-NMR spectrum of bovine pancreatic trypsin inhibitor, in aqueous solution, exhibits a very rich high resolution spectrum with an enormous number of sharp lines visible over a range of a few kHz (data courtesy of Dr. G.A. Gray, Varian Associates). (b) Spectrum of (a) on the scale of (c). (c) The ^{1}H-NMR spectrum of rhodopsin reconstituted in model membranes of dimyristoylphosphatidylcholine with perdeuterated acyl chains (DMPC-d_{54}) exhibits a broad featureless line approximately 25 kHz wide (MacKay et al., 1983). The high resolution lines are impossible to detect in this spectrum because they are masked by the relatively large proton-proton dipolar interactions, which are not averaged to zero by the slow anisotropic rotations of the protein in the membranes.

spin pairs in proteins produce a broad featureless line (MacKay et al., 1983) which, inevitably, masks the spectral features which are characteristic of the different amino acids in the primary sequence (Fig. 1c).

2. DEUTERIUM NUCLEAR MAGNETIC RESONANCE (^{2}H-NMR) PROVIDES WELL DEFINED, QUANTITATIVE, LOCAL STRUCTURAL INFORMATION ON BILAYER SYSTEMS

The quadrupolar interactions of ^{2}H nuclei are always much greater than their dipolar interactions, and the spectral shapes are determined by an average over orientations of a particular nuclear site rather than by interactions between nuclei at different sites. Thus, the measurement of the magnitude of the quadrupolar splitting of the ^{2}H-NMR spectrum of phosholipid molecules labelled at a particular molecular position provides a direct measure of the orientational order parameter S_{CD} of the vector joining the carbon and hydrogen atoms at that position. The ^{2}H-NMR spectrum of a lipid bilayer *powder* sample is a very broad superposition of doublets, but unlike the featureless dipolar broadened ^{1}H-NMR spectra, the sharp spectral ('Pake doublet') features of ^{2}H-NMR powder spectra provide a ready measurement of S_{CD} as may be seen from Fig. 2. Measurements of S_{CD} in the polar head groups and acyl chains of phospholipid molecules have provided a distinctive map of orientational order as a function of depth in a variety of phospholipid bilayer model membranes. Excellent reviews are available on the results and interpretation of this quantitative measure of membrane organization, and of a variety of related properties, including the influence of chain unsaturation and the presence of other molecules such as cholesterol on orientational order in the fluid phase, and the study of the gel-liquid crystalline phase transition (Mantsch et al., 1977, Seelig, 1977; Seelig and Seelig, 1980; Jacobs and Oldfield, 1981; Griffin, 1981; Davis, 1983; Smith, 1983).

Since proteins are distributed randomly in most membranes, the study of lipid-protein interactions will, in general, require the interpretation of ^{2}H-NMR spectra from a distribution of inequivalent sites even for phospholipid or protein molecules in which specific hydrogen atoms have been isotopically labelled with deuterium. Superpositions of ^{2}H-NMR powder spectra can be difficult to deal with. Nevertheless, it is possible to extract useful, quantitative information from such inhomogeneously broadened spectra. Firstly, use of the quadrupolar echo Fourier Transform technique (Davis et al., 1976; Davis, 1983) can give nearly perfect ^{2}H-NMR spectra. These spectra can be analyzed quantitatively by simulation methods (Jacobs and Oldfield, 1981), moment analysis (Bloom et al., 1978; Davis, 1983; Jarrell et al., 1981) and the method of de-Pakeing (Bloom et al., 1981; Sternin et al., 1983). However, as has been pointed out in detail elsewhere (Paddy et al., 1981; Devaux 1983), any quantitative analysis based only on the separation of peaks in powder spectra is subject to systematic errors.

3. NUCLEAR MAGNETIC RESONANCE PROVIDES QUANTITATIVE INFORMATION ON MOLECULAR MOTION

Every spectroscopic technique has a natural characteristic time which is set by the spread of transition frequencies associated with different molecular configura-

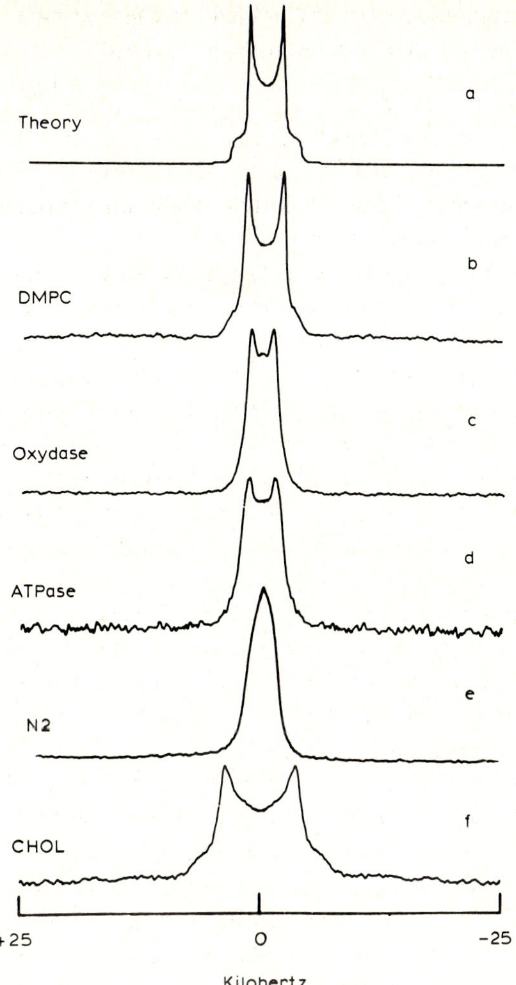

Fig. 2. Deuterium powder spectra (from Oldfield, 1982) with various influences thereon. (a) A hypothetical spectrum showing ideal line shape. The quadrupolar 'Pake doublet' allows a measure of the bond order parameter S_{CD}. (b) Experimental spectrum (34 MHz) of 1-myristoyl-2-(14,14,14-trideuterio)myristoyl-*sn*-glycero-3-phosphocholine (DMPC-d_3) in excess water at 30°C. (c) DMPC-d_3, as in (b), but containing 67 weight% cytochrome oxidase. (d) DPMC-d_3 containing 65 weight% of ATPase from sarcoplasmic reticulum. (e) DMPC-d_3, containing 67 weight% myelin proteolipid apoprotein from beef brain. (f) DMPC-d_3, containing 33 weight% cholesterol; note the large increase in quadrupolar splitting.

tions. A convenient measure is given by the second moment, M_2, of the spectrum. When the correlation time, τ_c, for molecular motion is short on the spectroscopic time scale, i.e. $M_2 \tau_c^2 \gg 1$, the corresponding experimental spectrum is 'motionally narrowed', such that the spectral frequencies are associated with the average values of the interactions over the motions. Since the ^{2}H-NMR spectrum is determined mainly by quadrupolar interactions whose characteristic time is a few microseconds, the determination of an ^{2}H-NMR spectrum, in itself, provides immediate information on whether certain molecular motions occur much faster or much more slowly than 10^6 s^{-1}.

More precise information on the faster motions is provided by measurement of the spin-lattice relaxation time, T_1, which is indicative of the 'spectral density' of the fluctuating, spin-dependent interactions at the NMR detection frequency of the spins (Davis, 1983; Brown, 1982; Brown 1983). Since ^{2}H-NMR experiments are usually carried out at frequencies between 10 and 80 MHz, T_1 measurements are sensitive to motions having correlation times in the range from 10^{-9} to 10^{-8} s, approximately.

It has been found empirically (Paddy et al., 1981; Seelig and Seelig, 1980; Jacobs and Oldfield, 1981; Davis, 1983) that a more useful relaxation measurement for obtaining information on lipid-protein interaction is the time constant for the decay of the echo, T_{2e}, in a quadrupolar echo experiment, which is sensitive to values of $\tau_c \gtrsim 10^{-7}$ s.

An extremely powerful technique for the study of the details of molecular motions having correlation times in this range is provided by the study of the ^{2}H-NMR line shape as a function of the separation of the two pulses in the quadrupolar echo Fourier transform method (see, e.g., Spiess and Sillescu, 1980; Rice et al., 1981; Griffin, 1981). This method has not, as yet, been applied systematically to the study of lipid-protein interactions but has great potential for advances in the immediate future. Similarly, changes in the shape of 'alignment spectra' associated with the Jeener-Broekaert three-pulse sequence can provide detailed information on ultra-slow motions (Lausch and Spiess, 1983). This powerful method, which has been so successful in the study of polymers, has not yet been applied to lipid-protein interactions.

II. Review of ^{2}H-NMR studies of the effect of lipid-protein interactions on the acyl chains of phospholipid molecules

1. THE STRATEGY UNDERLYING THE FIRST ATTEMPTS TO STUDY LIPID-PROTEIN INTERACTIONS USING ^{2}H-NMR

The early strategy, derived partly from successful ESR experiments (see Chapter 4), involved a search for two types of phospholipid sites in membranes *above the gel-liquid crystalline phase transition* corresponding to those phospholipid molecules in direct contact with proteins ('boundary' or 'annular' lipids) and those not in contact with proteins ('bulk' lipids). The reasoning was that, since S_{CD} is quite low ($\lesssim 0.2$) for the acyl chains of all pure phospholipid bilayer membranes in their liquid crystalline phase, direct contact with the relatively rigid protein structures should give rise to ^{2}H-NMR powder spectra characterized by measureably increased values of S_{CD}. That this is a reasonable expectation is evident from studies of the influence of cholesterol, which has a rigid steroid ring, on the values of S_{CD} of phospholipid acyl chains. As illustrated in Fig. 2F, the addition of cholesterol to phospholipid bilayer membranes produces a substantial increase in the values of S_{CD} (Stockton and Smith, 1976; Oldfield, 1982).

2. SUMMARY OF ^{2}H-NMR RESULTS IN THE LIQUID CRYSTALLINE REGION

To date, ^{2}H-NMR studies have been carried out on ^{2}H-labelled acyl chains of phospholipid molecules in a large variety of protein-lipid membrane combinations. The systems which had been studied up to 1982 are listed in Table 6 of the review by Devaux (1983). Systems studied thus far include membranes reconstituted with a variety of proteins, bacterial membranes into whose lipids ^{2}H-labelled nuclei have been incorporated biosynthetically, and erythrocyte membranes doped with ^{2}H-labelled dipalmitoyl phosphatidylcholine (DPPC-d_{62}) molecules using the PC exchange protein (Maraviglia et al., 1982). The results, which are illustrated in Figures 2 and 3, have been extensively interpreted, discussed and compared, not only by the authors themselves but also in several excellent review articles (Seelig and Seelig, 1980; Jacobs and Oldfield, 1981; Davis, 1983; Devaux, 1983). It is now generally agreed that the results are, by and large, consistent and may be summarized as follows for the influence of proteins on the lipid acyl chains in the *liquid crystalline phase* only.

(a) There is no evidence for spectra due to distinct boundary lipid and bulk lipid sites on the time scale ($\approx 10^{-5}$ s) of the ^{2}H-NMR measurements.

(b) The average value of S_{CD} is not changed appreciably by the proteins. This is true even for protein concentrations considerably larger than those found under normal physiological conditions.

(c) The presence of the proteins does not change T_1 by more than about 30%. In some cases, the change is less than 10% indicating that the relatively rapid component of the chain motions (i.e., $\tau_c \leqslant 10^{-8}$ s) is not influenced much by the proteins.

(d) The values of T_{2e} are always decreased markedly by the presence of even small concentrations of protein. Thus, the protein-lipid interaction introduces a new slow motion of the phospholipid molecules which is capable of modulating the quadrupolar interactions sufficiently to influence T_{2e} (Paddy et al., 1981), but which is too slow to affect T_1.

3. COMMENTS ON THE ^{2}H-NMR RESULTS IN THE LIQUID CRYSTALLINE REGION

We now comment on each of the four results quoted above, in turn.

(a) It should be noted that the amount of *quantitative* information provided by ^{2}H-NMR on boundary lipids in the liquid crystalline phase is sparse. One quantitative measurement is illustrated in Fig. 3c–f and described in the figure legend (see Bienvenue et al. (1982) for details). There, an upper limit of a few percent was placed on the ^{2}H-NMR spectral intensity in the wings of the powder pattern due to a hypothetical immobilized lipid acyl chain. Because an appreciable fraction of the spectral intensity from such an immobilized chain would overlap with the normal liquid crystalline ^{2}H-NMR spectrum, a measurement of this type gives an upper limit of the order of 10–20% (i.e., five to ten lipid molecules per protein) for the hypothetical immobilized lipid. This is considerably smaller than the number of lipids in direct contact with the protein (presumably more than 20 on the basis of area considerations and spin-label results as described in Chapter 4) but not sensitive enough to say for certain that there does not exist a few tight binding sites per protein.

Although there is no evidence for distinct boundary and bulk lipids on the ^{2}H-NMR time scale, the lipid-protein interaction does broaden the ^{2}H-NMR lines by a greater amount (in Hz) than $(\pi T_{2e})^{-1}$ (see Result **d**, above) in at least one case (Bienvenue et al., 1982). This observation was interpreted in terms of inhomogeneous broadening due to the small number of proteins 'sampled' by each lipid molecule as it diffuses laterally in the membrane during the ^{2}H-NMR time scale, which implies that the proteins are randomly distributed in the liquid crystalline membrane.

For sufficiently small lipid to protein (L/P) molar ratios in rhodopsin-DMPC reconstituted membranes (Bienvenue et al., 1982), it has been observed that the ^{2}H-NMR spectrum changed to that identified with a 'protein-associated' lipid phase, i.e., for $L/P \leqslant 30$, the ^{2}H-NMR spectrum became independent of temperature, exhibiting no evidence of a gel-liquid crystalline phase transition, e.g., compare Figs. 3d, e and 4d, e. As discussed further in Sections II.4 and II.5, this spectrum was identical to that observed *below* the phase transition for values of $L/P > 30$, under which conditions the lipids separated into a phase indistinguishable from the pure lipid gel phase, and a protein-associated lipid phase with $L/P = 30$. As emphasized elsewhere (Devaux, 1983), it is important not to confuse such a spatially separated protein-associated lipid phase with boundary lipids.

Boundary lipids are seen in electron spin resonance (ESR) experiments (see Chapter 4) and not in ^{2}H-NMR experiments. It has been pointed out in almost all papers referenced here that this is consistent with the fact that a single lipid-protein 'collision' as a result of lateral diffusion takes a time, short on the ^{2}H-NMR time scale, but not short on the ESR time scale. An especially detailed and insightful discussion of this point is given by Devaux (1983).

Biological membranes in which cholesterol is incorporated (Davis et al., 1980), or is present naturally (Maraviglia et al., 1982), exhibit relatively large values of S_{CD} in accord with the behaviour of model membranes containing cholesterol (Stockton and Smith, 1976; Rice et al., 1979).

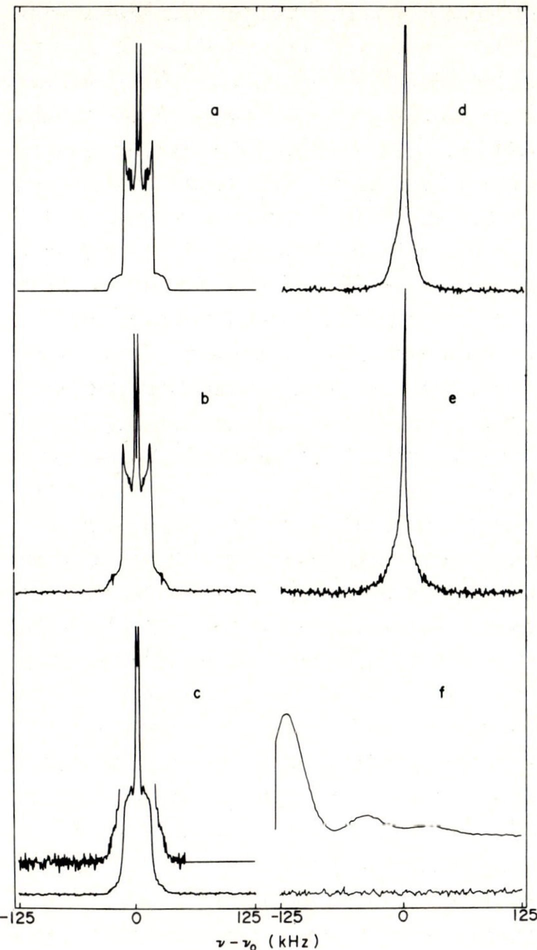

Fig. 3. (a–e) ^{2}H-NMR spectra of mixtures of DMPC-d_{54} and rhodopsin for different values of the molar lipid/protein ratio at 23°C. The pure DMPC-d_{54} sample has its gel-liquid crystalline phase transition at 20°C. (a) L/P = ∞, i.e., pure DMPC-d_{54}; (b) L/P = 150; (c) L/P = 50; (d) L/P = 30; (e) L/P = 12. The individual powder patterns of the pure lipid in (a) are broadened with the addition of protein, (b and c), but the average quadrupolar splitting is unchanged. The spectra for L/P $\leqslant$ 30 are different in shape but analysis shows that the average splitting is still unchanged. (f) The top trace shows the quadrupolar echo whose Fourier transform gave the spectrum for L/P = 50 in (c). The bottom trace shows the difference between this echo and that obtained from the Fourier transform of (c) with the wings of the spectrum zeroed as illustrated in the amplified spectrum shown in the top trace of (c). The fact that no residual difference signal was obtained in this manner within experimental error was used (Bienvenue et al., 1982) to place an upper limit of a few percent of the number of lipids immobilized by the rhodopsin in this sample (from Bienvenue et al., 1982).

(**b**) Although the failure to observe distinct boundary and bulk lipids is easy to understand in terms of exchange between the two types of sites, which is rapid on the ^{2}H-NMR time scale, further discussion is required in view of the observations that cholesterol incorporated in phospholipid bilayers causes a large increase in the *average value* of S_{CD} for a given acyl chain position, whereas proteins, which are also supposed to be rigid, produce relatively small changes. The suggestions for resolving this problem, which have been published up to now, are of a qualitative sort. One type of proposal (see, e.g., Seelig and Seelig, 1980; Jacobs and Oldfield, 1981) is that the protein-lipid interface on the hydrophobic region of the bilayer is rough because of the irregular nature of the amino acid side chains. Another suggestion is that the proteins are not really rigid but have a 'squishy' character (Bloom, 1979) and are mechanically matched to the lipids at their interface. We shall return to a discussion of some implications of the actual three-dimensional structure of integral membrane proteins concerning this problem later in Section IV.2.

(**c**) It is normally assumed implicitly within the context of these T_1 measurements that the rapid local orientations of acyl chains are associated with *trans-gauche* isomerization, and that it is reasonable for such motions not to be strongly affected by lipid-protein interactions. In model membranes, it has been noted (Jeffrey et al., 1979; Brown, 1982, 1983) that a substantial fraction of the spin-lattice relaxation rate has a frequency dependence characterized by $T_1^{-1} \propto \omega_0^{-1/2}$, where $\omega_0/2\pi$ is the nuclear spin transition frequency. A similar behaviour in nematic liquid crystals can be reliably attributed to collective molecular motions, which are called 'director fluctuations' in the jargon of liquid crystals (Doane, 1979). Brown (1982, 1983) has presented cogent arguments that the characteristic $\omega_0^{1/2}$ behaviour of T_1 in model membranes is also due to collective molecular motions. However, it should be noted that the mechanism of defect diffusion on the acyl chains of phospholipid molecules is also predicted to yield a $T_1 \propto \omega_0^{1/2}$ behaviour over a wide frequency range (Kimmich et al., 1983, see page 302). It seems to us, on intuitive grounds, that the introduction of proteins into membranes is much more likely to disrupt 'collective motions' than defect diffusion. Since the introduction of proteins into phospholipid bilayer membranes has been found *not* to affect the values of T_1 of ^{2}H-labelled nuclei on the acyl chains very much in the crystalline phase, this intuition suggests that defect diffusion is more likely to be the operative mechanism for the interpretation of T_1 on the basis of the available experimental evidence. Further experimental work is required on the frequency dependence of T_1 in the presence of proteins in order to clarify the spin-lattice relaxation mechanism in membranes.

(**d**) The nature of the slow motions of acyl chains induced by the lipid-protein interaction has not yet been established (Paddy et al., 1981). Some of the candidates are: exchange of lipids between boundary and bulk sites; rotation of proteins or motion of their secondary structural elements such as α-helices (Henderson, 1981; Unwin and Henderson, 1984) relative to each other; reorientational motions of lipids while bound to the protein surface; change in the average size or shape of the vesicles due to incorporation of proteins giving rise to broadening from lateral diffusion parallel to curved surfaces (Burnell et al., 1980).

4. SUMMARY OF ^{2}H-NMR RESULTS IN THE GEL REGION

There is general agreement among all workers in this field that the presence of proteins broadens the range of temperatures over which the gel-liquid crystalline phase transition takes place, and that the effect of the lipid-protein interaction is to reduce the orientational order parameters in the gel region from values obtained in the absence of proteins. Since the gel phase values of S_{CD} for acyl chains in pure phospholipid bilayers are always substantially greater than in the liquid crystalline phase, this result is often characterized as a 'fluidization' of the acyl chains by the lipid-protein interactions.

Most of the interpretations of gel phase data presented thus far have been of a qualitative or semi-quantitative nature. Quantitative analyses of line shapes and moments of line shapes for reconstituted membranes with rhodopsin (Bienvenue et al., 1982) and cytochrome *c* oxidase (Paddy et al., 1981; see also Deese et al., 1981 and Kang et al., 1979a) at temperatures below the gel-liquid crystalline phase transition temperatures have yielded two striking results.

(i) As shown in Figs. 4 and 5, the ^{2}H-NMR spectra in the gel phase region are often characteristic of a superposition of two distinct phases. One of these is indistinguishable from the pure phospholipid gel phase while the other gives a ^{2}H-NMR spectrum which is identical to that obtained at higher temp-

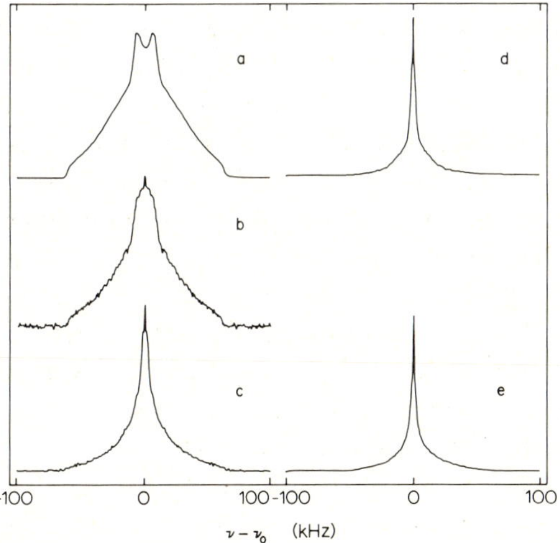

Fig. 4. ^{2}H-NMR spectra at 4°C for the same samples used to obtain the spectra of Fig. 3 (from Bienvenue et al., 1982).

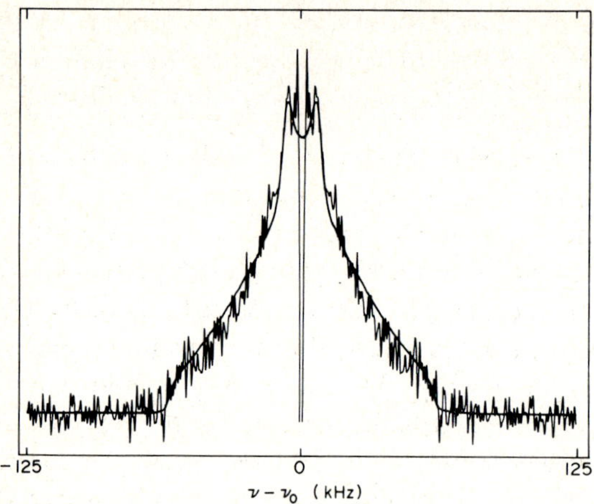

Fig. 5. Comparison of the normalized spectrum of pure DMPC-d_{54} at 4°C (Fig. 4a) with the difference between the normalized spectrum for L/P = 50 at 4°C (Fig. 4c) and 0.6 times the normalized spectrum for L/P = 12 at 4°C (Fig. 4e). The latter difference spectrum is the one with more noise. The sharp negative central peak in the difference spectrum is due to the different relative contributions to the spectrum from HDO in the different samples (from Bienvenue et al., 1982).

eratures for protein-associated lipids (see the second paragraph of Section II.3.a). For rhodopsin and DMPC, the number of lipids per protein molecule in the protein-associated lipid phase was found from ^{2}H-NMR experiments to correspond to $(L/P)_0 = 30$, while for cytochrome *c* oxidase and 1-(16,16,16-trideuteriopalmitoyl)-2-palmitoleoyl-sn-glycero-3-phosphocholine $((C^2H_3\text{-}P)(PO)PC)$, it was found that $(L/P)_0 = 35$ per protein monomer (Paddy et al., 1981). For values of $L/P \leqslant (L/P)_0$ only the lipid-associated ^{2}H-NMR spectrum is seen, while for $L/P > (L/P)_0$, the relative intensities of the two spectral components gives the same value of $(L/P)_0$ for all values of L/P.

(ii) The average values of the quadrupolar splittings of the ^{2}H-NMR spectra for the protein-associated lipid phase are, within experimental error, indistinguishable from those obtained in the liquid crystalline phase. It is important to keep in mind that the ^{2}H-NMR spectra of the protein-associated lipid phase and the pure phospholipid liquid crystalline phase are *not* identical in shape; it is only the average quadrupolar splittings which are identical. To the extent that the ^{2}H-NMR spectra reflect an axially symmetric environment, this result can be interpreted as implying that the average value of S_{CD} is the same for the lipids in the protein-associated phase as in the liquid crystalline phase.

5. COMMENTS ON THE ^{2}H-NMR RESULTS IN THE GEL REGION

The most striking result in the gel-phase region is that S_{CD} is reduced by the presence of the proteins. This implies that the acyl chains undergo a much greater degree of angular motion on the ^{2}H-NMR time scale. It should be emphasized, however, that this so called fluidization of the acyl chains does not necessarily imply that the physical state of the phospholipid molecules in the presence of proteins is similar to the liquid crystalline phase of the pure phospholipid bilayer. For example, no measurements of the phospholipid lateral diffusion constant have yet been reported for the protein-associated lipid phase. It would appear, from the similarity of the average S_{CD} in the protein-associated lipid phase to that of the protein-free liquid crystalline phase, that the structure of the protein-associated lipid phase probably resembles a bilayer, but whether or not all aspects of bilayer fluidity are retained remains to be established experimentally. It should also be emphasized that the values of $(L/P)_0$ given above do not correspond to boundary lipids. They correspond to the number of lipid molecules per protein molecule for a spatially separated phase of proteins and their associated lipids. This separated phase seems to co-exist in a stable form with the pure lipid phase below the temperature of the gel-liquid crystalline phase transition of the pure lipid. It is presumably responsible for the linear decrease with protein concentration of the enthalpy change associated with the gel-liquid crystalline phase transition observed for a variety of systems, e.g., see Alonso et al. (1982) for results on bacteriorhodopsin and for references to earlier calorimetric studies of other lipid-protein recombinants.

Finally, it is important to distinguish the superposition of ^{2}H-NMR spectra due to co-existing pure lipids in their gel state and protein-associated lipids, which were discussed above, and studies of co-existing gel and liquid crystalline states each with randomly dispersed proteins as has been well documented for a number of bacterial membranes (see, e.g., Davis et al., 1979, 1980; Kang et al., 1979b; Smith et al., 1980; Nichol et al., 1980). In addition, some especially detailed studies of co-existing gel and liquid crystalline phases have been carried out recently for mixtures of lipids and synthetic amphiphilic peptides (Davis et al., 1983; Huschilt et al., 1984). These studies are discussed in more detail in Section IV.3.

III. Review of ^{2}H-NMR studies of the effect of lipid-protein interactions on the polar headgroup region of lipid molecules

1. INTEGRAL MEMBRANE PROTEINS

The involvement of the head group region of lipids has been much less investigated than that of the acyl chains. In an early study, Gally et al. (1981) incorporated ^{2}H-labelled glycerol into mutants of *E. Coli*. The quadrupole splittings obtained from intact membranes, from aqueous dispersions of total lipid extract, and from dispersions of the purified phosphatidylethanolamine (PE) and phosphatidylglycerol (PG) were very similar. This suggested a minimal effect of membrane protein on the average conformation of the glycerol headgroup moiety. In the spectra of the membranes, however, the subcomponent resonances were of significantly different width. This suggested the onset of slower motions of the glycerol moiety within its ordered environment, due to the presence of protein. This behaviour is very similar to that observed for the acyl chains of the lipid molecules as discussed in Sections II.2.d and II.3.d.

These conclusions were confirmed in a later study on the same system by Borle and Seelig (1983), who also estimated spin-lattice relaxation times, T_1, for various segments of PG in the backbone and headgroup glycerol moieties in the native membranes, and compared them with those of the total lipid extract or the PG derived therefrom. In both membranes and isolated lipids, the T_1 values of the head group glycerol moiety were longer than those of the backbone moiety. In contrast to the quadrupole splittings, there were differences in the T_1 values from membranes and from isolated lipids. The shorter T_1 values obtained in the presence of the protein were taken to indicate a slowing down by the protein of those headgroup motions responsible for the spin-lattice (T_1) relaxation.

The influence of cytochrome oxidase on the headgroup properties of 1-palmitoyl-2-oleoyl-phosphatidylcholine (POPC) was investigated by Tamm and Seelig (1983) using ^{2}H-NMR of the deuterated choline residue and ^{31}P- and ^{14}N-NMR of the naturally occurring isotopes. This system has also been extensively studied by ^{2}H-NMR of labelled acyl chains, as discussed above. The ^{2}H- and ^{14}N spectra of the protein-lipid recombinants had resonances which were considerably broader than, but quadrupolar splittings very similar to, those of POPC alone. The ^{31}P-NMR spectra of both systems were essentially identical, but the ^{31}P T_1 values showed minima in their temperature dependencies at 25 and 15°C, respectively. The T_1 values were roughly 25% lower in the recombinants, as were the activation energies for rapid reorientation of the choline moiety. Using a fast exchange model for lipid near and removed from the protein, relaxation rates were estimated to be 30% greater for lipids in contact with protein.

The main conclusion from the above data is that, even at very low lipid to protein ratios, only small effects on head group ordering or mobility due to protein are evident. The authors took this to indicate an excellent fluid-like match between the outer surface of cytochrome *c* oxidase and its surrounding lipids as enunciated in the squishy protein picture (Bloom, 1979).

2. PERIPHERAL MEMBRANE PROTEINS

A number of peripheral proteins have been shown to bind preferentially to charged lipids. It is obviously of great interest to investigate whether such peripheral proteins perturb the ^{2}H-NMR spectra of the charged lipids to which they bind, particularly in the head group region. Recently, two such studies have been carried out, one involving the myelin basic protein in mixtures of DMPC and DMPG (Sixl et al., 1984) and another involving cytochrome *c* in mixtures of DMPC and DMPS (phosphatidylserine) (Devaux et al., 1984). Whereas the myelin basic protein produced significant decreases in the ^{2}H quadrupole splittings of all ^{2}H-labelled sites of the head group region of DMPG, with no change in those of DMPC implying some selective lipid-protein association, cytochrome *c* produced no appreciable changes in the head group quadrupole splittings of DMPS. Similarly, the myelin basic protein also produced observable changes in the chemical shift anisotropy of the ^{31}P-NMR powder pattern while the cytochrome *c* study showed no such effects.

Each of these studies showed effects of the protein on T_1 and T_2, but in view of their recent nature and the subtleties involved in the interpretation of relaxation time measurements (see, e.g., Section II.3), we do not discuss them here. The results obtained for the myelin basic protein offer great promise for the study of molecular aspects of the binding of peripheral proteins to charged lipids. It is tempting to speculate at this early stage as P.F. Devaux (private communication) has done, that the binding mechanism of the myelin basic protein involves much deeper penetration of the membrane surface than does that of cytochrome *c*. Indeed, Sixl et al. (1984) invoke, as an explanation of their data, previous studies of the binding of myelin basic protein to acidic phospholipids, which indicate such partial penetration.

IV. Relationship between the three-dimensional structure of integral membrane proteins and ^{2}H-NMR results on lipids

1. THE α-HELIX IS A FUNDAMENTAL *TRANS*-MEMBRANE SECONDARY STRUCTURAL ELEMENT OF INTEGRAL MEMBRANE PROTEINS

Bacteriorhodopsin, whose structure is better known than that of any other integral membrane protein, consists of seven *trans*-membrane α-helices joined together by relatively short turns and bends in the aqueous medium outside the bilayer (Henderson and Unwin, 1975; Henderson, 1981; Unwin and Henderson, 1984). The hydrophobic nature of the interior of a lipid bilayer imposes strong constraints on the segments of the protein located within the bilayer. These constraints, which we list below, represent the most obvious and direct influence of lipid-protein interactions on protein conformation. They must be taken into account in interpreting any measurements involving integral membrane proteins.

(a) The amino acid side chains must be predominantly hydrophobic.

(**b**) Charged or polar residues are likely to be sequestered inside the three-dimensional structure out of direct contact with the acyl chains of the lipid molecules (Engelman and Zaccai, 1980) and geometrically arranged so as to stabilize the three-dimensional structure via Coulombic interactions, (i.e., a positively charged residue on one α-helix is likely to be paired with a negatively charged residue on another).

(**c**) Hydrogen bonds are likely to be completed *within* a secondary structural element as in the case of the α-helix or between neighbouring elements as in the case of β-sheets (Henderson, 1981; Unwin and Henderson, 1984).

An implicit consequence of (**a**) is that the length of the hydrophobic region of the protein would be expected to match, approximately, the thickness of the hydrophobic region of the lipid bilayer, as is observed to be the case for bacteriorhodopsin in its natural membrane. In the light of the experience gained with bacteriorhodopsin, the primary sequences of integral membrane proteins can be scanned for one or more series of predominantly hydrophobic residues which would form an α-helix of length capable of spanning the lipid bilayer (Engelman et al., 1982). This type of search has been quite successful and, in addition to bacteriorhodopsin, has been applied to various integral membrane proteins including glycophorin and fd phage coat protein (see, e.g., Henderson, 1981), rhodopsin (Ovchinnikov, 1982) and fumarate reductase (Wiener et al., 1984). It is interesting to ask whether ^{2}H-NMR measurements are able to provide insight into this type of hydrophobic matching.

2. A SIMPLE GEOMETRICAL INTERPRETATION OF ORIENTATIONAL ORDER IN MEMBRANES

A detailed quantitative theoretical analysis of the effects of mismatch of the hydrophobic regions of lipids and proteins has been carried out recently by Mouritsen and Bloom (1984) in their ‘mattress model of lipid-protein interactions’. Their paper includes a detailed review of previous theoretical approaches to lipid-protein interactions. They point out that if the lipid molecules exchange rapidly on the ^{2}H-NMR time scale between sites on and off the protein surface, the average value, $\langle S_{CD} \rangle$, of the orientational order parameters *for the entire acyl chain* should be correlated with the hydrophobic thickness. Thus, the general observation that proteins do not change $\langle S_{CD} \rangle$ in the fluid phase can be interpreted as a straightforward consequence of the matching of the hydrophobic thicknesses of the lipid bilayers to those of the proteins for the lipid-protein combinations investigated up to now. The origin of this matching may lie in the strategy employed by the experimenters since there has been a natural tendency to study combinations of proteins with lipids having acyl chain compositions typical of those found in the natural membrane source of the protein. Such lipids and proteins would be expected to be matched hydrophobically. With this interpretation, the implication of the ^{2}H-NMR results is that the lack of change of $\langle S_{CD} \rangle$ due to the lipid-protein interactions is a manifestation of a 'structure-function relationship' involving lipids and proteins.

Furthermore, the observation discussed in Sections II.4 and II.5 that *protein-associated lipids* have the same values of $\langle S_{CD} \rangle$ even in the gel phase region, as does the pure lipid bilayer in the fluid region, may be interpreted as a consequence of the constraint on the lipid chains imposed by hydrophobic matching with the proteins.

3. EXPERIMENTS WITH SYNTHETIC AMPHIPHILIC POLYPEPTIDE MOLECULES

Most ^{2}H-NMR experiments on lipid-protein interactions have been carried out on either biological membranes in which the ^{2}H probes were introduced biosynthetically or reconstituted membranes in which natural proteins were incorporated into a lipid bilayer of selected composition. There are difficulties in the interpretation of such experiments due to the complexity of most proteins and the lack of detailed information on their three-dimensional structure. An attempt has been made recently to isolate specific questions associated with the interactions between lipid molecules and individual *trans*-membrane polypeptides which simulate the *trans*-membrane α-helical secondary structural elements of integral membrane proteins (Davis et al., 1983). The polypeptide:

$$\text{Lys}_2\text{-Gly-Leu}_n\text{-Lys}_2\text{-Ala-amide} \qquad (\text{K}_2\text{GL}_n\text{K}_2\text{A})$$

was synthesized, for values of n = 16, 20 and 24, by the solid phase peptide synthesis technique and incorporated into aqueous dispersions of two bilayer systems, potassium palmitate and dipalmitoyl phosphatidylcholine (DPPC). The lipid system contained perdeuterated acyl chains on which ^{2}H-NMR experiments were carried out.

The amphiphilic nature of the peptide, the lysines and the terminal amino group being positively charged and the leucines being hydrophobic, makes it well suited to examine the effects of mismatching of the hydrophobic regions of the lipid and peptide molecules. As anticipated, circular dichroism measurements showed that the ($K_2GL_nK_2A$) peptide molecule adopted a very tight α-helical conformation when incorporated in a lipid bilayer. The effect of mismatch was very profound in the potassium palmitate sample as may be seen from Fig. 6. The addition of peptide molecules with n = 24 to this soap system produced large increases in the values of S_{CD}. X-ray diffraction measurements showed that this increase in $\langle S_{CD} \rangle$ was accompanied by a corresponding increase in bilayer thickness. Although the influence of the peptide on the DPPC order parameters and bilayer thickness was less pronounced, the DPPC/peptide/water system exhibited an interesting modification of its phase behaviour in the vicinity of the gel-liquid

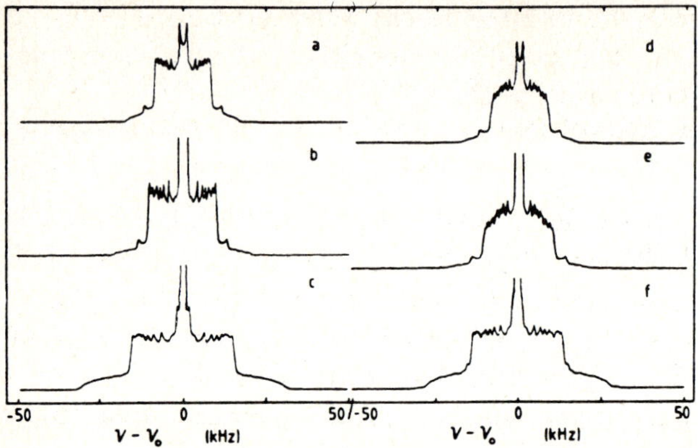

Fig. 6. ^{2}H-NMR spectra of mixtures of perdeuterated potassium palmitate and the polypeptide $K_2GL_{24}K_2A$ in 30% water by weight. (a) L/P = ∞ at 49°C; (b) L/P = 200 at 49°C; (c) L/P = 100 at 49°C; (d) L/P = ∞ at 65°C; (e) L/P = 200 at 65°C; (f) L/P = 100 at 65°C. (from Davis et al., 1983).

crystalline phase transition temperature (Davis et al., 1983; Huschilt et al., 1984). In the study of Davis et al (1983), the moments of the ^{2}H-NMR line were analyzed to determine the fraction of the DPPC-d_{62} in each of the co-existing gel and liquid crystalline (fluid) phases. The results were checked by a method of spectral subtraction similar to that demonstrated in Figs. 4 and 5 (see also Section II.2 and Bienvenue et al., 1982). The latter method has been considerably refined by Huschilt et al. (1984). Several samples having different lipid-to-peptide ratios, L/P, were studied by spectral subtraction methods. At a given temperature in the coexistence region, each sample gave a spectrum which was a linear superposition of two spectra, one corresponding to the value $(L/P)_f$ in the fluid phase at that temperature and the other corresponding to the value $(L/P)_g$ in the gel phase at the same temperature. The two spectra so obtained did not change with L/P; only the relative intensities of the spectra associated with $(L/P)_f$ and $(L/P)_g$ changed as L/P was varied between these two limiting values.

The use of synthetic amphiphilic polypeptides to study lipid-protein interactions is thus very promising. One can envisage the tailoring of a variety of amphiphilic peptide molecules to answer specific questions of importance in natural membranes.

V. Review of ^{2}H-NMR measurements on integral membrane proteins and polypeptides

1. THE STUDY OF ^{2}H-LABELLED AMINO ACIDS INCORPORATED BIOSYNTHETICALLY INTO BACTERIORHODOPSIN

Oldfield and co-workers have studied bacteriorhodopsin samples, enriched biosynthetically with ^{2}H-labelled amino acids. A summary and novel interpretation of their data was presented recently by Keniry et al. (1984). As discussed in Section IV.1, bacteriorhodopsin, the unique protein of the purple membrane of *H. halobium* and *H. cutirubrum*, has been shown to consist of seven helices embedded in a lipid matrix (Henderson and Unwin, 1975). Various more detailed models of its secondary structure have appeared, one of which is shown in Fig. 7 (Engelman et al., 1982). The ^{2}H-NMR spectra of bacteriorhodopsin, labelled in a variety of amino acids, always exhibited a very narrow central component as well as a broad component with a quadrupole splitting (Fig. 8). The relative intensities of the two spectral components depended upon the nature of the labelled amino acid. Keniry et al. (1984) postulated that an aqueous interface of the lipid-protein complex could be defined (lines B in Fig. 7) such that the narrow spectral components were due to residues outside the two B lines while the broad components arose from residues between the two B lines. This assumed that the reorientation of amino acids in the aqueous region would be sufficiently unhindered to average the ^{2}H quadrupolar splitting almost completely, while those within the lipid region would be constrained in their motions by the lipids and/or

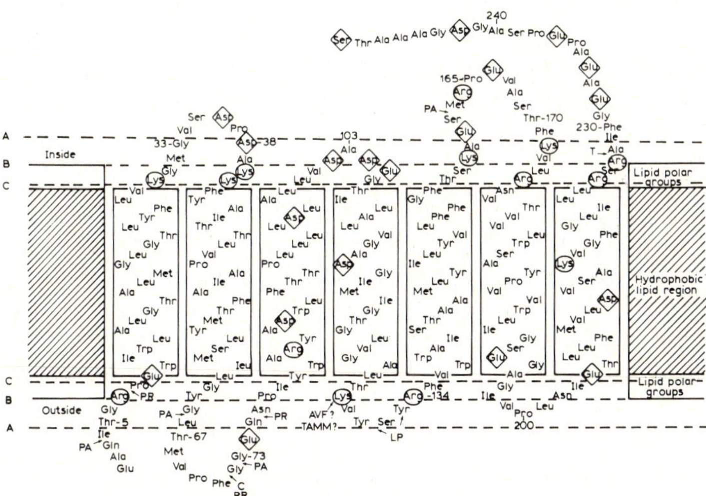

Fig. 7. A proposed model (Engelman et al., 1982) for the secondary structure of bacteriorhodopsin showing various definitions of the membrane surface (lines A, B or C). Taken from Keniry et al. (1984).

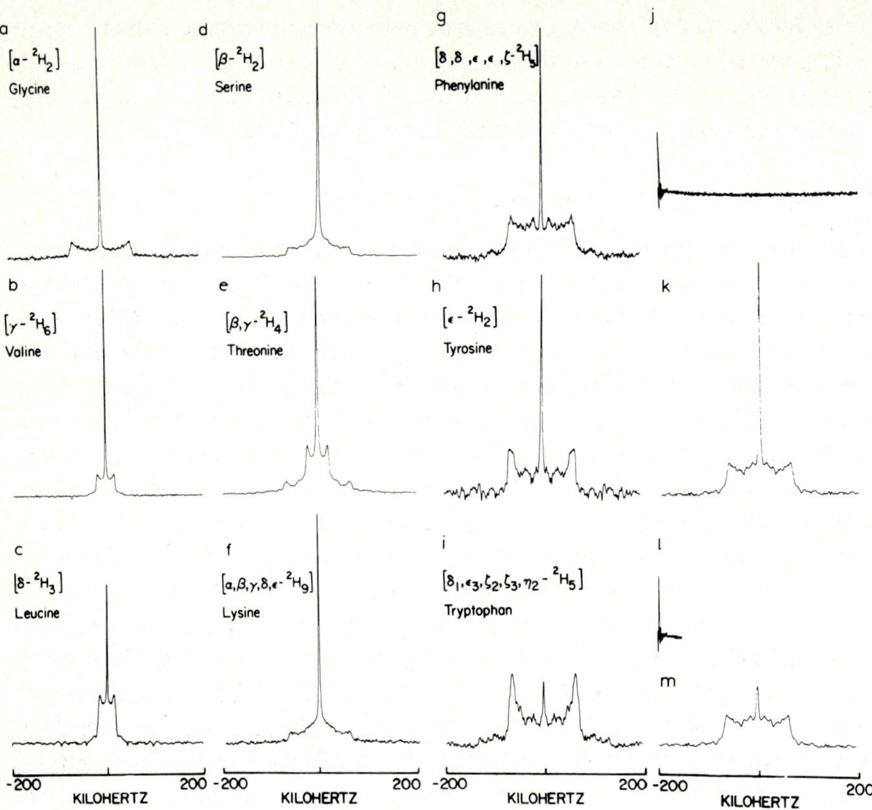

Fig. 8. Deuterium NMR spectra (55.3 MHz, 4°C) of ^{2}H-labelled bacteriorhodopsin in membranes of *Halobacterium halobium*. In spectra a–i, the indicated amino acids were labelled. Spectra j–m are time- and frequency-domain pairs for incorporated [$^2H_5$]-phenylalanine with different numbers of accumulated data points (j,k, 4096; l,m, 512) to show the discrimination that may be made between the broad and narrow spectral components. Taken from Keniry et al. (1984).

other protein helices leading to large quadrupolar splittings. Keniry et al. (1984) considered various models for the protein secondary structure and concluded that the model shown in Fig. 7 gave good agreement with the ratios of the narrow to broad components for all the amino acids. Displacement of the B lines by ± 2 units in the primary sequence, corresponding to line pairs A and C in Fig. 7, were each found to give much poorer agreement with the basic hypothesis. A further interesting result was that mild tryptic cleavage of C-terminal residues resulted in the loss of intensity in the central component of the spectra even when, as in the case of valine-γ-d_6, the deuterated residues are not removed by the process. This was interpreted as indicating "that the motional states of membrane surface residues are highly sensitive to the nature of the surroundings."

This paper is of great interest both for the details of the results obtained (i.e., it is by no means obvious that the residues outside B in Fig. 7 should have enough angular freedom to average out the quadrupolar splittings so completely) and for the promise it gives ^{2}H-NMR spectroscopists of ultimately making structure- and function-sensitive measurements on integral membrane proteins.

2. ^{2}H-NMR STUDIES OF EXCHANGEABLE HYDROGEN SITES

In the previous sections, we reviewed a ^{2}H-NMR study of proteins in membranes using ^{2}H-labelled amino acids incorporated biosynthetically into the integral membrane proteins. An alternative method of incorporating ^{2}H labels into such positions is via exchange with deuterated solvents such as ^2H_2O and CH_3O^2H. Such a study has recently been carried out (Pauls et al., 1984) on the synthetic peptide $K_2GL_{24}K_2A$ discussed in Section IV.3. In that system, there are 15 exchangeable hydrogens associated with the five NH_3^+ groups and, in the α-helical conformation, approximately 30 associated with the hydrogen bonding of the CO group of each amino acid to the NH of the residue four ahead in the linear sequence. The latter hydrogens are of special interest for ^{2}H-NMR studies since they provide a means of studying motion associated with the peptide backbone.

The ^{2}H-NMR spectrum of the exchangeable hydrogens of $K_2GL_{24}K_2A$ incorporated into DPPC membranes is shown in Fig. 9a at 10°C, and Fig. 9b at 42°C, corresponding to the gel and liquid crystalline phases of DPPC, respectively. These spectra exhibit the narrower $N^2H_3^+$ spectrum associated with rapid rotation of the N^2H_3 group superimposed on the broader N-^{2}H spectrum. In the gel phase, the latter spectrum is characteristic of hydrogen bonding sites in solids with little molecular motion (Hunt and MacKay, 1974, 1976). It is a characteristic ^{2}H-NMR powder pattern for nuclei at sites corresponding to a non-axially symmetric charge distribution. In this case, the 'asymmetry parameter' was determined to be η = 0.14 (Pauls et al., 1984) which is quite typical of this type of hydrogen bonding site (Hunt and MacKay, 1974, 1976). On the other hand, the liquid crystalline spectrum of Fig. 9b can be explained in terms of rapid rotation of the peptide about its long axis (Pauls et al., 1984), being an axially symmetric ($\eta = 0$) powder pattern. A quantitative analysis of the relationship between the spectra of Fig. 9a and b (Pauls et al., 1984) gives a value of about 19° between the axis of rotation of $K_2GL_{24}K_2A$ and the N-H bond direction, a result which is compatible with the geometry of the α-helix (Pauling and Corey, 1951).

One of the striking observations in this study was that as the rate of rotation increased with increasing temperature in the gel phase, T_{2e} decreased going through a minimum, as predicted by Pauls et al. (1984), at the temperature for which the rate of rotation matched (approximately) the change in the average quadrupolar splitting due to the motion. This is illustrated in Fig. 10. In the vicinity of the minimum, the shape of the spectrum as determined by the quadrupolar echo technique is distorted; as discussed in Section I.3, an analysis of this

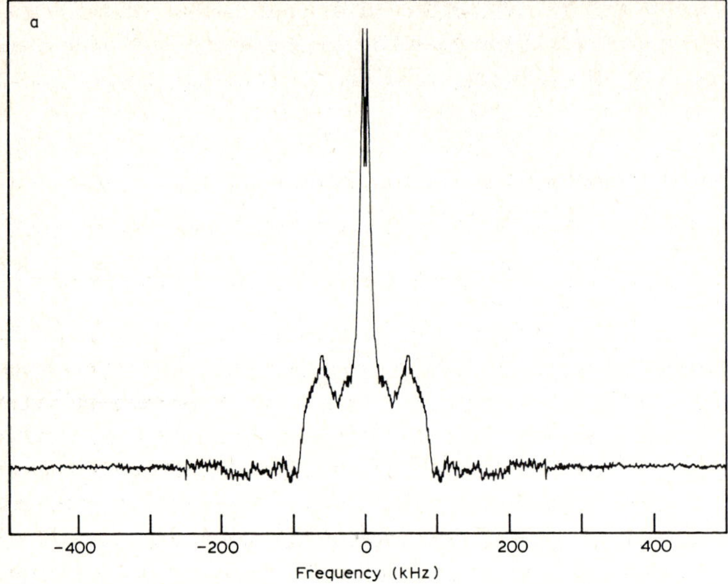

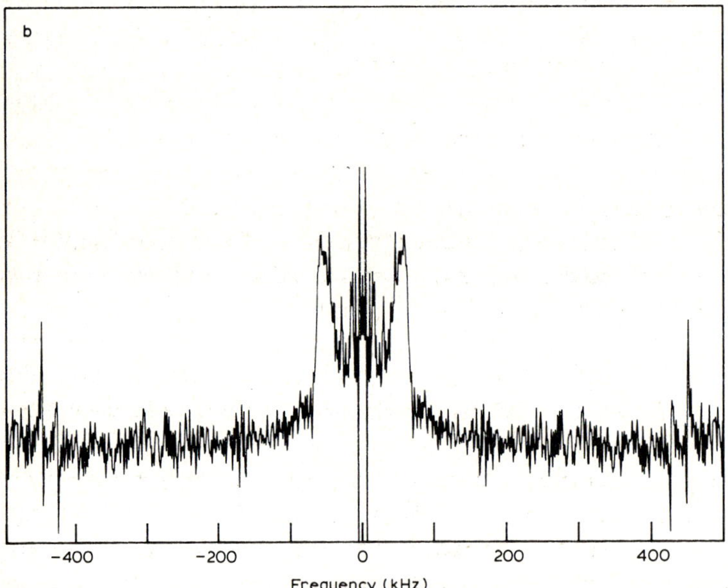

Fig. 9. ^{2}H-NMR spectra of the exchangeable hydrogen atoms in the polypeptide $K_2GL_{24}K_2A$-amide incorporated into DPPC bilayers. (a) The spectrum at 10°C, in the gel phase of DPPC, consists of two components. The broad component is characteristic of the N^2H ---- 0 hydrogen bonding sites of the immobilized α-helical backbone of the peptide while the narrower spectrum arises from rotating $N^2H_3^+$ groups. (b) The spectrum from the same sample at 42°C. It is characteristic of the α-helical peptide molecules rotating about their long axes.

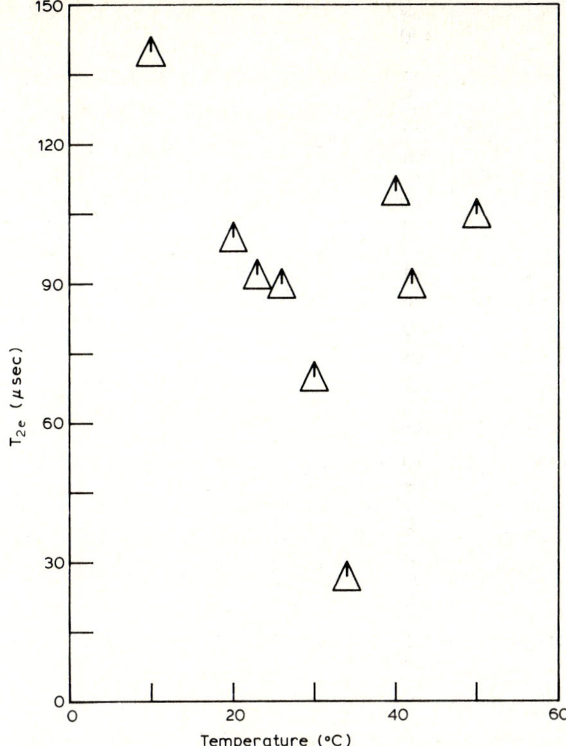

Fig. 10. Plot of T_{2e} versus temperatures for the ^{2}H-NMR signals shown in Fig. 9.

modified spectrum is capable of providing detailed information on molecular motion (Spiess and Sillescu, 1980). It seems that measurements such as these of exchangeable hydrogens in proteins are capable of providing important information on the dynamical properties of the protein backbones.

VI. Future prospects

In closing this review, we discuss briefly a few representative examples of experiments and techniques which are capable of providing new insights into lipid-protein interactions in the immediate future. We express these in relation to some of the well known limitations of the ^{2}H-NMR method.

1. LIMITATIONS OF THE ^{2}H-NMR METHOD SENSITIVITY, TIME SCALE, SPECIFICITY

As discussed in Section I, ^{2}H-NMR is essentially a 'broad line' technique. As a result, its *sensitivity* is relatively low. Still, as may be seen from Fig. 8 in Section V, the sensitivity in the most favourable case (bacteriorhodopsin) is sufficient to detect ^{2}H-NMR signals from a few amino acids – perhaps, even a single amino acid in some cases.

Specificity as to the origin of the ^{2}H-NMR spectral features has thus far come only from the labelling process and not from any chemical influence on the spectrum, as is the case in high resolution proton NMR where the chemical shift enables one to relate the origin of NMR signals from proteins to specific amino acid residues and thence to positions in the primary sequence. Up to now, no evidence for specific lipid-protein binding effects has been obtained because, apparently, most lipids exchange rapidly between sites on and off the protein surface on the ^{2}H-NMR *time scale* of a few microseconds. As discussed in Section II.3.a, there is a need for greater sensitivity in this type of measurement.

2. THERMODYNAMIC STUDIES

Though the relatively long ^{2}H-NMR time scale makes it difficult to observe lipid-protein binding effects, it has made it possible for ^{2}H-NMR to provide well defined thermodynamic information. As discussed in Sections II.2.a, II.5 and IV.3, partition of labelled lipids between different thermodynamic phases can now be determined quite precisely (Bienvenue et al., 1982; Davis et al., 1983; Huschilt et al., 1984), and the ^{2}H-NMR spectral properties can provide unique information on the physical properties of each phase. Such measurements, though they must be done with care, are straightforward. They should, in future, be combined systematically with calorimetry and other macroscopic measurements on membrane systems.

3. STRATEGIC APPROACH TO THE SEARCH FOR SELECTIVE LIPID-PROTEIN BINDING

It would be of great interest to investigate specificity in lipid-protein interactions using ^{2}H-NMR. So far, there is no direct evidence for bound lipid-protein complexes having lifetimes long on the ^{2}H-NMR time scale. However, as discussed in section II.3.a, there is still scope for improvement in the sensitivity for this type of experiment. We suggest that the time is ripe for more selective experiments to be done, making use of the wide range of experience gained in the study of lipid-protein systems using ^{2}H-NMR. One approach is to continue along the lines recently begun in studies of the interactions between lipids and peripheral proteins (Section III.2). If possible, those lipids which are believed to bind to specific proteins (such as negatively charged lipids interacting with proteins having positively charged binding sites) should be selectively labelled and diluted in lipids which are not expected to bind. The number of labelled lipids should not exceed appreciably the number of anticipated binding sites since the signal from unbound lipids partly obscures the hypothetical signal of interest. This dilution approach has been used successfully by Sixl et al. (1984) in studies of the myelin basic protein in mixtures of PG and PC, and has also been attempted by Devaux et al. (1984), as discussed briefly in Section III.2. Insofar, as we know, this approach has not yet been attempted in other systems in a systematic manner.

4. MORE NMR TRICKS SHOULD BE USED

We close by noting that only a limited range of the presently available sophisticated battery of NMR techniques for treating broad line NMR spectra has been utilized in membrane systems. These techniques include coherent averaging of dipolar, quadrupolar and anisotropic chemical shift interactions (Haeberlin, 1976; Mehring, 1976; Abragam and Goldman, 1982, see chapter 2), two-dimensional Fourier transform NMR (Bax, 1982) and multiple quantum NMR (Weitekamp, 1983), as well as some of the relaxation methods described in section I.3. Up to now, NMR spectroscopists have avoided using more sophisticated methods in favour of more direct and easier to interpret simple spectroscopic techniques. Part of the reason for this is that the relaxation parameters in membranes are often not ideal for many of the sophisticated methods to be easily exploited. Another reason is that membranes are complex systems and it has been difficult to formulate questions sufficiently well defined to use any but the simplest NMR methods. We feel that results obtained using ^{2}H-NMR during the past few years now enable us to pose some good questions concerning lipid-protein interactions.

Acknowledgements

Wc wish to thank the following persons for providing us with help in the form of Figures, data in advance of publication and/or helpful comments and criticism: F.W. Dahlquist, J.H. Davis, P.F. Devaux, G. Gray, G. Hoatson, A. MacKay, E. Oldfield, M. Paddy, K.P. Pauls and A. Watts. The research of one of us (M.B.) was supported by Natural Sciences and Engineering Research Council.

References

Abragam, A. and Goldman, M. (1982) *Nuclear Magnetism: Order and Disorder*. Clarendon Press, Oxford.

Alonso, A., Restall, C.J., Turner, M., Gomez-Fernandez, J.C., Goni, F.M. and Chapman, D. (1982) Protein-lipid interactions and differential scanning colorimetric studies of bacteriorhodopsin reconstituted lipid-water systems. Biochim. Biophys. Acta 689, 283–289.

Bax, A. (1982) *Two-dimensional Nuclear Magnetic Resonance in Liquids*. Delft University Press, Delft and D. Riedel Publishing Co., Dordrecht, Holland.

Bienvenue, A., Bloom, M., Davis, J.H. and Devaux, P.F. (1982) Evidence for protein-associated lipids from deuterium nuclear magnetic resonance studies of rhodopsin-dimyristoylphosphatidylcholine recombinants. J. Biol. Chem. 257, 3032–3038.

Bloom, M. (1979) Squishy proteins in fluid membranes. Can. J. Phys. 57, 2227–2230.

Bloom, M., Davis, J.H. and Dahlquist, F.W. (1978) Determination of orientational order in bilayer system using moments of deuterium magnetic resonance spectra. 20th Ampere Congress, Tallinn, Estonia, p. 551. Editors: E. Lippmaa and T. Saluvere. Springer, Berlin.

Bloom, M., Davis, J.H. and MacKay, A.L. (1981). Direct determination of the oriented sample NMR spectrum from the powder spectrum for systems with local axial symmetry. Chem. Phys. Lett. 80, 198–202.

Borle, F., and Seelig, J. (1983) Structure of *E. coli* membranes. Deuterium magnetic resonance studies of the phosphoglycerol head group in intact cells and model membranes. Biochemistry 22, 5536–5544.

Brown, M.F. (1982) Theory of spin-lattice relaxation in lipid bilayers and biological membranes. ^{2}H and ^{14}N relaxation. J. Chem. Phys. 77, 1576–1599.

Brown, M.F. (1983) New view of lipid bilayer dynamics from ^{2}H and ^{13}C NMR relaxation time measurements. Proc. Natl. Acad. Sci. U.S.A. 80, 4325–4329.

Burnell, E.E., Cullis, P.R. and de Kruijff, B. (1980) Effects of tumbling and lateral diffusion on phosphatidylcholine model membrane P-NMR lineshapes. Biochim. Biophys. Acta 603, 63–69.

Davis, J.H. (1983) The description of membrane lipid conformation, order and dynamics by ^{2}H NMR. Biochim. Biophys. Acta 737, 117–171.

Davis, J.H., Bloom, M., Butler, K.W. and Smith, I.C.P. (1980) The temperature dependence of molecular order and the influence of cholesterol in *Acholeplasma laidlawii* membrane. Biochim. Biophys. Acta 597, 477–491.

Davis, J.H., Clare, D.M., Hodges, R.S. and Bloom, M. (1983) Interaction of a synthetic amphiphilic polypeptide and lipids in a bilayer structure. Biochemistry 22, 5298– 5305.

Davis, J.H., Nichol, C.P., Weeks, G. and Bloom, M. (1979) Study of the cytoplasmic and outer membrane of *Escherichia coli* by deuterium magnetic resonance. Biochemistry 18, 2103–2112.

Deese, A.J., Dratz, E.A., Dahlquist, F.W. and Paddy, M.R. (1981) Interaction of rhodopsin with two unsaturated phosphatidylcholines: A deuterium nuclear magnetic resonance study. Biochemistry 20, 6420–6427.

Devaux, P.F. (1983) ESR and NMR studies of lipid-protein interactions in membranes. In: *Biological Magnetic Resonance*, Vol. V, pp. 183–299. Editors L.J. Berliner and J. Reuben, Plenum Press, New York.

Devaux, P.F., Hoatson, G., Fauvre, E., Fellman, P., Farren, B., MacKay, A. and Bloom, M. (1984) Study of the binding of cytochrome *c* to membranes containing negatively charged lipids. Manuscript in preparation.

Doane, J.W. (1979) NMR of liquid crystals. In: *Magnetic Resonance of Phase Transitions*, pp. 171–242. Editors F.J. Owens, C.P. Poole, Jr. and H.A. Farach. Academic Press, New York.

Engelman, D.M. and Zaccai, G. (1980) Bacteriorhodopsin is an inside-out protein. Proc. Natl. Acad. Sci. U.S.A. 77, 5894–5898.

Engelman, D.M., Goldman, A. and Steitz, T.A. (1982) The identification of helical segments in the polypeptide chain of bacteriorhodopsin. Methods Enzymol. 88, 81–88.

Gally, H.U., Pluschke, G., Overath, P. and Seelig, J. (1981) Structure of *E. coli* membranes. Glycerol auxotrophs as a tool for the analysis of the phospholipid head group region by deuterium magnetic resonance. Biochemistry 20, 1826–1831.

Griffin, R.-G. (1981) Solid state nuclear magnetic resonance in lipid bilayers. Methods Enzymol. 72, 108–174.

Haeberlen, U. (1976) High resolution NMR in solids. Selective averaging. Adv. in Magn. Reson. (Suppl. 1) – .

Henderson, R. (1981) Membrane protein structure. In: *Membranes and Intercellular Communications*: Les Houches Session XXXIII, July 30–August 30, 1979, pp. 229–249. Editors: R. Balian, M. Chabre and P.F. Devaux. Elsevier/North Holland, Amsterdam.

Henderson, R. and Unwin, P.N.T. (1975) Three dimensional model of purple membrane obtained by electron microscopy. Nature 257, 28–32.

Hunt, M.J. and MacKay, A.L. (1974) Deuterium and nitrogen pure quadrupole resonance in deuterated amino acids. J. Magn. Reson. 15, 402–414.

Hunt, M.J. and MacKay, A.L. (1976) Deuterium and nitrogen pure quadrupole resonance in amino acids II. J. Magn. Reson. 22: 295–301.

Huschilt, J.C., Hodges, R.S. and Davis, J.H. (1984) Phase equilibria in an amphiphilic peptide-phospholipid model membrane by ^{2}H NMR difference spectroscopes. Biochemistry (in press).

Jacobs, R.E. and Oldfield, E. (1981) NMR of Membranes. Progr. NMR Spectrosc. 14, 113–136.

Jarrell, H.C., Byrd, R.A. and Smith, I.C.P. (1981) Analysis of the composition of mixed lipid phases by the moments of the ^{2}H NMR spctra. Biophys. J. 34, 451–463.

Jeffrey, K.R., Wong, T.C., Burnell, E.E., Thompson, M.J., Higgs, T.P. and Chapman, N.R. (1979) Molecular motion in the lyotropic liquid crystal containing potassium palmitate: a study of proton spin-lattice relaxation times. J. Magn. Reson. 36, 151–171.

Kang, S.Y., Gutowsky, H.S., Hsung, J.C., Jacobs, R., King, T.E., Rice, D. and Oldfield, E. (1979a) Nuclear magnetic resonance investigation of the cytochrome oxidase-phospholipid interaction: A new model for boundary lipid. Biochemistry 18, 3257–3267.

Kang, S.Y., Gutowsky, H.S. and Oldfield, E. (1979b) Spectroscopic studies of specifically deuterium labelled membrane systems. Nuclear magnetic resonance investigation of protein-lipid interactions in *Escherichia coli* membranes. Biochemistry 18, 3268–3271.

Keniry, M.A., Gutowsky, H.S. and Oldfield, E. (1984) Surface dynamics of the integral membrane protein bacteriorhodopsin. Nature 307, 383–386.

Kimmich, R., Schuur, G. and Scheurmann, A. (1983) Spin-lattice relaxation and lineshape parameters in nuclear magnetic resonance of lamellar lipid systems: Fluctuation spectroscopy of disordering mechanisms. Chem. Phys. Lipids 32, 271–322.

Lausch, M. and Spiess, H.W. (1983) Deuteron spin alignment spectra of powders in presence of ultraslow motions. J. Magn. Reson. 54, 466–479.

MacKay, A.L., Burnell, E.E., Bienvenue, A., Devaux, P.F. and Bloom, M. (1983) Flexibility of membrane proteins by broad-line proton magnetic resonance. Biochim. Biophys. Acta 728, 460–462.

Mantsch, H.H., Saito, H. and Smith, I.C.P. (1977) Deuterium magnetic resonance. Applications in chemistry, physics and biology. Progr. NMR Spectrosc. 11, 211–272.

Maraviglia, B., Davis, J.H., Bloom, M., Westerman, J. and Wirtz, K.W.A. (1982) Human erythrocyte membranes are fluid down to −5°C. Biochim. Biophys. Acta 686, 137–140.

Mehring, M. (1976) *High Resolution NMR in Solids*. Springer, Berlin.

Mouritsen, O.G. and Bloom, M. (1984) Mattress model of lipid-protein interactions in membranes. Biophys. J. 46, 141–153.

Nichol, C.P., Davis, J.H., Weeks, G.W. and Bloom, M. (1980) Quantitative study of the fluidity of *Escherichia coli* membranes using deuterium magnetic resonance. Biochemistry 19, 451–457.

Oldfield, E. (1982) NMR of protein-lipid interactions in model and biological membrane systems. In: *Membranes and Transport*, Vol. 1, pp. 115–123. Editor: E.N. Martonosi. Plenum Press, New York.

Ovchinnikov, Yu.A. (1982) Rhodopsin and bacteriorhodopsin: structure-function relationship. FEBS Lett. 148, 179–191.

Paddy, M.R., Dahlquist, F.W., Davis, J.H. and Bloom, M. (1981) Dynamical and temperature dependent effects of lipid-protein interactions. Application of deuterium nuclear magnetic resonance and electron paramagnetic resonance spectroscopy to the same reconstitutions of cytochrome *c* oxidase. Biochemistry 20, 3152–3162.

Pauling, L. and Cory, R.B. (1951) The structure of synthetic polypeptides. Proc. Natl. Acad. Sci. U.S.A. 37, 241–250.

Pauls, K.P., MacKay, A.L., Söderman, O., Bloom, M. and Hodges, R.S. (1984) Dynamic properties of an integral membrane polypeptide backbone measured by ^{2}H NMR, to be published.

Rice, D.M., Meadows, M.D., Scheinman, A.O., Goni, F.M., Gomez-Fernandez, J.C., Moscarello, M.A., Chapman, D. and Oldfield, E. (1979) Protein-lipid interactions. A nuclear magnetic resonance study of sarcoplasmic reticulum Ca^{2+}, Mg^{2+}-ATPase, lipophilin and proteolipid apoprotein-lecithin systems and a comparison with the effects of cholesterol. Biochemistry 18, 5893–5903.

Rice, D.M., Wittebort, R.J., Griffin, R.G., Meirovitch, E., Stimson, E.R., Meinwald, Y.C., Freed, J.H. and Scheraga, H.A. (1981) Rotational jumps of the tyrosine side chain in crystalline enkephalin. ^{2}H NMR line shapes for aromatic ring motion in solids. J. Am. Chem. Soc. 103, 7707–7710.

Schulz, G.E. and Schirmer, R.H. (1979) *Principles of Protein Structure*. Springer, New York, Heidelberg.

Seelig, J. (1977) Deuterium magnetic resonance: theory and application to lipid membranes. Q. Rev. Biophys. 10, 353–418.

Seelig, J. and Seelig, A. (1980) Lipid conformation in model membranes and biological membranes. Q. Rev. Biophys. 13, 191–61.

Sixl, F., Brophy, P.J. and Watts, A. (1984) Selective protein-lipid interactions at membrane surfaces: a deuterium and phosphorus nuclear magnetic resonance study of the association of myelin basic protein with the bilayer head groups of dimyristoylphosphatidyl-choline and dimyristoylphosphatidyl-glycerol. Biochemistry 23, 2032–2039.

Smith, I.C.P. (1983) Deuterium NMR. In: *NMR of Newly Accessible Nuclei*, Volume 2, pp. 1–26. Editor. P. Laszlo. Academic Press, New York.

Smith, I.C.P., Butler, K.W., Tulloch, A.P., Davis, J.H. and Bloom, M. (1979) The properties of gel state lipid in membranes of *Acholeplasma laidlawii* as observed by ^{2}H NMR. FEBS Lett. 100, 57–61.

Spiess, H.W. and Sillescu, H. (1980) Solid echoes in the slow motion regime. J. Magn. Reson. 42, 381–389.

Sternin, E., Bloom, M. and MacKay, A.L. (1983) De-Pake-ing of NMR spectra. J. Magn. Reson. 55, 274–282.

Stockton, G.W. and Smith, I.C.P. (1976) A deuterium magnetic resonance study of the condensing effect of cholesterol on egg phosphatidylcholine bilayer membranes I. Perdeuterated fatty acid probes. Chem. Phys. Lipids 17, 251–263.

Tamm, L.K. and Seelig, J. (1983) Lipid solvation of cytochrome *c* oxidase. Deuterium, nitrogen-14 and phosphorus-31 nuclear magnetic resonance studies on the phosphocholine head group and on *cis*-unsaturated fatty acyl chains. Biochemistry 22, 1474–1483.

Unwin, P.N.T. and Henderson, R. (1984) The structure of proteins in biological membranes. Sci. Am. 250, 78–94.

Weitekamp, D.P. (1983) Time-domain multiple-quantum NMR. Adv. Magn. Reson. 11, 111–274 (Editor: J.S. Waugh).

Wiener, J.H., Lemire, B.D., Jones, R.W., Anderson, W. and Scraba, D.G. (1984) A model for the structure of fumarate reductase in the cytoplasmic membrane of *Escherichia Coli*. J. Cell. Biochem. 24, 207–216.

Watts/De Pont (Eds.)
Progress in Protein-Lipid Interactions

CHAPTER 3

Modulation of lipid polymorphism by lipid-protein interactions

B. DE KRUIJFF[a], P.R. CULLIS[b], A.J. VERKLEIJ[a], M.J. HOPE[b], C.J.A. VAN ECHTELD[c], T.F. TARASCHI[d], P. VAN HOOGEVEST[e,*], J.A. KILLIAN[e], A. RIETVELD[e] and A.T.M. VAN DER STEEN[e,**]

[a]*Institute of Molecular Biology, State University of Utrecht, Padualaan 8, 3584 CH Utrecht, The Netherlands,* [b]*Department of Biochemistry, University of British Columbia, Vancouver BC, V6T 1W5, Canada,* [c]*Department of Cardiology, University Hospital, State University of Utrecht, Catharijnesingel 101, Utrecht, The Netherlands,* [d]*Hahnemann Medical College, Department of Pathology, 230 N Broad St., Philadelphia, PA 19102, U.S.A. and* [e]*Department of Biochemistry, State University of Utrecht, Padualaan 8, 3584 CH Utrecht, The Netherlands*

I. Introduction

The functioning of biological membranes is determined to a large extent by the chemical, structural and dynamical properties of its main components, e.g., lipids and proteins, and by the interactions among and between these molecules.

Concerning the lipid backbone of the membrane it is now evident that the bilayer is the basic structure which is ideally suited for the main general function of a membrane, namely to act as a semi-permeable barrier. Proteins are supposed to provide the membrane with its specific and selective functions such as signal transmission, pumping and other transport activities and energy conservation.

In the last decade it has become clear that the above structural view of the lipid part of a membrane is too simple. In particular the realization that each

---

* Present address: CIBA-Geigy Ltd., Agricultural Division, R & D Seeds/Biotechnology, Schwarzwaldallee 211, CH-4002 Basle, Switzerland.

** Present address: Organon Teknika, Veedijk 58, B-2300 Turnhout, Belgium.

Abbreviations: CL, cardiolipin; DEPE, dielaidoylphosphatidylethanolamine; DOPE, dioleoylphosphatidylethanolamine; LPC, lysophosphatidylcholine; NMR, nuclear magnetic resonance; PA, phosphatidic acid; PC, phosphatidylcholine; PE, phosphatidylethanolamine; PG, phosphatidylglycerol; PS, phosphatidylserine; SRP, signal recognition particle; T_{BH}, bilayer → hexagonal H_{II} transition; WGA, wheat-germ agglutinin.

membrane contains substantial amounts of lipids which upon isolation and dispersion in aqueous buffers do not organize themselves in a bilayer but in a non-bilayer configuration has suggested that lipids might play additional structural and functional roles (Cullis and De Kruijff, 1979; Cullis et al., 1983; De Kruijff et al., 1984). The most commonly encountered non-bilayer configurations adopted by lipids in model membrane systems are the hexagonal H_{II} phase and the related lipidic particles. These non-bilayer structures can greatly affect the functional behavior of the model membranes. They might be intermediates in vesicle fusion, they appear to be involved in lipid flip-flop, and might act as carriers for polar compounds.

Although unambiguous proof for the existence of non-bilayer lipid structures in biological membranes is still lacking, there are now several morphological and functional indications which suggest that these structures can be formed in a variety of membranes (Kachar and Reese, 1982; Pinto da Silva and Kachar, 1982; Corless and Costello, 1981; Crowe and Crowe, 1982; Simpson, 1978; Larsson et al., 1980; Chi and Lagunoff, 1978; Van Venetië and Verkleij, 1982). From the functional point of view, it is clear that mechanisms should exist by which the structure of lipids can be locally and/or transiently modulated to allow specific processes to occur at discrete sites. Lipid-protein interactions are among the most likely candidates for such regulatory roles. It is the aim of this Chapter to provide a comprehensive review of the various ways lipid polymorphism can be modulated by lipid-protein interactions.

II. Lipid polymorphism

1. PHASE PREFERENCES OF MEMBRANE LIPIDS: TECHNICAL ASPECTS

The pioneering and original studies of the group of Luzatti on the structure of lipid/water systems (Luzatti, 1968; Luzatti et al., 1968) has led the basis for most of our present knowledge on lipid polymorphism. This group established that virtually every lipid showed polymorphism in that it could adopt different phases depending on the experimental conditions.

One lipid could well be organized in ten or more different phases which often were found to be specific for the type of lipid under investigation. The structural assignment of these phases in some cases has been very difficult and despite the enormous progress made in this area still the exact molecular architecture of many phases remains obscure.

With this overwhelming complexity of lipid polymorphism in mind, it is reassuring that when restricted to conditions of excess aqueous buffer at ‘physiological’ pH, ionic strength and temperature, membrane lipids adopt only a limited number of phases of which the bilayer and the hexagonal H_{II} phase are the most relevant (for general reviews of membrane lipid polymorphism see Cullis and De

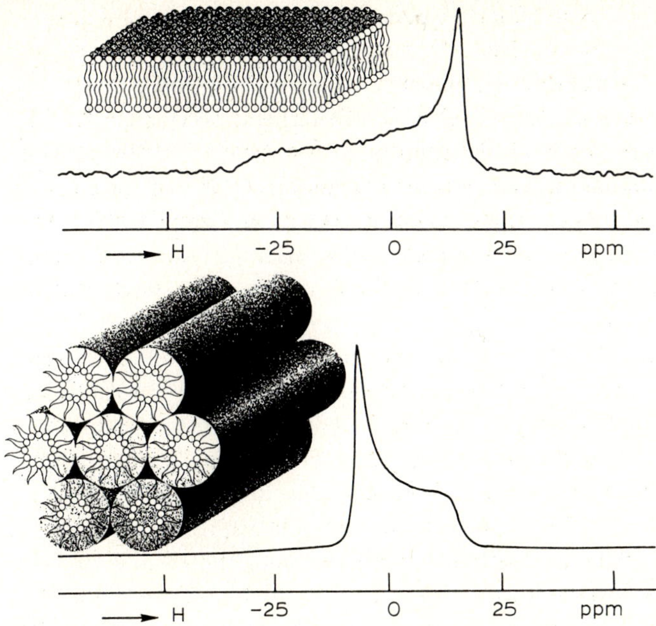

Fig. 1. Schematic representation of the bilayer and the hexagonal H_{II} phase and the corresponding characteristic ^{31}P-NMR spectra of aqueous dispersion of respectively $22{:}1_c/22{:}1_c$-PC (upper) and $18{:}1_c/18{:}1_c$-PE (lower) at 20°C.

Kruijff, 1979; Cullis et al., 1983; De Kruijff et al., 1984). Schematic representations of these structures are presented in Fig. 1.

In the hexagonal H_{II} phase the lipid molecules are organized in long tubes with their polar headgroups surrounding the aqueous channel present within these tubes. The individual tubes are hexagonally organized and the subscript II is used to distinguish it from the less familiar hexagonal H_I phase. In the hexagonal H_I phase, which is, for instance, adopted by lysophospholipids at low water content, the headgroups cover the surface of the tubes which now have a hydrophobic interior.

Compared to the bilayer organization, the hexagonal H_{II} phase is hydrophobic in nature due to the inverted orientation of the lipid molecules in the cylinders. It is in particular this characteristic, together with the low hydration, which forms the basis for the functional significance of H_{II} type of lipids in membrane function. Spectroscopic techniques have indicated that the acyl chains are progressively more disordered towards the terminal methyls, which is consistent with this inverted orientation and the strong curvature of the lipid monolayers surrounding the aqueous channels (Cullis and De Kruijff, 1979; Cullis et al., 1983; De Kruijff et al., 1984). The tube diameters range from 50 to 80 Å, depending on the type of lipid or ionic composition of the medium (Verkleij, 1984). In excess buffer phase separation occurs between the hexagonal H_{II} phase and the excess aqueous phase. In these cases the phase boundary along the tubes is formed by a lipid monolayer (Van Venetië and Verkleij, 1981).

Although classification of membrane lipids according to their phase preference is somewhat arbitrary, as will become clear in subsequent sections, it is now evident that under physiological conditions of excess water (> 50% by weight) and ambient (0–40°C) temperatures phosphatidylcholines (PC), sphingomyelins, diglucosyl- and digalactosyldiglycerides and phosphatidylinositols isolated from various membranes, prefer the bilayer organization, whereas phosphatidylethanolamines (PE) and monoglucosyl- and monogalactosyldiglycerides prefer the hexagonal H_{II} phase (see, for review, De Kruijff et al., 1984). The negatively charged phospholipids (phosphatidylserine (PS), phosphatidylglycerol (PG), phosphatidic acid (PA) and cardiolipin (CL)) can adopt both phases depending on the ionic composition and pH of the medium (Cullis et al., 1983). The quantitatively less abundant lysophospholipids and glycolipids such as gangliosides, prefer micellar organizations under these conditions.

For a fruitful discussion of the relationship between lipid-protein interactions and lipid polymorphism it is useful to briefly describe the experimental methods used to characterize the various possible lipid structures.

Small angle X-ray diffraction is the classical technique to elucidate the macroscopic organization of lipid aggregates (Luzatti, 1968). In systems that posses long range order, analyses of the relationship between first and higher order reflections can provide an unambiguous means for lipid phase determination. In multi-bilayer systems the spacings of the first and higher order reflections relate as $1:\frac{1}{2}:\frac{1}{3}:\ \ldots$. The higher order reflections are often weak and in some multilamellar systems even absent, which can make a conclusive assignment of the phase rather difficult. The first order reflection corresponds to the lamellar repeat distance, which is the sum of the bilayer thickness and the thickness of the water layer separating subsequent bilayers.

For lipids organized into the hexagonal H_{II} phase a series of sharp reflections which relate as $1:\frac{1}{\sqrt{3}}:\frac{1}{2}:\frac{1}{\sqrt{7}}:\ldots$ are observed which are typical for the two-dimensional periodicity of hexagonally packed cylinders. The $\frac{1}{\sqrt{7}}$ and higher order reflections are very weak and often not observed. The repeat distance d is determined by the first order reflection of the hexagonally H_{II} phase and related as $d = \frac{\sqrt{3}}{2}\,a$ to the diameter, a, of the lipidic cylinders. The occurrence of a second order reflection with a $\frac{1}{\sqrt{3}}$ relationship to the first order reflection is the most clear indication for the presence of a hexagonal phase.

When three-dimensional periodicity is present in a lipid preparation often many reflections can be found which in some cases can be interpreted in terms of the lattice type and space group to which the structure belongs.

In particular, in mixtures of bilayer and hexagonal H_{II} types of membrane lipids sometimes cubic symmetry is found. The molecular architecture of these cubic phases has yet not been satisfactorily resolved, but appears to be related to the so-called lipidic particles, which will be discussed in subsequent sections.

When the long range order is distorted or not present at all, the sharp Bragg reflections are replaced by continuous scattering profiles that are much more difficult to interpret (Blaurock, 1982). This particularly complicates the analysis of the phase state in samples that contain different phases, of which only one is ordered. In this case X-ray can overlook the presence of disordered structures.

Among the electron microscopic techniques to study lipid polymorphism, freeze-fracturing has been shown to be the most valuable method. This success can be attributed to: (1) the physical fixation by fast-freezing; (2) the replication method which excludes beam damage during electron microscopical imaging, and (3) the resolution power of about 30 Å. Moreover, this technique, as with any microscopic method, provides direct visualization of the macromolecular structure in contrast to the spectroscopic and diffraction methods which give averaged information about the entire sample. Figure 2 shows the characteristic smooth fracture faces of lipid bilayers organized in a multilamellar fashion. The hexagonal H_{II} phase gives rise to distinct fracture faces composed of long parallel lines, which occur along at least two fracture planes at angles of approximately 120° to each other (Deamer et al., 1979). One of the most important contributions of the freeze-fracture electron microscopy to lipid polymorphism has been the discovery

Fig. 2. Freeze-fracture morphology of (A) multilamellar egg-PC liposomes, (b) $18{:}1_c/18{:}1_c$-PE in H_{II} phase (20°C), and (C) lipidic particles observed in $18{:}1_c/18{:}1_c$-PC/CL (1:1) liposomes upon the addition of Ca^{2+} (Verkleij et al., 1979). Original magnification 100 000 ×, reduced by a factor of 65/100 by the publisher.

of the so-called lipidic particles (Verkleij et al., 1979). These are small (60–120 Å) spherical entities found on the fracture of bilayers of many different lipid systems in which at least one of the components present favours hexagonal H_{II} phase formation. In some cases the lipidic particles can be one-, two- or three-dimensionally organized (Fig. 2). Despite considerable discussion on their molecular organization, there is a growing consensus that they represent intrabilayer inverted micelles. These structures might be present within one bilayer or lie at the nexus of intersecting bilayers (Fig. 3). In some cases these lipidic particles are aligned with the tubes of which the hexagonal H_{II} phase is formed (Van Venetië and Verkleij, 1981; Verkleij et al., 1980), which is the most clear evidence for a structural relationship between these phases. For a recent review on lipidic particles see Verkleij (1984).

It has to be noted that, even using fast freezing devices in the case of temperature-dependent bilayer → hexagonal H_{II} phase transitions, the high temperature phase (which always is the hexagonal H_{II} phase) may not be preserved during quenching of the sample. To avoid such phase changes the use of ultra-rapid freezing devices is highly recommended, in particular since these methods can be applied without the use of cryoprotectants, which can affect the phase state of the lipids.

The introduction of nuclear magnetic resonance (NMR) techniques, in particular ^{31}P-NMR, has greatly contributed to our present knowledge of membrane lipid polymorphism (Cullis and De Kruijff, 1979). They allow for a convenient and quantitative discrimination between the most important phases found for

Fig. 3. Two possible modes of inverted micelle integration in a bilayer system.

hydrated membrane lipids. Samples can be heterogeneous, long range order is not required and these techniques can be equally well applied to model and biological membranes. The most important drawbacks are the low sensitivity and the indirect way of structure assignment, due to the fact that the NMR line-shape is affected by both structural and dynamic parameters.

In the case of ^{31}P-NMR the NMR line-shape of phospholipids is determined by the chemical shift anisotropy (CSA) of the phosphate group and the dipolar interaction between the phosphorus and the neighbouring methylene protons (for reviews see Cullis and De Kruijff, 1979; Seelig, 1978). This dipolar interaction complicates the interpretation of the spectra and therefore is reduced or eliminated by applying high power proton decoupling. The electron density around the phosphorus atom is anisotropic and is determined by the bonding pattern. Different orientations of the phosphate group in the magnetic field of the spectrometer will give rise to different shielding of the nucleus and, hence, will result in different resonant frequencies. The way this CSA is averaged by different motions of the lipid phosphates, forms the basis for the use of ^{31}P-NMR in the study of lipid polymorphism. In the case of an organization in extended bilayers, the fast axial rotation of the phospholipid molecules gives rise to the characteristic asymmetrical ^{31}P-NMR line-shape depicted in the top part of Fig. 1. The distance between the high field peak and the low field shoulder ($\Delta\sigma$) is approximately 40 ppm for most membrane phospholipids.

This diagnostic 'bilayer' line-shape is only observed for lipids organized in relatively large membrane systems. When the membrane vesicles become smaller than about 1000 Å radius (Burnell et al., 1980), lateral diffusion of the lipid molecules and vesicle tumbling will (partially) average the CSA leading to a gradual change to an isotropic NMR signal, such as depicted in Fig. 4.

In the case of phospholipids organized in cylinders, such as found in the hexagonal H_{II} phase, the lateral diffusion of the lipid molecules around the pipes provides additional averaging of the CSA resulting in a ^{31}P-NMR line shape with a reversed shape and a reduction of $\Delta\sigma$ by a factor two (Fig. 1, bottom). For lipid molecules organized in structures allowing isotropic motion, such as micelles, small vesicles, inverted micelles or the highly curved interwoven bilayers present in one of the cubic phases, an isotropic ^{31}P-NMR line-shape is observed (Fig. 4). It is clear that ^{31}P-NMR alone cannot give conclusive information on the molecular structure of such systems.

From a comparison between Figs. 1 and 4, it can be seen that the dominant spectral components of the bilayer, 'hexagonal H_{II}', and 'isotropic' ^{31}P-NMR spectra occur at characteristic chemical shift positions, which is a very useful diagnostic tool to characterize lipid structure under conditions where more than one phase is present. It should be realized that in theory it is possible by changing the headgroup structure to generate hexagonal H_{II} and isotropic ^{31}P-NMR line-shapes for phospholipids organized in bilayers. However, there is now consider-

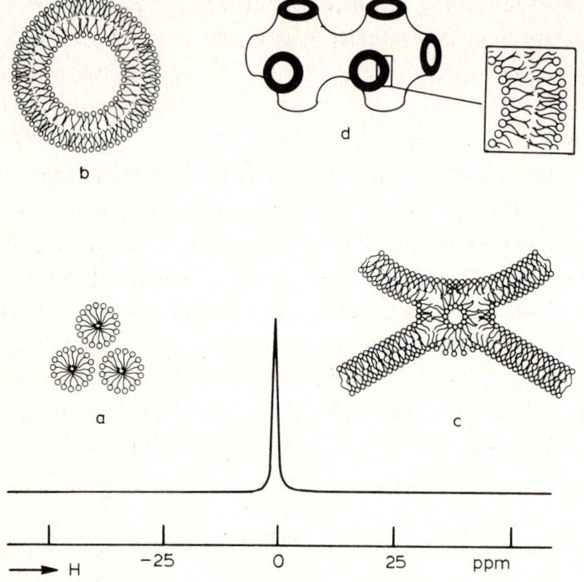

Fig. 4. Isotropic ^{31}P-NMR signals can originate from (a) micelles, (b) small unilamellar vesicles, (c) inverted micelles and (d) a cubic phase (Larssen et al., 1980).

able evidence that the phosphate segment of a lipid headgroup has a stable structure which is similar for a large variety of different phospholipids. Suggestions (Thayer and Kohler, 1981) that changes in headgroup structure could give rise to erroneous phase assignments remain speculative in the absence of experimental evidence.

^{2}H-NMR on specifically deuterated lipids is potentially a useful technique to study lipid polymorphism since it allows the detection of the structure of specific lipids in complex mixtures (Tilcock et al., 1982). The way the quadrupolar moment of the deuteron is motionally averaged in different macroscopic structures resembles the way the CSA of the ^{31}P nucleus is affected by motion of the lipid molecules (for review see Seelig, 1977). In the absence of changes in local order a bilayer to hexagonal H_{II} transition will result in a decrease in the quadrupolar splitting by a factor of 2 (Gally et al., 1980).

From this summary of the experimental methods used to elucidate lipid structures it is clear that, whenever possible, a combination of the above-mentioned methods should be used to get a full description of the phase state of lipids in any particular system.

2. MODULATION OF LIPID STRUCTURE

Before discussing the relationship between lipid-protein interactions and lipid polymorphism, it is essential to briefly summarize the parameters which determine the phase state of hydrated membrane lipids in the absence of protein.

Aqueous dispersions of PE, monoglucosyl- and monogalactosyldiglycerides show, at neutral pH, a temperature- and fatty acid composition-dependent polymorphism, in that at low temperatures the bilayer and at higher temperatures the hexagonal H_{II} phase is preferred. In the case of PE, the bilayer → hexagonal H_{II} transition is an interbilayer fusion event (Cullis et al., 1980), which requires less than 1 kcal/mol (Seddon et al., 1983; Cullis et al., 1978), and which can occur very rapidly (in the order of 10^{-3} s) (Cullis and De Kruijff, 1978). The acyl chains in the H_{II} phase are always liquid-crystalline which means that this transition always occurs at higher temperatures than the gel → liquid-crystalline phase transition. In some cases the temperature interval between these transitions can be very small (Cullis and De Kruijff, 1978; Marsh and Seddon, 1982). The temperature of the bilayer → hexagonal H_{II} transition (T_{BH}) is highly dependent on the acyl chain composition. Table 1 summarizes values of T_{BH} for various synthetic and natural PEs. Increasing unsaturation and chain length promotes H_{II} phase formation. It is evident that most natural PEs prefer the H_{II} phase at physiological temperatures. The phase state of PE is also dependent on the charge of the headgroup and the ionic strength of the medium. Deprotonation of the $-NH_3^+$ group results in bilayer stabilization. This effect is already noticeable at slightly alkaline pH (Seddon et al., 1983; Cullis and De Kruijff, 1978; Hardman, 1982). In contrast, protonation of the phosphate group will result in bilayer destabilization (Seddon et al., 1983). Increasing the ionic strength will promote H_{II} phase formation (Harlos and Eibl, 1981). Some of these effects might be related to headgroup hydration, which is known to be important for determining the structure adopted by the lipids. A low hydration will increase the tendency to form a hexagonal H_{II} phase (Seddon et al., 1983). Unfortunately, there is only limited information on headgroup hydration of lipids in excess water.

The isothermal modulation of lipid structure is probably biologically most interesting. In particular, changes in lipid structure induced by electrostatic interactions are relevant. The phase state of negatively charged phospholipids is very sensitive to changes in the ionic composition of the aqueous medium. Whereas all major naturally occurring negatively charged membrane phospholipids in 100 mM NaCl at pH 7 prefer a lamellar organization, changes in pH and salt concentration and the presence of divalent cations can trigger a transition to the hexagonal H_{II} phase. In the case of PG, PA, CL and PS, protonation of the phosphate or carboxyl group (in the case of PS) results in hexagonal H_{II} phase formation (for review see Cullis et al., 1983). Low concentrations of divalent cations induce hexagonal H_{II} phase formation in PA (at pH values between 6 and 8) (Verkleij et al., 1982; Farren et al., 1983) and in CL (Rand and Sengupta, 1972; Cullis et al., 1978; De Kruijff et al., 1982; Seddon et al., 1983). This latter system has been most thoroughly studied. In the case of the sodium salt of the highly unsaturated beef heart CL (which in all eukaryotic cells is localized exclusively in the

TABLE 1
Bilayer-hexagonal H_{II} transition temperatures of PEs

Product	Temperature (°C)	Remarks
Synthetic lipids		
Saturated (diester)		
18:0/18:0-PE	105[a]	1 M NaCl
Saturated (diether)		
1,2-dihexadecyl-*sn*-glycero-3-phosphoethanolamine	87[a], 88[b]	1 M NaCl
1,2-dietetradecyl-*rac*-glycero-3-phosphoethanolamine	93.5[a]	1 M NaCl
Saturated (ether-ester)		
1-hexadecyl-2-palmitoyl-*sn*--glycero-3-phosphoethanolamine	102[b]	
Unsaturated (ester)		
$16{:}1_c/16{:}1_c$-PE	~0[c]	
$18{:}1_c/18{:}1_c$-PE	10[d]	
$18{:}1_t/18{:}1_t$-PE	55[d], 60–63[e], 65[f]	
$18{:}2_{cc}/18{:}2_{cc}$-PE	0–25[g], < −15[e]	
$18{:}3_{ccc}/18{:}3_{ccc}$-PE	0–30[g]	
20:4/20:4-PE	<−30[g]	
22:6/22:6-PE	<−30[g]	
Mixed species		
16:0/18:1-PE	75[g]	
Natural species		
Egg	25–30[d], 32–45[b], 28[h], 28[i]	
Egg (from egg-PC via phospholipase D)	50[i], 63[b]	
E. coli	55–60[d]	Wild-type
Endoplasmic reticulum	7[j]	Rat liver
Sarcoplasmic reticulum	−10[k]	Rabbit muscle
Inner mitochondrial membrane	10[l]	Rat liver
Erythrocyte membrane	8[4]	Human
Soya	−10[m]	
Soya (from soya-PC, via phospholipase D)	0–20[n]	
Bovine white matter phospholipid	18[b]	Containing plasmalogen

[a] Harlos and Eibl (1981); [b] Boggs et al. (1982); [c] Van Dijck et al. (1976); [d] Cullis and De Kruijff (1978a); [e] Tilcock and Gullis (1982); [f] Ghosh and Seelig (1982); [g] Dekker et al. (1983); [h] Hardman (1982); [i] Mantsch et al. (1981); [j] De Kruijff et al. (1980a); [k] Cullis et al. (1983); [l] Cullis et al. (1980b); [m] Cullis and De Kruijff (1978b); [n] Cullis and Hope (1980).

inner mitochondrial membrane), the bilayer phase is preferred at neutral pH in the absence of divalent cations. Due to the negative charge on the lipids the dispersion swells considerably, resulting in a large interbilayer distance or even in the formation of large unilamellar vesicles.

The presence of 1–3 mM Ca^{2+} results in the precipitation of CL in a (Ca^{2+}-CL) complex, which above 0°C is organized in the hexagonal H_{II} phase (De Kruijff et al., 1982). This Ca^{2+}-induced bilayer → hexagonal H_{II} transition is endothermic with $\Delta H = 1.8$ kcal/mol and is accompanied by a marked increase in Ca^{2+} binding from a maximum of 0.35 Ca^{2+}/CL in the bilayer phase to the stoichiometric value of 1.0 in the hexagonal H_{II} phase (De Kruijff et al., 1982).

Other divalent cations such as Mg^{2+}, Mn^{2+} and Ba^{2+} also interact strongly with CL, resulting in the formation of the respective salts which show temperature-dependent lamellar → hexagonal H_{II} phase transitions (Van Venetië and Verkleij, 1981; Rand and Sengupta, 1972; Vasilenko et al., 1982). That a specific divalent cation/CL complex is essential for H_{II} phase formation in this system is unlikely, since high concentrations of NaCl and low pH also induce the hexagonal H_{II} phase (Seddon et al., 1983).

Biological membranes contain a large variety of different lipids. Therefore, an understanding of the phase behaviour of lipid mixtures is a prerequisite to understand the possible phase state of lipids in biomembranes. Mixtures of lipids which form the most common lipid structures, e.g., the bilayer and H_{II} phase are clearly most interesting. One general observation is that incorporation of increasing amounts of bilayer-forming lipids in a dispersion of H_{II} type lipids will cause a bilayer stabilization of the latter lipid (Cullis et al., 1983). Figure 5 illustrates this for a dispersion of soya-PE, which is titrated with egg-PC. At an equimolar concentration, a bilayer ^{31}P-NMR line-shape is observed for all the lipids present in the mixture. Another general finding is that, in such systems at intermediate concentrations, a new ^{31}P-NMR spectral component is observed (De Kruijff et al., 1980b). This component has the characteristic chemical shift position and narrow line width for phospholipids undergoing rapid isotropic motion. Such motion was apparently not possible in either the hexagonal H_{II} phase or the extended bilayers which are formed by the PC alone. Freeze-fracturing has revealed that in these mixtures the so-called lipidic particles (Fig. 2) are present, which most likely represent intrabilayer inverted micelles (Fig. 3). The isotropic ^{31}P-NMR signal will contain contributions from lipids organized in these particles but might also be caused by lipid molecules diffusing laterally along the highly curved bilayers with which these particles are associated.

As expected from the phase behaviour of the pure components, the formation of isotropic structures and associated lipidic particles in lipid mixtures can be modulated by factors such as lipid composition, temperature and electrostatic interactions. One intriguing result from such studies has been that, in lipid mixtures in which part of the lipids adopt a non-bilayer phase (isotropic or hexagonal

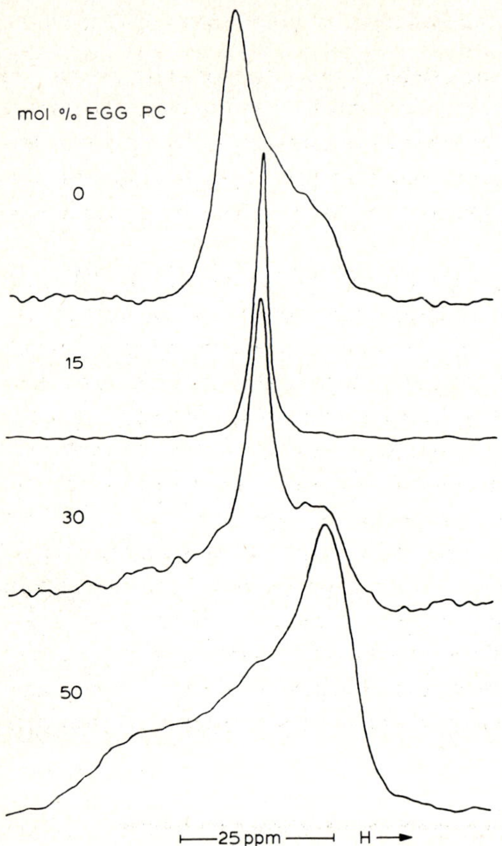

Fig. 5. ^{31}P-NMR spectra of soya-PE dispersions containing increasing amounts of egg-PC. Reproduced with permission from Cullis and De Kruijff (1978b).

H_{II}), the bilayer-preferring lipid is proportionally equally abundant in the non-bilayer as in the bilayer phase (Tilcock et al., 1982; De Kruijff et al., 1979). This most likely is the result of the low energy barrier between bilayer and non-bilayer lipid structures. Another interesting aspect is that the modulation of bilayer → hexagonal H_{II} phase transitions sometimes follows other rules, as established in the regulation of bilayer fluidity, e.g., modulation of the gel → liquid-crystalline phase transition. For instance, increasing chain length increases the gel → liquid-crystalline phase transition, but decreases the bilayer → hexagonal H_{II} phase transition temperature (Seddon et al., 1983). Similarly, cholesterol incorporation decreases membrane fluidity but promotes H_{II} phase formation (Tilcock et al., 1982; Cullis et al., 1978).

3. THE MOLECULAR SHAPE CONCEPT: A RATIONALE FOR LIPID POLYMORPHISM

From the foregoing it will be clear that a detailed unifying molecular description of all aspects of membrane lipid polymorphism will be very difficult. Any theory will have to deal with the relative importance of factors such as headgroup hydration, dipolar, electrostatic, hydrophobic and Van der Waals interactions, and molecular dynamics in determining the way the lipid molecules aggregate into defined macroscopic structures.

As a working model we (Cullis and De Kruijff, 1979; Cullis et al., 1983; De Kruijff et al., 1984) commonly use the shape-structure concept. In this concept it is proposed that the 'shape' of a lipid molecule is the primary factor which determines the way the molecules will aggregate into larger structures. The word shape is placed in inverted commas to indicate that, in this concept, shape is meant to be more than the geometric space occupied by the isolated molecule. In our view, it includes headgroup hydration, intra- and intermolecular interactions, and lipid dynamics. In this concept lipid molecules with a relatively large hydrophilic moiety have an inverted cone shape which is easily accommodated into micellar structures (Fig. 6). When the size of the polar part is more in balance with the hydrophobic portion of the molecule, the shape will be more cylindrical, which is more compatible with a bilayer organization. In the case when the headgroup is relatively small, the molecule will have a cone shape which is ideally suited to accommodate the molecules into inverted structures

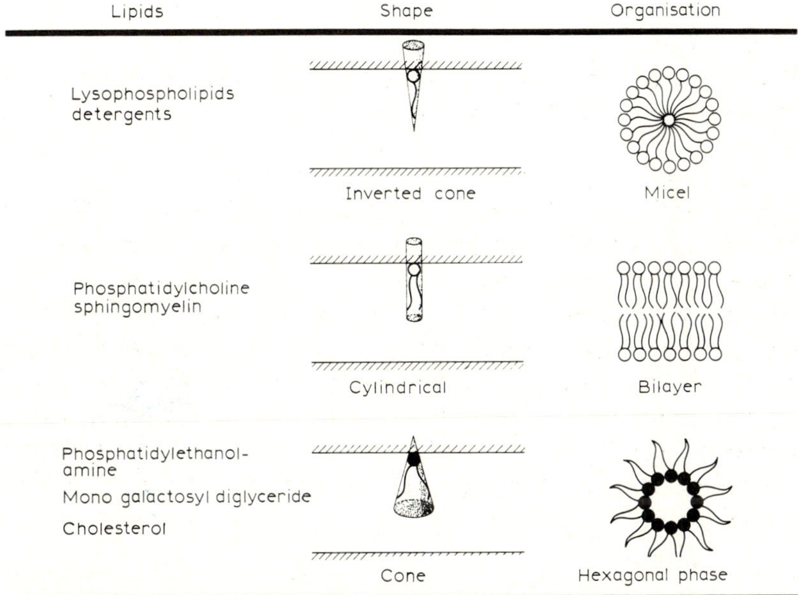

Fig. 6. The molecular shape concept of lipid polymorphism.

such as the hexagonal H_{II} phase or inverted micelles. With this concept most of the available data on polymorphism of membrane lipids can be understood. It explains the effect of fatty acid composition and temperature on the polymorphism of PEs and describes the effect of electrostatic interactions on the structure of dispersions of negatively charged lipids rather well. One of the most convincing examples of the shape-structure relationship is the observation of a lamellar phase for an equimolar mixture of an inverted cone-shaped lipid such as lyso-PC and the cone-shaped unsaturated PE (Madden and Cullis, 1982).

III. Lipid-protein interactions and lipid polymorphism

In general, two classes of proteins can be recognized in biological membranes. The first class is made up by proteins which are located at the exterior of the membrane which they might partially penetrate, and which are interacting with the membrane mainly via electrostatic interactions, possibly in combination with a hydrophobic interaction (extrinsic proteins). These proteins can often be dissociated from the membrane by washing with salt solutions. The second class comprises of proteins which are deeply buried in the hydrophobic interior of the membrane and which often span the bilayer (intrinsic membrane proteins). To solubilize these proteins, detergents are often needed. This classification of membrane proteins is followed when discussing lipid polymorphism in relation to lipid-protein interactions.

1. EXTRINSIC MEMBRANE PROTEINS AND POLYPEPTIDES

a. Poly-L-lysine

Since the availability of membrane proteins which interact only electrostatically with lipids is limited, poly-L-lysine is commonly used as a model for such proteins. This highly basic linear polymer of lysine has a random coil structure at neutral pH and interacts strongly with negatively charged phospholipids, resulting in an increase in temperature and heat content of the gel → liquid-crystalline phase transition (Papahadjopoulos et al., 1975). Further, in mixed systems, it can induce lateral phase separations, resulting in the formation of rigid poly-L-lysine/lipid domains (Hartmann and Galla, 1978).

Studies on the influence of poly-L-lysine on lipid polymorphism have been directed towards an answering of two important questions. Firstly, does the electrostatic polypeptide-lipid interaction affect the macroscopic lipid structure in a one component lipid system? Secondly, can the ability of poly-L-lysine to induce phase separations in mixed lipid systems result in bilayer → non-bilayer phase changes?

Addition of poly-L-lysine (degree of polymerization, $N = 200$) to a CL dispersion in 100 mM NaCl, pH 7.0, results in an immediate precipitation of the lipids in the form of a stoichiometric poly-L-lysine/CL complex (one lysine per CL phosphate) (De Kruijff and Cullis, 1980). During this interaction, a transient disruption of the bilayers must occur, since all CL molecules become available for interaction with poly-L-lysine. The resulting complex is organized in a multilamellar structure as evidenced by freeze-fracturing, ^{31}P-NMR (Fig. 7) and X-ray diffraction (De Kruijff and Cullis, 1980). The interlamellar repeat distance is 53 Å. Peptides with a decreasing degree of polymerization down to $N = 5$, give very similar binding patterns, and structural characteristics of the poly-L-lysine/CL complex have been observed (Vaandrager et al., unpublished observations). Lowering N to 3 and 2, resulted in a weaker binding but still lamellar phases were present, even in the presence of a large excess of the peptide.

Thus, in contrast to the induction of a hexagonal H_{II} phase by Ca^{2+}, high salt or low pH, the charge neutralization by poly-L-lysine does result in a stable bilayer organization of CL. A model for the poly-L-lysine/CL complex which accommodates all the present knowledge on this system is shown in Fig. 8. The poly-L-lysine chain is sandwiched between two bilayers, and the various lysine residues can stoichiometrically interact in an alternating manner with CL present in both bilayers.

The close apposition of the bilayers in this complex, as well as the transient disrupture of the bilayers during the interaction with the liposomes, might explain

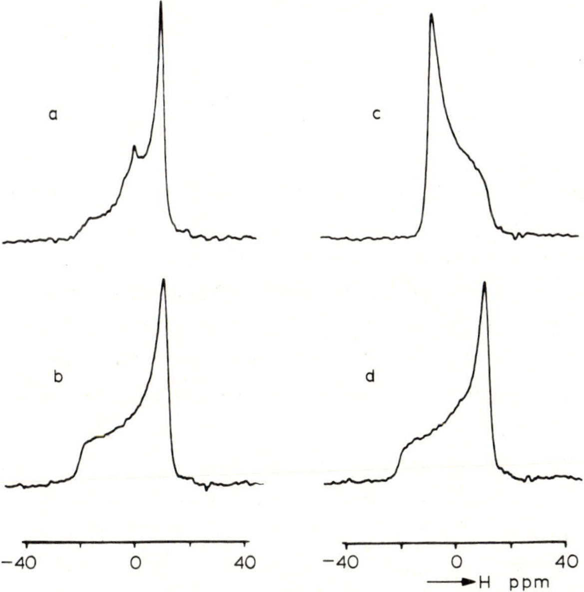

Fig. 7. ^{31}P-NMR spectra at 30°C of (A) 50 μmol of CL in 1.0 ml buffer (B) 50 μmol of CL in 1.0 ml buffer, 5 min after the addition of 40 mg poly-L-lysine, and (C, D) as in (A, B) after addition of 100 μl 1 M $CaCl_2$. Reproduced with permission from De Kruijff and Cullis (1980a).

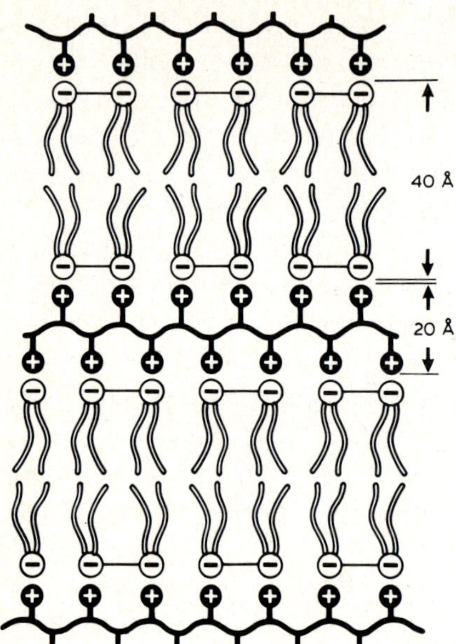

Fig. 8. Schematic representation of the poly-L-lysine CL (1:1, lysine/P_i) multilamellar complex.

the high lipid vesicle fusion potency (Eytan and Almary, 1983) of poly-L-lysine. The structural stability of the poly-L-lysine/CL complex is further indicated in competition experiments with Ca^{2+}. In the absence of the polypeptide, Ca^{2+} addition results in H_{II} phase formation. However, in the presence of poly-L-lysine, addition of a large molar excess of Ca^{2+} does not affect the lamellar structure nor the poly-L-lysine/CL binding (Fig. 7) (De Kruijff and Cullis, 1980). Even the reverse is true; poly-L-lysine addition to the hexagonally organized CL/Ca^{2+} complex results in dissociation of Ca^{2+} and the formation of the poly-L-lysine/CL lamellar complex (Vaandrager et al., unpublished observations). Apparently the strong cooperativity in binding of the individual monovalent lysine residues in the polypeptide plays a determinant role in the stability of the complex. For the other major negatively charged membrane lipids PS and PG, poly-L-lysine addition also results in the formation of a peptide-lipid complex in which the phospholipids are organized in a multi-lamellar phase (De Kruijff and Cullis, 1980).

The second question concerning the ability of poly-L-lysine to trigger bilayer → non-bilayer transitions in mixed lipid systems is addressed by the results shown in Fig. 9. Incorporation of 33 mol% beef heart CL in soya-PE model membranes stabilizes the bilayer structure of the PE which, in the absence of CL, prefers a hexagonal H_{II} phase at temperatures above 10°C. The origin of the small isotropic ^{31}P-NMR signal is unknown (Fig. 9A). Addition of poly-L-lysine at 30°C results in ^{31}P-NMR spectra in which at the low field side of the isotropic peak signal

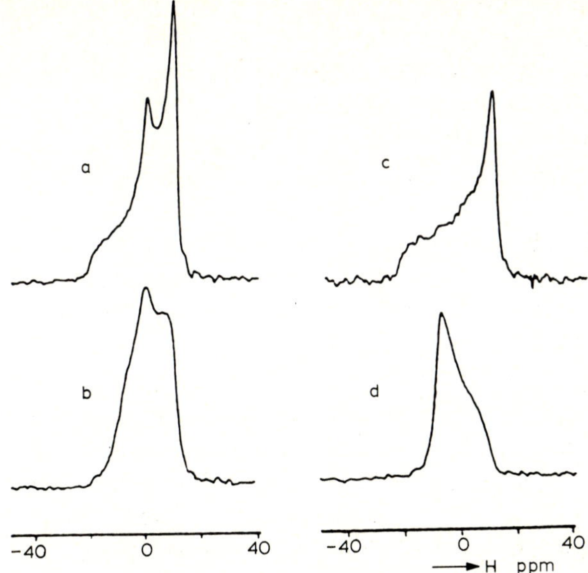

Fig. 9. ^{31}P-NMR spectra of a dispersion of 50 μmol soya-PE CL (2:1) in 1.0 ml buffer (A) at 30°C, (B) 5 min after the addition of 40 mg poly-L-lysine, (C) as (B) except recorded at 0°C, (D) as (B) except that 100 μl 1 M $CaCl_2$ was added. Reproduced with permission from De Kruijff and Cullis (1980a).

intensity is present from lipid molecules organized in cylindrical structures (Fig. 9B). Freeze-fracturing revealed the presence of extended areas of hexagonal H_{II} phase (De Kruijff and Cullis, 1980a). Addition of Ca^{2+} instead of poly-L-lysine caused a similar but much more dramatic effect (Fig. 9D). Since the hexagonal H_{II} spectral component induced by poly-L-lysine vanished at lower temperatures (Fig. 9C), it most likely originates predominantly from the PE component in the mixture. This result can be interpreted as follows. In the starting situation all lipid molecules are randomly dispersed in the bilayer. As a consequence of the poly-L-lysine/CL interaction phase separation occurs between a lamellar poly-L-lysine/CL complex and a hexagonally organized PE component. Since all lipid molecules are organized in a liquid-crystalline phase (as indicated by the ^{31}P-NMR line-shape), this experiment demonstrates that electrostatic peptide-lipid interactions can induce separation and polymorphic phase transitions in a fluid membrane system. Since the asymmetric lipid composition of biological membranes is such that negatively charged lipids are often preferentially located together with the hexagonal H_{II} preferring PE in the inner leaflet of the membrane (Op den Kamp, 1979), this type of peptide-lipid interaction could be of importance for the (local) regulation of lipid structure on that side of the membrane.

b. *Small amphipathic peptides*

Many extrinsic membrane proteins also (partially) interact hydrophobically with the membrane lipids. Model peptides for such proteins are the small basic peptides, mellitin and the cardiotoxins. These peptides, which are isolated from venoms, have a strong potency to interact electrostatically with negatively charged lipids. Several studies have indicated that subsequently the acyl chain packing is strongly perturbed and that a penetration of the peptide in the model membrane occurs (Habermann, 1980; Dufourq et al., 1982). There is little known about the effect of these peptides on the macroscopic organization of the model membranes. This is even more surprising in view of the fact that high concentrations of these peptides can totally disrupt the lipid bilayer.

For cardiotoxin IV (isolated from snake venom) it has been shown that, in a complex natural lipid mixture, addition of the peptide resulted in the formation of non-bilayer lipid structures such as the lipidic particles and the hexagonal H_{II} phase (Gulik-Krzywicki et al., 1981). Interpretation of these results should be made with great care in view of the presence of trace amounts of phospholipases in these preparations. However, these initial experiments are promising and should be extended to better defined lipid systems to substantiate the suggestion that the combination of charge neutralization and peptide penetration can result in the formation of non-bilayer lipid structures.

c. *Cytochrome* c *and apocytochrome* c

Cytochrome *c* is a basic (eight net positive charges per molecule at pH 7.0) protein component of the outer leaflet of the inner mitochondrial membrane where it participates in the electron transport through the terminal part of the respiratory chain. It is imported into the mitochondrion as a precursor molecule, the haem-free apocytochrome *c*, which is synthesized on free ribosomes in the cytosol (Schatz and Mason, 1974). The polypeptide chains of both proteins are identical, but their secondary and tertiary structure is greatly different. Cytochrome *c* is a highly structure, nearly spherical protein, whereas apocytochrome *c* has virtually no structure and adopts a random coil configuration (Fisher et al., 1973). During import the precusor is converted into the mature protein (Hennig and Neupert, 1981).

Cytochrome *c*/lipid interactions have been extensively investigated (for a review see Nicholls, 1974). The protein binds electrostatically to negatively charged phospholipids, penetrates lipid mono- and bilayers (Papahadjopoulos et al., 1975; Gulik-Krzywicki et al., 1969), enchances the cation permeability of lipid bilayers (Kimelberg and Papahadjopoulos, 1971), it decreases the temperature and energy content of the gel → crystalline phase transition (Papahadjopoulos et al., 1975) and can induce lipid phase separations (Brown and Wüthrich, 1977; Birrell and Griffith, 1976). The protein does not interact with vesicles made of zwitterionic lipids.

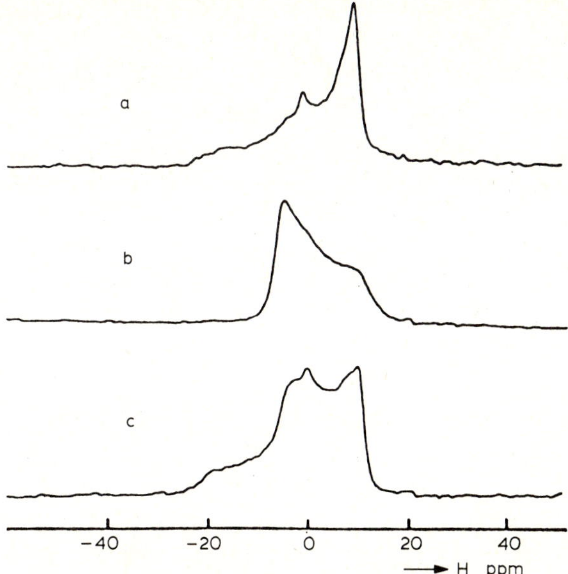

Fig. 10. ^{31}P-NMR spectra at 30°C of an aqueous dispersion of (A) CL in the presence of (B) Ca^{2+} and (C) cytochrome *c*. Beef heart CL (50 μmol) was dispersed in 1.0 ml 100 mM NaCl, 10 mM Tris/HCl, 0.2 mM EDTA, pH 7.0. In (B) 0.1 ml 1 M $CaCl_2$ and in (C) 0.2 ml buffer containing 36 mg cytochrome *c* was added. For details see De Kruijff and Cullis (1980b).

In the case of beef heart CL, cytochrome *c* addition results in the formation of a hexagonal H_{II} phase for part of the lipid, as well as the formation of a structure in which the lipid molecules undergo rapid isotropic motion (De Kruijff and Cullis, 1980b; Gulik et al., 1969) (Fig. 10). On the fracture face small particles are present, which must reflect the protein or protein-lipid complexes. These structural changes are specific for CL (De Kruijff and Cullis, 1980b). Addition of cytochrome *c* to PS or PG also results in a strong lipid-protein interaction, but now the lipids are organized in multilamellar system with smooth fracture faces. The formation of the hexagonal H_{II} phase by cytochrome *c* suggests that the protein actually (partially) resides within the aqueous cylinders, which is compatible with the dimensions of the protein and the water channel. The specificity of the cytochrome *c*-induced change in structure for CL, together with the notion that CL appears to be essential for the functioning of the cytochrome *c* oxidase, opens the possibility that an intrabilayer localization of cytochrome *c* caused by the interaction with CL might facilitate the flow of electrons to the oxidase (De Kruijff et al., 1981).

Apocytochrome *c* affects lipid structure in a quite different way. This protein, also upon interaction, appears to penetrate the bilayer and affects the macroscopic organization of the lipids but now the structural changes are of a different nature and specificity. Freeze-fracture electron microscopy shows particulate

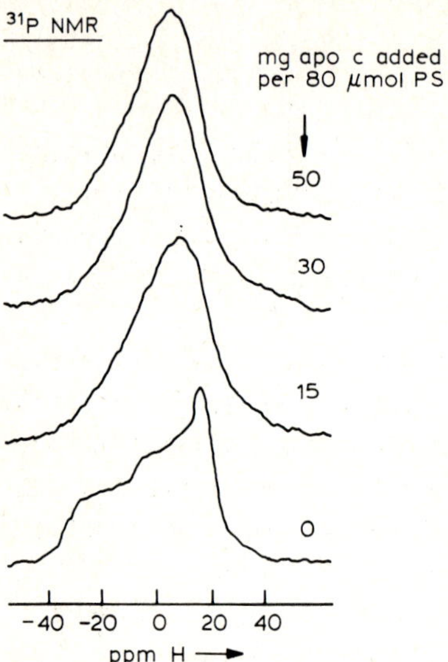

Fig. 11. ^{31}P-NMR spectra of apocytochrome *c*-PS recombinants. For details see Rietveld et al. (1983).

structures on the lipid bilayers of both CL and PS following the addition of the protein (Rietveld et al., 1983). In addition, the typical asymmetrical bilayer line-shape of the ^{31}P-NMR spectrum of both lipids is narrowed, suggesting additional motional averaging of the CSA. No hexagonal H_{II} phase is observed. This is shown in Fig. 11 for the apocytochrome *c*/PS system. The change of the spectrum to a more isotropic line-shape suggests increased motional freedom of the lipid molecules which was not observed for cytochrome *c* and poly-L-lysine.

The exact molecular structure of the apocytochrome *c* negatively charged lipid recombinant awaits much more detailed studies. The present results, however, are intriguing in the light of the import of apocytochrome *c* into the mitochondrion. In this process a specific receptor for apocytochrome *c* located in the outer membrane plays an important role (Hennig and Neupert, 1981). The question of how the protein is translocated across the membrane is totally unclear. Since PS is a negatively charged lipid in the outer membrane, a PS-apocytochrome *c* interaction followed by a local change in lipid structure could possibly provide a pathway for the translocation of the protein across the membrane. Very recent model membrane experiments indeed demonstrated that this protein can get (partially) across a PS bilayer (Rietveld and De Kruijff, 1984).

That the three basic proteins poly-L-lysine, cytochrome *c* and apocytochrome *c* affect lipid structures in CL systems in a protein-specific way, suggests that the overall protein structure, the spatial distribution of the positively charged amino acids, and the hydrophobicity of the protein must play a determinant role in the final structure of the protein-lipid recombinant.

2. INTRINSIC MEMBRANE PROTEINS AND PEPTIDES

a. Gramicidin

The pentadecapeptide gramicidin has been a popular model for intrinsic membrane proteins in studies on lipid-protein interactions in model membrane systems. It is easily available and its chemical structure is known. This extremely hydrophobic peptide can from *trans*-membrane channels through which small molecules can pass the membrane.

In the channel conformation the gramicidin molecule is most likely present in the form of a dimer which is held together by hydrogen bonding. There is still considerable discussion and confusion concerning the actual structure of the dimer. One group (Urry et al., 1983) is strongly in favor of the $-NH_2$ terminal to $-NH_2$ terminal (L,D) model (Figs. 12 and 17). Other groups proposed antiparallel – and parallel – double helices or hybrid structures (Veatch et al., 1974; Ivanov, 1982). Even studies on chemically prepared dimers have not yet unambiguously resolved this matter (Ivanov, 1982). Since it is well known that gramicidin can adopt different conformations in different solvents, it is possible that its conformation might be dependent on the type of membrane system studied. In lamellar lipid systems formed by saturated PCs, gramicidin broadens the gel to liquid-crystalline phase transition and reduces its energy content (Chapman et al., 1977). Below the transition temperature it decreases chain order. In the liquid-crystalline state for gramicidin/lipid < 1:15 m/m the chains are more ordered, whereas at higher concentrations the chains are more disordered. Similarly, a recent difference infra-red spectroscopic study revealed that, at high

Fig. 12. Schematic representation of the π_6 (L, D) N-N dimer of gramicidin in a bilayer. The sequence of gramicidin is: HCO-L-Val-Gly-L-Ala-D-Leu-L-Ala-D-Val-L-Val-D-Val-L-Trp-D-Leu-L-Trp-D-Leu-L-Trp-D-Leu-L-Trp-NHCH$_2$-CH$_2$OH.

concentrations, gramicidin causes an increase in lipid chain *gauche* isomers above the transition temperature of the lipid (Lee et al., 1984).

Studies on the effect of gramicidin on lipid polymorphism have been exciting as they have revealed the dramatic way the macroscopic structure adopted by lipids can be affected by lipid-protein interactions.

A good insight into the way a particular compound can affect lipid polymorphism can be obtained by studying the effect of this molecule on the temperature-dependent bilayer to hexagonal H_{II} transition of a PE. This is shown in Fig. 13 for $18:1_t/18:1_t$-PE-gramicidin lipid mixtures which were obtained by hydrating a mixed lipid-peptide film with 100 mM NaCl, 10 mM Tris/HCl, pH 7.0 (Van Echteld et al., 1981). Using ^{31}P-NMR, the fraction of lipid molecules giving rise to a bilayer or hexagonal line-shape was determined as a function of temperature. For the pure lipid a sharp transition occurs at approximately 58°C. Incorporation of increasing amounts of gramicidin broadens the transition and shifts it to lower temperatures. The molecular efficiency of these effects is strong. The midpoint of the transition shifts by approximately 5°C per mol% gramicidin incorporated (see insert, Fig. 13). From the shape of the temperature profile it can be suggested that gramicidin has two effects. In the first place gramicidin incorporation appears to result in the formation of a DEPE-gramicidin 'complex' which is organized in a liquid-crystalline hexagonal H_{II} phase even at temperatures where the PE would otherwise be organized in a lamellar gel phase (e.g., below 35°C). The structure of this complex is rather temperature independent, the amount formed being proportional to the amount of gramicidin present. The second effect appears to be that the presence of this complex lowers and broadens the bilayer to hexagonal H_{II} transition temperature of the excess 'free' PE molecules. Similar results have

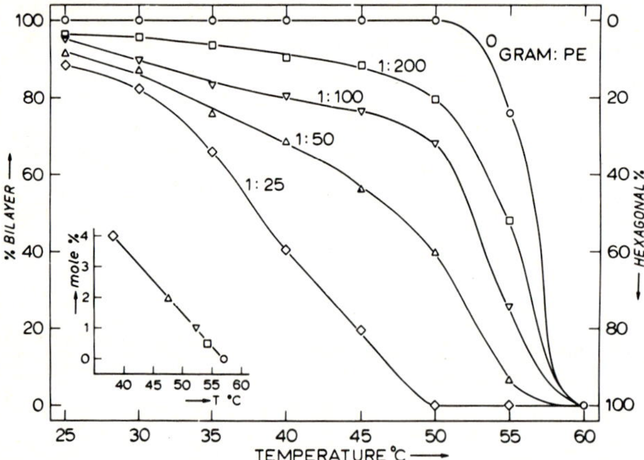

Fig. 13. Effect of gramicidin on the bilayer → H_{II} transition of $18:1_t/18:1_t$-PE. The insert shows the relationship between gramicidin concentration (mol%) and the temperature at which 50% of the lipid is organized in the H_{II} phase. Reproduced with permission from Van Echteld et al. (1981).

been obtained for $18{:}1_c/18{:}1_c$-PE-gramicidin systems (Van Echteld et al., 1981).

One direct implication of these results is that gramicidin is associated with the hexagonal H_{II} phase. The most likely organization is that of gramicidin present in the hydrophobic core and oriented with the long axis of the molecule parallel to the fatty acid chain, or as dimers spanning the hydrophobic domain between adjacent water cylinders.

The strong bilayer destabilizing effect of gramicidin is even apparent in PC systems which are typical bilayer-forming lipids. X-ray, freeze-fracture electron microscopy and ^{31}P-NMR techniques demonstrated that incorporation of gramicidin in $18{:}1_c/18{:}1_c$-PC dispersions resulted in the formation of a hexagonal H_{II} phase for part of the lipids (Van Echteld et al., 1981). For a 1:10 molar gramicidin/lipid ratio already some 30% of the lipids are organized in a hexagonal H_{II} phase as can be derived from the ^{31}P-NMR line-shape (Figs. 14 and 15). The amount of H_{II} phase induced by gramicidin was found to be virtually independent on temperature (Van Echteld et al., 1982). Very similar results have been observed when gramicidin was added from an ethanolic solution to preformed $18{:}1_c/18{:}1_c$-PC liposomes (Killian et al., 1984).

Further insight in the nature of the phase change has been obtained from

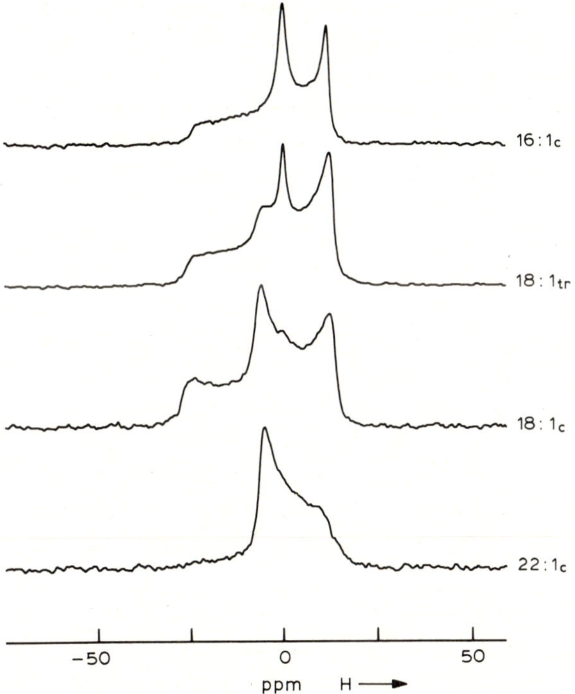

Fig. 14. 81.0 MHz ^{31}P-NMR spectra of various mixed PC-gramicidin (10:1; mol/mol) dispersions. Reproduced with permission from Van Echteld et al. (1982).

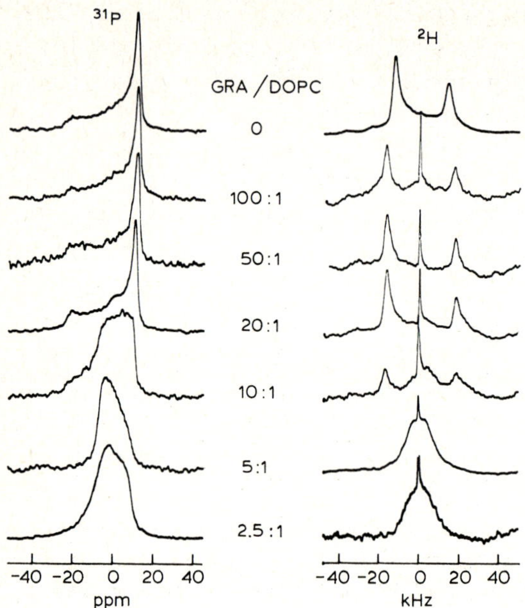

Fig. 15. 81.0 MHz ^{31}P-NMR and 30.7 MHz ^{2}H-NMR spectra of 1,2-(11,11-dideutero), $18{:}1_c/18{:}1_c$-PC gramicidin dispersions. A dry film of 50 μmol of the ^{2}H-labelled PC (Tilcock et al., 1982) containing the appropriate amount of gramicidin was hydrated with 1.0 ml 100 mM NaCl, 10 mM Tris/HCl, pH 7.0, whereafter the ^{2}H-spectrum was recorded as described in Tilcock et al. (1982).

^{2}H-NMR studies on 1,2-(11,11-dideutero), $18{:}1_c/18{:}1_c$-PC-gramicidin systems. In the absence of the peptide a quadrupolar splitting of 8 kHz is observed for this lipid at 30°C (Fig. 15). Upon incorporation of gramicidin this splitting is gradually replaced by one of 2–3 kHz width. Theoretically a change from a lamellar to a hexagonal H_{II} phase would result in a decrease of quadrupolar splitting by a factor of 2, due to rapid reorientation of the molecules around the cylinders of the H_{II} phase (Seelig, 1977). The larger reduction in quadrupolar splitting demonstrates that, in addition, the local order of acyl chains is decreased in the H_{II} phase as compared to the bilayer organization. Such differences in acyl packing between lamellar and H_{II} phase have been observed in other lipid systems (Tilcock et al., 1982; Gally et al., 1980; Hardman, 1982), and further substantiate the inverted nature of the hexagonal H_{II} phase. However, it should be realized that also, in the hexagonal H_I phase formed by detergent PC mixtures at low water content, the acyl chain order of the PC molecule is decreased when compared to the bilayer situation (Beyer, 1983; Ulmius et al., 1982). The hexagonal H_{II} promoting ability of gramicidin is also manifest in mixed lipid systems. In the absence of gramicidin an equimolar mixture of $18{:}1_c/18{:}1_c$-PE and $18{:}1_c/18{:}1_c$-PC is organized in a lamellar phase (Fig. 16). Incorporation of increasing amounts of gramicidin induces the formation of an H_{II} phase for part of the lipids.

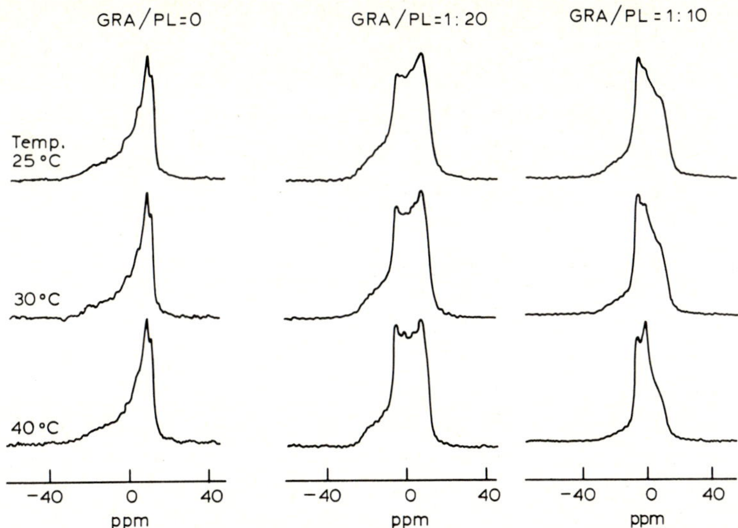

Fig. 16. 81.0 MHz ^{31}P-NMR spectra of $18:1_c/18:1_c$-PE-$18:1_c/18:1_c$-PC (1:1)-gramicidin dispersions at various temperatures. Mixed films of 50 mol $18:1_c/18:1_c$-PE and $18:1_c/18:1_c$-PC containing the appropriate amount of gramicidin were hydrated with 150 mM NaCl, 10 mM Tris/HCl, pH 7.0, whereafter the ^{31}P-NMR spectrum was recorded as described in Van Echteld et al. (1981).

One intriguing question which will have to be resolved in the future is whether, under such conditions, the H_{II} phase is preferentially composed of one of the two lipid species present. This question touches the heart of lipid-protein interactions and translates into: can a membrane protein interact preferentially with one lipid species in a mixture, and does this give rise to special structural and functional features of the system?

The efficiency of gramicidin to induce the H_{II} phase in PC systems is highly dependent on the nature of the acyl chains. In the case of unsaturated species the amount of H_{II} phase induced by gramicidin increases with increasing chain length (Fig. 14). The *trans*-unsaturated species being less effective than the *cis*-unsaturated one. In $16:1_c/16:1_c$-PC-gramicidin mixtures, a structure is formed which allows the lipid molecules to reorient themselves rapidly in all directions. Freeze-fracturing showed that such samples consisted of extended networks of interwoven lipid bilayers and ice-fracture planes. These types of structures, which resemble cubic phases (Verkleij, 1984), have been observed in a number of cases in which the system was in an intermediate state between the lamellar and the hexagonal H_{II} phase. The inverted micellar lipidic particles have not yet been observed in any gramicidin-containing lipid system.

In natural PC species gramicidin again induces the hexagonal H_{II} phase for part of the lipids (Van Echteld et al., 1982). The same was observed for the disaturated 18:0/18:0-PC-gramicidin system. In contrast, reducing the chain length to

16, 14 or 12 carbon atoms resulted in lamellar complexes only (Van Echteld et al., 1982). This might explain why the ability of gramicidin to affect lipid polymorphism was only recently discovered. Virtually all previous studies (Chapman et al., 1977; Rice and Oldfield 1979; Lee et al., 1984) on gramicidin-lipid systems had been conducted on model membranes of 14:0/14:0-PC and 16:0/16:0-PC, which remained organized in bilayer organization upon gramicidin incorporation. This shows that one should be careful in generalizing findings obtained for one particular lipid system.

How can we understand the hexagonal H_{II} phase-inducing ability of gramicidin? To answer this question we have to consider a number of possible parameters which are relevant to lipid polymorphism. These include interfacial hydration, the length of the gramicidin dimer as compared to the bilayer thickness, lipid packing and finally the shape of the gramicidin molecule. We will discuss these factors in turn.

It is well-known that headgroup hydration and, consequently, interbilayer attractive forces are very important in lipid polymorphism (Seddon et al., 1983). Lipids with a low headgroup hydration favour the H_{II} phase. This can be understood with the molecular shape concept and, alternatively, can be rationalized by realizing that the bilayer → hexagonal H_{II} phase transition is an interbilayer fusion event (Cullis et al., 1980), which most readily occurs when the opposing bilayers are in the closest proximity which, among other factors, can be achieved by a low interfacial hydration. Since gramicidin favours H_{II} phase formation, this would imply that gramicidin would dehydrate the surface of the lipid bilayer. In principle, this could be due to at least two effects: (i) a direct influence of the peptide on the hydration properties of the headgroups of the surrounding lipid molecules; (ii) attractive forces between gramicidin dimers in adjacent bilayers. Interbilayer head-to-head or tail-to-tail associations for instance via hydrogen bonding could result in a decrease in the thickness of the water layers separating the various bilayers in these multilayered systems. Even without such specific interactions it can be expected that the surface of the bilayer will become relatively hydrophobic due to a (partial) exposure of the hydrophobic residues of gramicidin present on the entry of the membrane channel.

In favour of a 'dehydration' contribution to the hexagonal H_{II} promoting ability of gramicidin is the visual observation that the hydration and swelling of lipids in buffer is strongly reduced upon increasing the gramicidin concentration. At high concentrations (molar ratios > 1:10, gramicidin/lipid) the lipids do not disperse at all anymore. Furthermore, the tube-to-tube distance of the hexagonal H_{II} phase induced by gramicidin in $22{:}1_c/22{:}1_c$-PC systems is virtually identical to the interlamellar repeat distance of the dispersions of this lipid in the absence of gramicidin (Van Echteld et al., 1982). For temperature-dependent bilayer → hexagonal H_{II} transitions in PE systems the tube-to-tube distance is ca. 15 Å larger than the interbilayer repeat distance (De Kruijff et al., 1984). This increase in repeat distance in the H_{II} phase must be predominantly due to the fact that the thickness of the aqueous cylinders in the H_{II} phase is ca. 15 Å larger than the thickness of the water layer between the lipid bilayers. From the geometry of the phases it can be estimated that these repeat distances are compatible with a similar water content of both phases. The absence of an increase in repeat distance of the bilayer → H_{II} phase transition induced by gramicidin in $22{:}1_c/22{:}1_c$-PC, thus can be interpreted as a (partial) dehydration of the lipids during (or prior) to H_{II} phase formation.

The strong fatty acid composition dependence of the induction of the H_{II} phase by gramicidin in PC systems is difficult to understand in this dehydration model for H_{II} phase formation, as at first sight one would not expect the hydration to be dependent on the acyl chain composition. However, it cannot be excluded that the surface structure of the PC-gramicidin bilayer is, for instance, dependent on the length of the fatty acids. For 16:0/16:0-PC in the liquid-crystalline state the thickness of the hydrophobic part of the bilayer can be estimated as 30–31 Å (Büldt et al., 1978). This is remarkably similar to the length of the gramicidin dimer present in a lipid environment which is approximately 30 Å (Wallace et al., 1981). Since the hexagonal H_{II} phase formation occurred with PC species which had acyl chain lengths exceeding 16 carbon atoms, it is tempting to suggest that a mismatch in length of the dimer and the hydrophobic part of the bilayer is involved in the H_{II} phase formation. This could be caused, for instance, by dimple formation around the entrance of the channel, which would result in a local curvature consistent with that found in the H_{II} phase or, alternatively, this could cause a change in hydration properties of the interphase. However, it should be realized that, in general, increasing the acyl chain length of a lipid promotes H_{II} phase formation, since lipid molecules with larger acyl chains can be more easily accommodated into a hexagonal H_{II} phase (Sheddon et al., 1983).

Lipid packing itself is also an important factor in the H_{II} phase formation induced by gramicidin. For instance in the gel state of PCs, gramicidin cannot induce the H_{II} phase, most likely because gel state lipids can be properly accommodated only in a lamellar structure. Furthermore, the difference in the amount of H_{II} phase induced by gramicidin in $18{:}1_c/18{:}1_c$-PC as compared to $18{:}1_t/18{:}1_t$-PC (Fig. 14) most likely is the result of the looser packing of the *cis* analogue of the PC. The area per molecule of these lipids at the air-water interface is 61 and 68Å/mol (surface pressure 30 mN/m^2) for the *trans* and *cis* compounds, respectively (Demel, unpublished observations). Thus, the $18{:}1_t/18{:}1_t$-PC is more cylindrical than $18{:}1_c/18{:}1_c$-PC. This means that this latter lipid would fit more easily into the hexagonal H_{II} phase. The observation (Rice and Oldfield, 1979; Lee et al., 1984) of a decreased chain order upon incorporation of high amounts of gramicidin favour such a model on gramicidin-induced H_{II} phase formation. It must be emphasized that the polymorphism of gramicidin-lipid systems cannot only be explained by lipid packing or bilayer 'fluidity'. For instance, gramicidin induces the hexagonal H_{II} phase into 18:0/18:0-PC but not in the more fluid $16{:}1_c/16{:}1_c$ liposomes.

A final point of importance might be the shape of gramicidin (or its dimer). As predicted from the shape concept (Fig. 6), gramicidin (or the dimer) should be conical to explain the hexagonal H_{II} phase formation as induced by the peptide. From the side-view of the space-filling model of gramicidin (Fig. 17), it can be seen that indeed the monomer has a pronounced cone shape due to the location of the four bulky tryptophan residues all at the C-terminal side of the molecule.

If the shape of the molecule is important for determining the macroscopic structure of lipids then we can make the prediction that the cone-shaped gramicidin should form bilayers with the inverted cone-shaped lyso-PC as is schematically represented in Fig. 18. As can be seen from the ^{31}P-NMR line-shapes of various lyso-PC/gramicidin mixtures this is indeed the case (Fig. 19). The isotropic signal originating from pure lyso-PC micelles is gradually replaced by an axial symmetric typical bilayer line-shape upon increasing the gramicidin content. These data demonstrate that gramicidin interacts with 16:0-lyso-PC to form a 1:4 molar lamellar complex which, in the case of an excess of lyso-PC, is in equilibrium with lyso-PC micelles (Killian et al., 1983). Qualitatively similar results were reported by Pasquali-Ronchetti et al. (1983).

Thus, it appears that indeed the shape of gramicidin is of importance in the lipid-protein interaction. However, there is one problem in that there is now

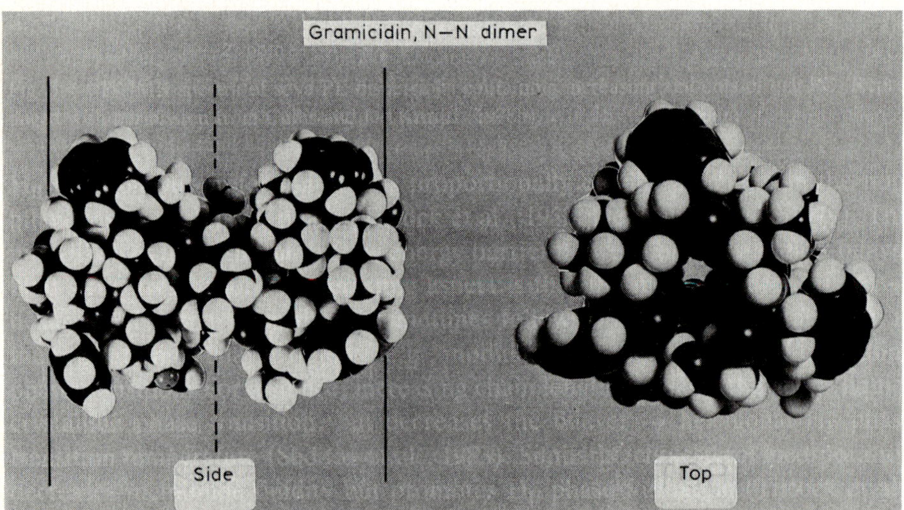

Fig. 17. Side and top view of space-filling models of the π(L, D), N-N-dimer of gramicidin. The dark lines represent the two bilayer interfaces. The dotted line indicates the middle of the bilayer and the site of association of the two gramicidin monomers. In the top view of the model the aqueous channel is visible.

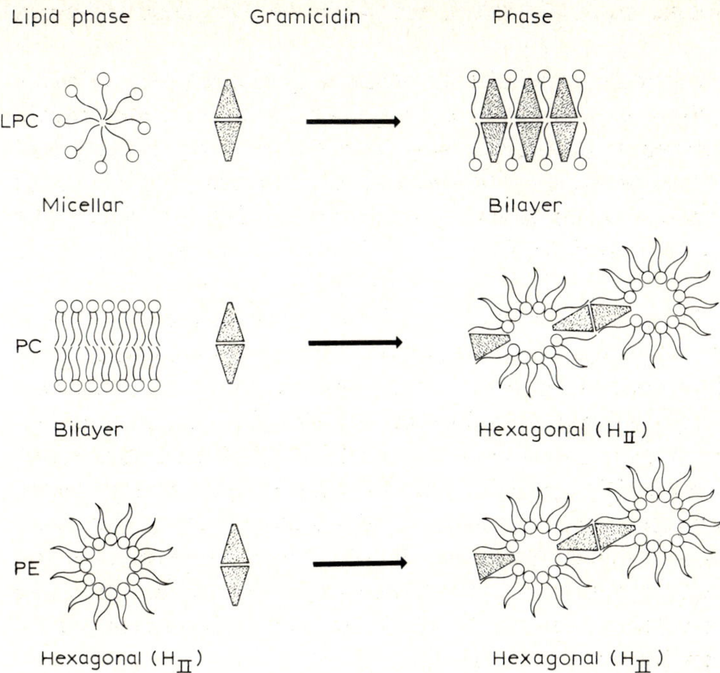

Fig. 18. Shape-structure relationships in gramicidin-lipid systems. The gramicidin is supposed to be organized as a C-C dimer.

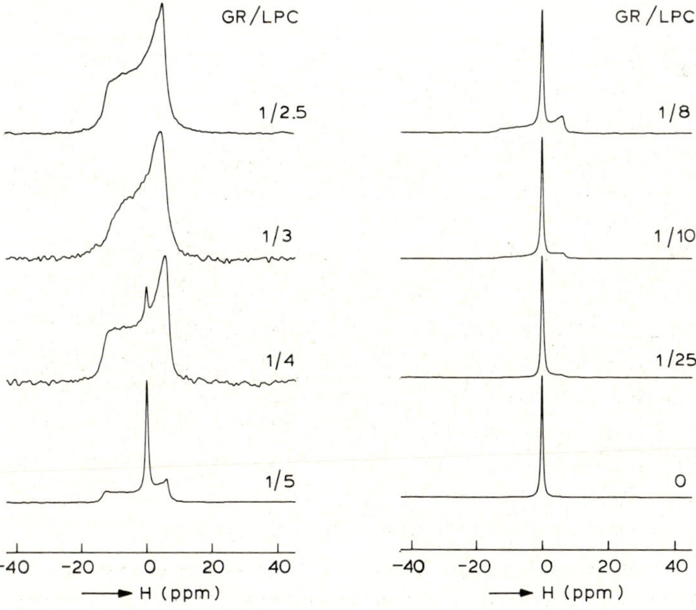

Fig. 19. 81.0 MHz ^{31}P-NMR spectra of various 1–16:0-LPC/gramicidin mixtures in aqueous buffer at 25°C. Reproduced with permission from Killian et al., 1983.

considerable support for the suggestion that, at least in some systems, gramicidin is present in the form of a dimer linked N by N (see Fig. 17). In this case the tryptophans are at the membrane-water interface which is opposite to the orientation inferred from the studies on lipid structure (Fig. 18).

To solve this apparent discrepancy and to further verify the importance of the shape of a protein for the structure of lipid-protein systems, will require chemical synthesis of gramicidin analogues including the covalently bonded N to N and C to C dimers.

b. Glycophorin

Although there have been many studies on the local structure and dynamics of lipids in recombinants of intrinsic membrane proteins, in only a limited number of cases has the question been addressed to whether this type of lipid-protein interaction affects lipid polymorphism. The best studied protein in that respect is glycophorin. Before discussing these studies it is useful to briefly review the properties of this major sialoglycoprotein of the human erythrocyte membrane. The primary structure and the topology of this protein in the erythrocyte membrane are known (Tomita and Marchesi, 1975) (Figs. 20 and 21). The large N-terminal carbohydrate-bearing portion of the protein is located at the outside of the membrane which is spanned by a 23 hydrophobic amino acid-long peptide, presumably organized in an α-helix. At the membrane-cytoplasm interface some positively charged amino acids are located (positions 96–100), which might be of importance for the proper positioning of the protein in the membrane. The smaller C-terminal part is exposed to the cytoplasm. The protein is highly negatively charged due to the presence of some 32 sialic residues present on the carbohydrate headgroup of the protein. These sugars act as receptors for several plant lectins such as WGA (Verpoorte, 1975). The protein can be readily purified and incorporated into model membranes by a variety of procedures. One particularly useful method is that of McDonald and McDonald (1975) which avoids the use of detergents. Protein and lipid are dissolved in an organic solvent mixture and, after evaporation, the mixed film is hydrated in buffer which results in the formation of lipid-protein recombinants. In the case of PCs large unilamellar glycophorin-containing vesicles, in which the protein spans the bilayer in a right side out orientation, can be obtained after differential centrifugation (Van Zoelen et al., 1978a). In such vesicles, glycophorin (present as small aggregates) perturbs the

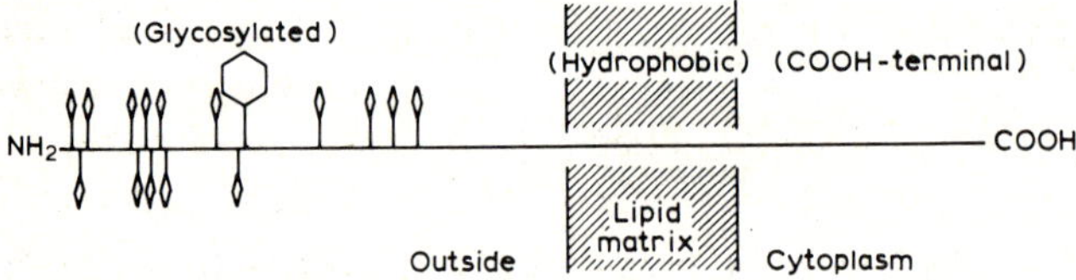

Fig. 20. Topology of glycophorin in the erythrocyte membrane.

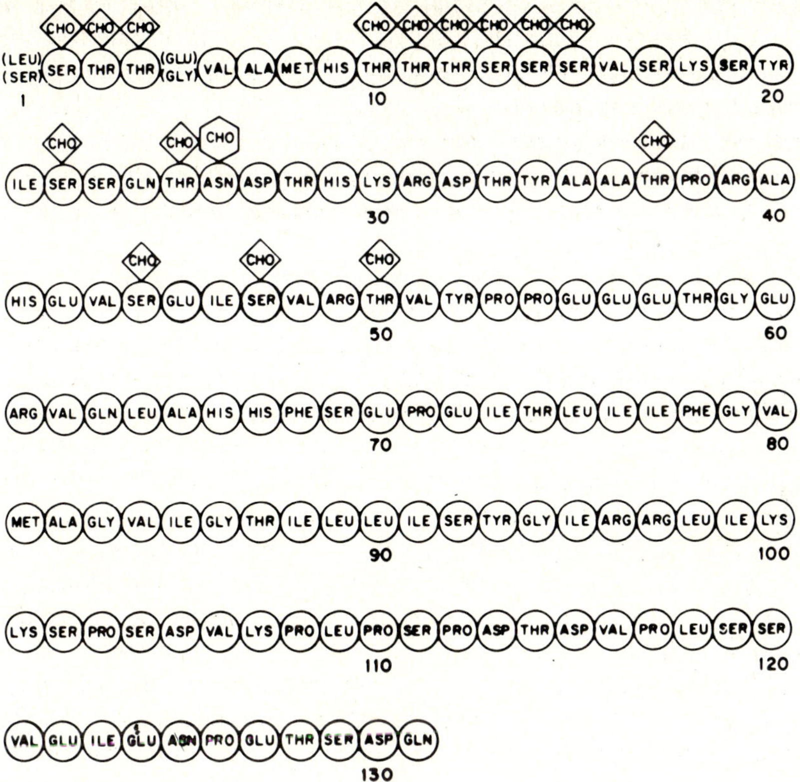

Fig. 21. Primary structure of human erythrocyte glycophorin A (Tomita and Marchesi, 1975).

lipid packing (Taraschi and Mendelsohn, 1980; Rüppel et al., 1982) and it reduces the heat content of the gel to liquid-crystalline phase transition (Van Zoelen et al., 1978a). The barrier function of these vesicles is dramatically affected by the presence of the protein. Phospholipid flip-flop is greatly enhanced (Van Zoelen et al., 1978b; De Kruijff et al., 1978) as is also the permeation of molecules with molecular weights of up to 900 (Van der Steen et al., 1982; Van Hoogevest et al., 1983). Packing defects at the protein-lipid interface and/or hydrophilic channels within protein aggregates are the most likely candidates for these transport pathways.

When the McDonald reconstitution procedure was applied to unsaturated PEs which prefer a hexagonal H_{II} organization, several interesting observations were made (Taraschi et al., 1982). Firstly, the mixed film dispersed readily in aqueous buffers, which is unusual for PE. Secondly, independent of the original lipid-protein ratio, two types of structures were formed: unilamellar vesicles with a high glycophorin/PE ratio (1:25, mol/mol) and H_{II} organized PE with a low glycophorin content. The relative amounts of both structures, which can be separated by centrifugation, was dependent on the initial lipid/protein ratio. The strong bilayer stabilization of glycophorin in this system is also evident from ^{31}P-NMR studies (Fig. 22). The glycophorin-rich PE vesicles at low temperatures displayed bilayer type line-shapes which changed to more isotropic line-shapes at higher temperatures. In no case were hexagonal type line-shapes observed. Fast reorientation of PE molecules by lateral diffusion or vesicle tumbling are the most likely motions which average the chemical shift anisotropy resulting in the isotropic line-shapes.

Using sonication methods, DOPE-glycophorin recombinants could be prepared with a varying lipid-to-protein ratio (Taraschi et al., 1982). For these vesicles it could be established that one glycophorin molecule can stabilize a bilayer organization of about 200 DOPE molecules (Taraschi et al., 1982). Taking into account that glycophorin appeared to be highly aggregated in PE bilayers, as

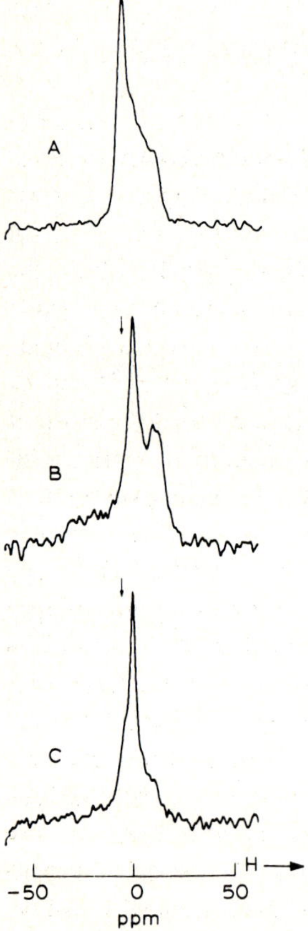

Fig. 22. 36.4 MHz ^{31}P-NMR spectra of $18{:}1_c/18{:}1_c$PE in the absence (0°C) (A) and presence (B) (0°C) and (C) 25°C of glycophorin. Arrow indicates the position of the main spectral component of $18{:}1_c/18{:}1_c$-PE in the H_{II} phase. Reproduced with permission from Taraschi et al. (1982a).

evidenced by freeze-fracture electron microscopy and rotational diffusion measurements (Van Hoogevest et al., 1984), it can be suggested that the molecular efficiency of the bilayer stabilization by glycophorin is even larger. Recently, a similar strong bilayer stabilization of PE systems by the incorporation of an intrinsic membrane protein was reported for DOPE model membranes containing the Ca^{2+}-ATPase from sarcoplasmic reticulum (Navarro et al., 1984). Interestingly, in these vesicles there was a maximal coupling of ATP hydrolysis and Ca^{2+} translocation. Similarly, chlorophyllase, a glycoprotein of chloroplast membranes, stabilizes the H_{II}-prefering monogalactosyldiglyceride into a bilayer configuration (Lambers et al., 1984). The glycophorin-induced bilayer stabilization is not confined to PE systems. Also, in the case of the isothermal induction of the H_{II} phase by Ca^{2+} in CL systems, glycophorin shows a similar behavior (Taraschi et al., 1983). There are several possible factors which may contribute to this effect. The large, charged hydrophilic headgroup of glycophorin could, for instance, be involved in either preventing an interbilayer fusion event, which precedes H_{II} phase formation, or because this headgroup cannot fit into the aqueous channels, prevents the H_{II} phase from forming.

Figure 23 gives some insight in this matter. Sonicated unilamellar DOPE-glycophorin (200:1, mol/mol) vesicles give rise to an isotropic ^{31}P-NMR signal due to the small size of the vesicles. Addition of neuraminidase, which removes the exposed sialic acid residues from the protein, does not result in a change in ^{31}P-NMR line-shape or vesicle morphology (Taraschi et al., 1982). Thus, the negative charge of the glycophorin headgroup does not appear to be responsible for the bilayer stabilization. In contrast, removal of the entire carbohydrate bearing headgroup by trypsin resulted in the formation of large amounts of H_{II} phase (Fig. 23). Apparently, the large headgroup of glycophorin prevents the fusion of the PE vesicles, which precedes H_{II} phase formation. Alternatively, a direct interaction between the sugar moiety of glycophorin and the PE headgroups could also play a role in the bilayer stabilization. Evidence for such interactions has been obtained for the glycophorin-PC system (Rüppel et al., 1982). That the membrane spanning part of glycophorin is involved in stabilizing the bilayer structure of PE, can be inferred from experiments in which sonicated DOPE-glycophorin (25:1, mol/mol) vesicles are treated with trypsin. Freeze-fracturing also shows that, in these protein-rich vesicles, fusion occurs but that the bilayer structure is maintained (Taraschi et al., 1982). Similarly, ^{31}P-NMR reveals that, in these vesicles, trypsin does not result in spectral components indicative of the H_{II} phase (Fig. 23). These results are schematically represented in Fig. 24.

The importance of the bilayer concentration of glycophorin for the structure of the PE-glycophorin vesicles is further emphasized in experiments with wheat-germ agglutinin (WGA). This multivalent lectin can bind to the carbohydrate parts of several glycophorin molecules at the same time, resulting in an aggregation of glycophorin in the bilayer and in clustering of the vesicles.

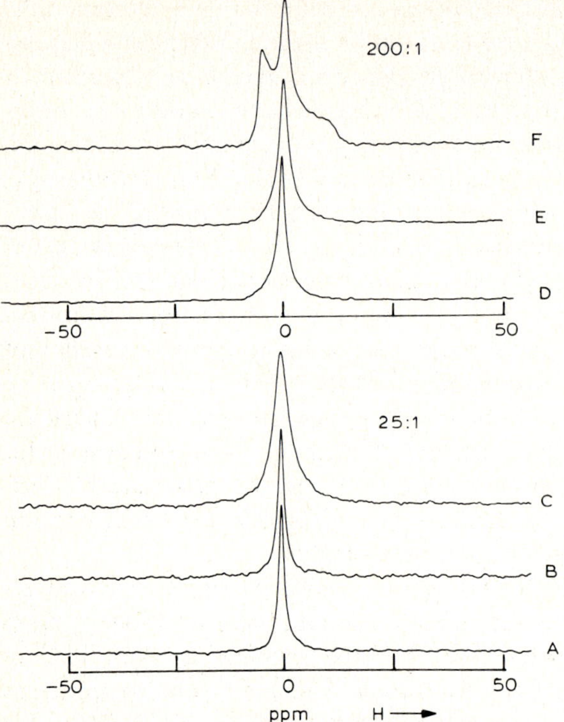

Fig. 23. Effect of neuraminidase and trypsin treatment on the 81.0 MHz ^{31}P-NMR spectra of cosonicated $18{:}1_c/18{:}1_c$-PE/glycophorin vesicles. The molar ratios of $18{:}1_c/18{:}1_c$-PE/glycophorin are indicated in the figure. Vesicles were treated with (B, E) neuraminidase (50 units/mg of protein, 2 h, 37°C), followed by (C, F) trypsin treatment (5%, w/w, with respect to glycophorin, 2 h, 37°C). Reproduced with permission from Taraschi et al. (1982b).

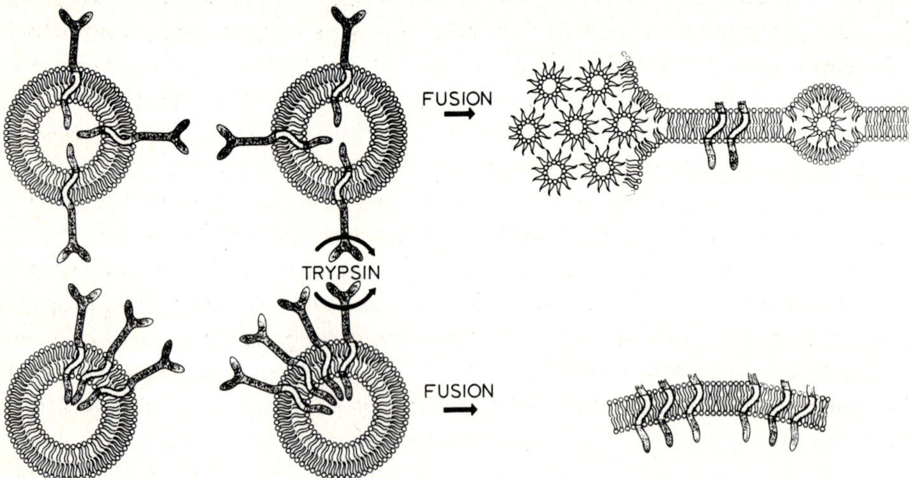

Fig. 24. Schematic representation of the structural consequences of the removal of the headgroup of glycophorin by trypsin in $18{:}1_c/18{:}1_c$-PE vesicles. Upper panel, $18{:}1_c/18{:}1_c$-PE/glycophorin 200:1 (mol/mol); lower panel, $18{:}1_c/18{:}1_c$-PE/glycophorin 25:1 (mol/mol).

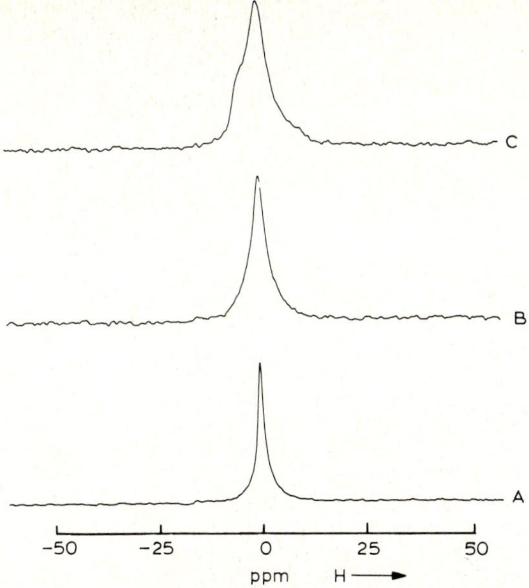

Fig. 25. 81.0 MHz ^{31}P-NMR spectra at 30°C of cosonicated $18{:}1_c/18{:}1_c$-PE/glycophorin vesicles (200:1, mol/mol) in the absence (A) and presence (B, C) of WGA. The molar ratios WGA:glycophorin were 0.5 and 1.0 in (B and C), respectively. For further details see Taraschi et al. (1982b).

Addition of excess WGA to DOPE-glycophorin 200:1 (mol/mol) vesicles immediately results in the formation of the H_{II} phase and lipidic particles (Figs. 25 and 26). This process, which can be completely inhibited by an excess of glucosamine, a sugar which competes with glycophorin for the binding to WGA, can be interpreted as being due to the local increase in concentration of glycophorin as a result in the WGA-glycophorin interaction, thereby removing the bilayer stabilization of the excess DOPE molecules (Taraschi et al., 1982). In fact, no such transition to an H_{II} phase is observed when WGA clusters the glycophorin molecules in DOPE-glycophorin 25:1 (mol/mol) vesicles. A schematic representation of the effect of WGA on the structure of $18{:}1_c/18{:}1_c$-PE/glycophorin recombinants is given in Fig. 27.

This result has intriguing functional consequences in particular in the area of ligand-receptor interactions. Clustering of the receptor caused by the addition of the ligand often results in uptake of the receptor-ligand complex via an endocytotic event involving membrane fusion. The experiment with WGA and glycophorin demonstrates that aggregation of the glycoprotein in the bilayer can result in the formation of non-bilayer lipid structures which might be intermediates in membrane fusion.

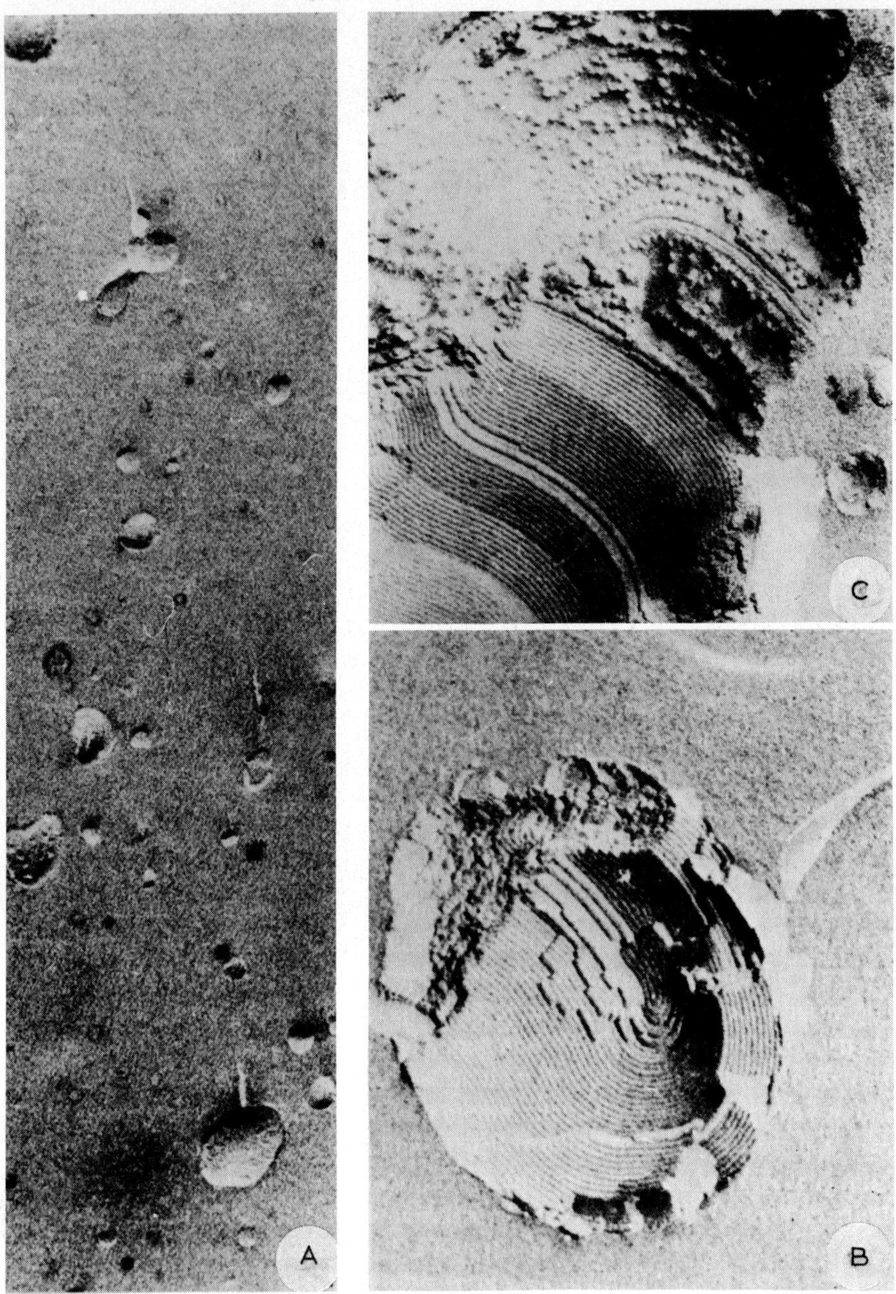

Fig. 26. Freeze-fracture electron micrographs of cosonicated $18{:}1_c/18{:}1_c$-PE/glycophorin vesicles (200:1, mol/mol) in the absence (A) and presence (B, C) of 200 nmol WGA (WGA/glycophorin, 1.0). Final magnification 100 000 ×. Reproduced with permission from Taraschi et al. (1982b).

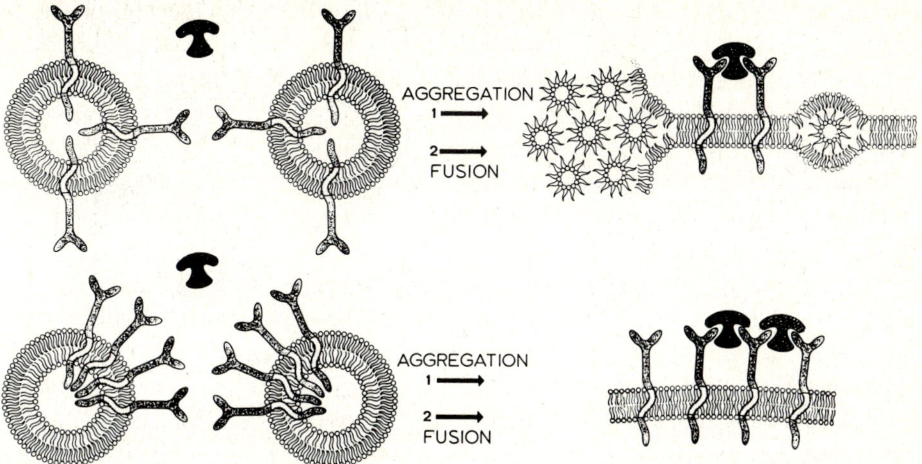

Fig. 27. Schematic representation of the effect of WGA-glycophorin interaction on the structure of $18{:}1_c/18{:}1_c$-PE/glycophorin vesicles. Upper panel: $18{:}1_c/18{:}1_c$-PE-glycophorin (200:1, mol/mol); lower panel: $18{:}1_c/18{:}1_c$-PE/glycophorin (25:1, mol/mol). In this model WGA is proposed to laterally aggregate glycophorin molecules.

IV. Functional aspects

The ability of membrane lipids to adopt non-bilayer structures opens interesting possibilities in explaining membrane functions such as fusion and transport, which we will now summarize.

1. FUSION

Virtually every biological membrane at some stage is involved in fusion with another membrane. This proces is very fast and strictly regulated. The existence of different types of fusion (such as endocytosis, exocytosis, cell division) and the large variations in membrane composition has hampered the formation of a universal fusion mechanism.

In several cases it has now been demonstrated that proteins play essential rôles, such as in recognition and initiation, in the fusion event. At the same time, it is obvious that the fusion event itself involves lipids, since these provide the membrane with its essential barrier properties. Whatever mechanism is operating, it is clear that transiently the membrane lipids will have to leave the bilayer organization. This forms the basis for the suggestion that H_{II}-preferring lipids, by virtue of their ability to adopt non-bilayer structures, are involved in the fusion process. This idea is now supported by many observations. For instance, addition of fusogenic lipids to erythrocytes causes fusion and the formation of the H_{II} phase (Cullis and Hope, 1978; Hope and Cullis, 1981). Furthermore, vesicles containing lipids which can adopt the H_{II} phase fuse under conditions in which this phase preference is expressed.

In the fused vesicles lipidic particles can be observed which often appear to be localized at the site of fusion (Verkleij et al., 1979, 1980). In Fig. 28 a schematic model is presented in which inverted non-bilayer lipid structures act as intermediates in the fusion process. Tentative drawings of membrane fusion intermediates and their possible corresponding features as visualized by freeze-fracturing are shown in Fig. 29.

Besides the possible formation of non-bilayer structures by H_{II} phase-preferring lipids, these lipids could also be involved in membrane fusion as a result of their hydration properties. One essential stage in the fusion process is the close apposition of the two membranes prior to fusion. In order to avoid strong repulsive forces, the membranes should be dehydrated at the site of fusion. The H_{II} phase-preferring lipids have in general a low headgroup hydration and therefore it would be energetically favourable if these lipids would be present at the site of fusion. An evaluation of the relative importance of both effects is, as yet, premature due to lack of experimental data. If the headgroup hydration of an H_{II}-preferring lipid is independent of the acyl chain composition, then a careful compari-

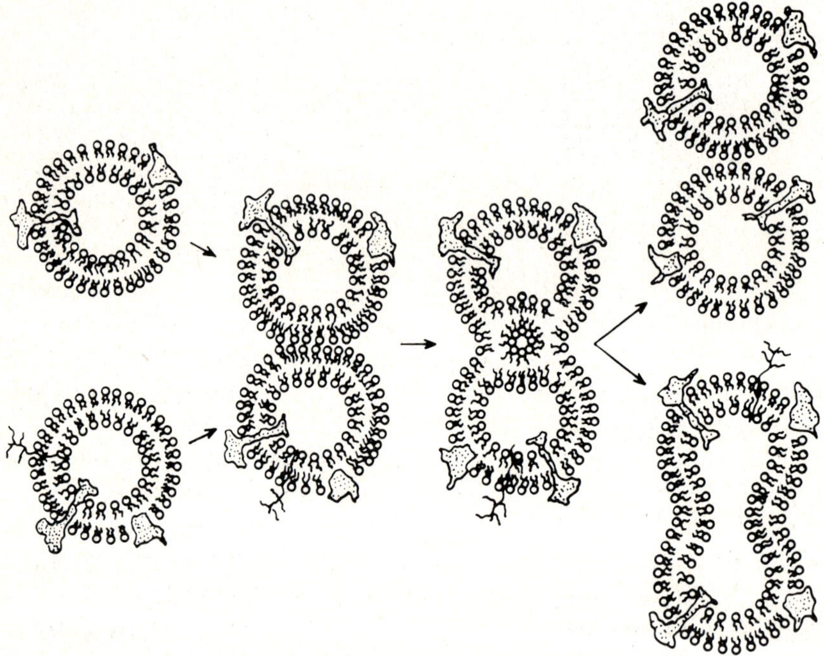

Fig. 28. Proposed mechanism of membrane fusion proceeding via an inverted cylinder or inverted micellar intermediate. The process whereby the membranes come into close apposition is possibly protein mediated, whereas the fusion event itself is proposed to involve formation of an 'inverted' fusion intermediate.

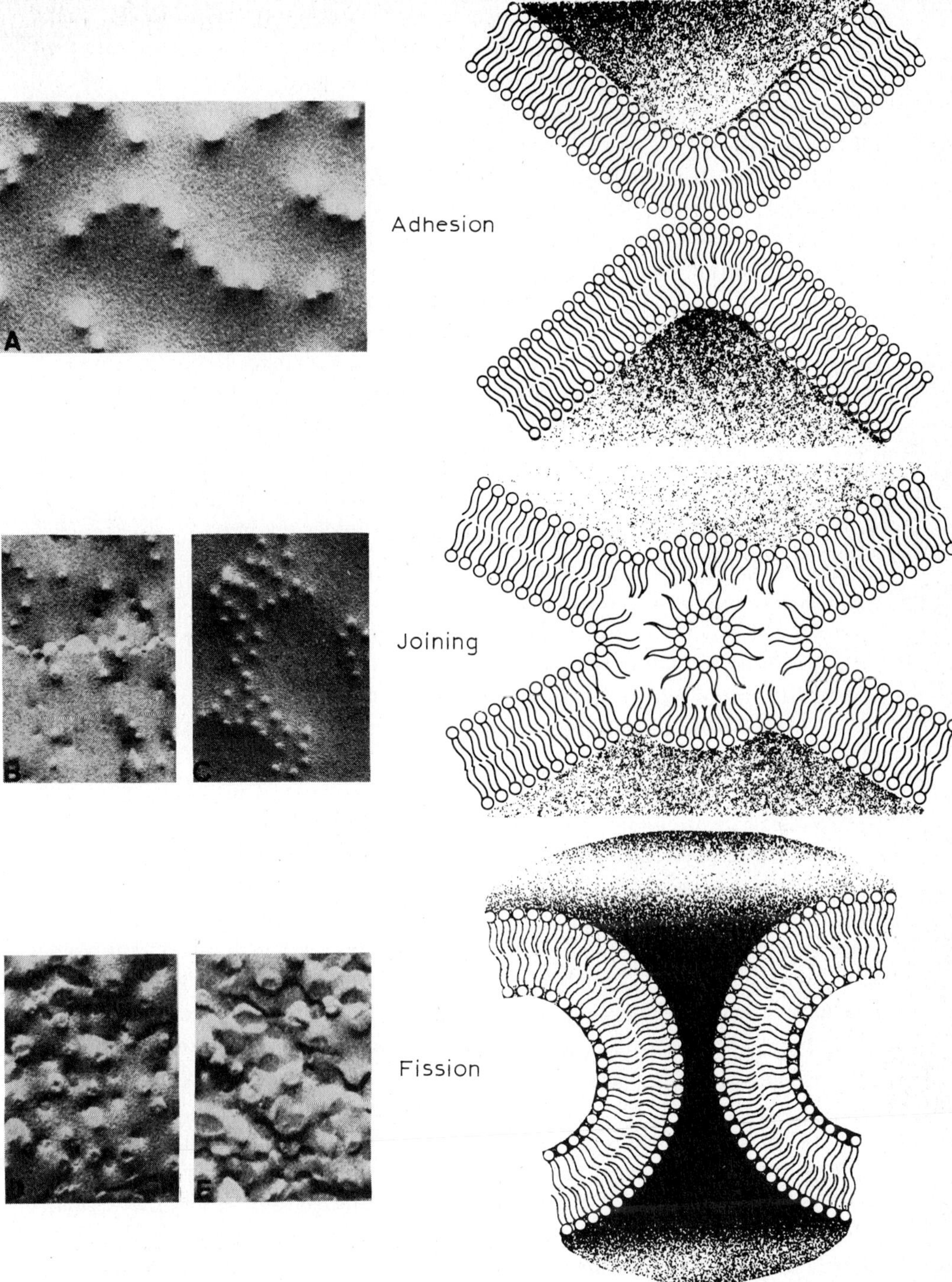

Fig. 29. Tentative drawings of membrane fusion intermediates and their corresponding features as visualized by freeze fracturing. A, B, C, D and E are micrographs taken from the sample DOPE/DOPC/cholesterol (molar ratio 3:1:2) heated up to 60°C for 10 min, cooled down to 4°C. (× 100 000, reduced by a factor of 90/100 by the publisher). Reproduced with permission from Verkleij (1984).

son of the phase state and fusion capacity of the liquid-crystalline model membranes with a different degree of acyl unsaturation of the H_{II} phase-preferring lipid could give an insight in this matter.

Within the above concepts, the involvement of lipid-protein interactions in membrane fusion can be envisaged in two ways.

(1) In biological membranes the ability of non-bilayer lipids to induce fusion is suppressed by lipid-protein interactions. Perturbation of these interactions locally and temporally allows fusion. Such perturbations might be proteolysis and/or the aggregation of receptors.

(2) Fusogenic peptides like gramicidin could directly destabilize bilayer structure and allow membrane fusion to occur. One intriguing question is whether the fusogenic proteins found in viruses can trigger phase changes in the lipid part of the target cell membrane (for a recent hypothesis on the involvement of hydrophobic polypeptides in membrane fusion, see Lucy, 1984).

2. TRANSPORT

The most important function of a membrane is to act as a selective permeability barrier separating two aqueous compartments of specific chemical composition. This is achieved by a delicate interplay between passive diffusion through the lipid part of the membrane, and active and/or facilitated transport processes often involving membrane proteins or peptides. Molecular details of most transport phenomena remain obscure. This is still the situation even for the apparently simple case of the biologically very important fast diffusion of water through a lipid bilayer.

The notion that lipids can readily adopt inverted structures opens attractive possibilities concerning the passage of polar solutes across a lipid bilayer. In Fig. 30 one of these possibilities is depicted. Transient formation of inverted micelles in a lipid bilayer could lead to translocation of: (1) the agent promoting the formation of such structures (thus giving an ionophoric activity to the lipid); (2) pockets of water; and (3) lipid molecules (flip-flop). Although recent considerations suggest that the probability of forming inverted micelles in an isolated bilayer is low (Siegel, 1984), there are several experimental results available which suggest that the presence of non-bilayer lipid structures could facilitate the transport of H_{II} phase-triggering compounds, such as divalent cations.

Using a two phase water/organic solvent system it was clearly demonstrated that H_{II} phase-preferring negatively charged lipids could translocate $^{45}Ca^{2+}$ into an organic phase (Tyson et al., 1976). Interestingly, the presence of ruthenium red, the classical inhibitor of the inner mitochondrial Ca^{2+} transport system, blocked this uptake by CL and in addition inhibited the formation of non-bilayer lipid structures by Ca^{2+} in liposomes formed by this lipid (Cullis et al., 1980).

PA, which can adopt inverted structures in the presence of Ca^{2+}, incorporated into liposomes was reported to act as a Ca^{2+} ionophore (Serhan et al., 1981).

This result, which was recently questioned (Holmes and Voss, 1983), is biologically interesting since during many cellular responses there is a transient increase in Ca^{2+} permeability of the plasma membrane, which is accompanied by a rapid turnover of several phospholipids including PA (Salmon and Honeyman, 1980; Putney et al., 1980).

Whether water permeability is increased by the presence of non-bilayer structures is not known. The transbilayer involvement of lipids is greatly increased by the presence of non-bilayer lipid structures (Gerritsen et al., 1980; Noordam et al., 1981; Hoekstra and Martin, 1982). Since, in such systems, the majority of the lipid molecules in the model membrane become accessible for exchange proteins or chemical labelling, it can be concluded that transport of lipids from the inside of the honeycomb type (Cullis et al., 1980) of liposomal structures to the outer monolayer also occurs, probably by a combination of lateral diffusion and lipid flip-flop.

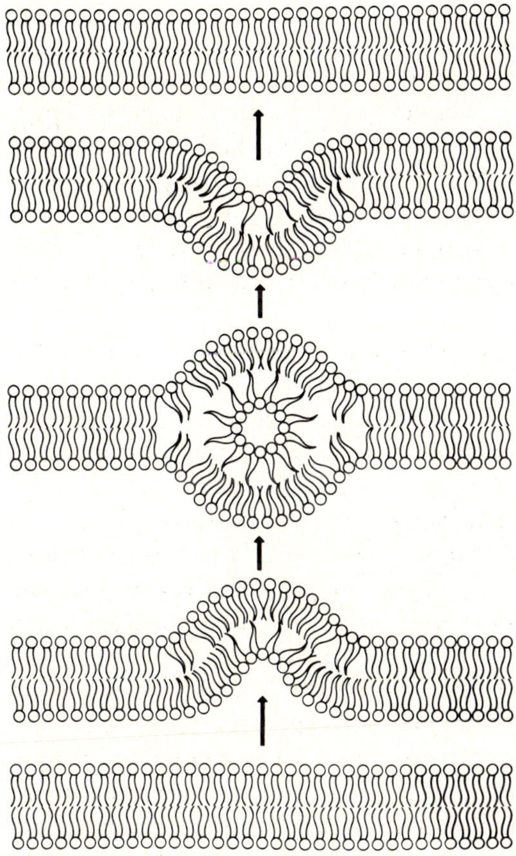

Fig. 30. Dynamic formation of inverted micelles provides a permeation pathway for lipids and polar solutes.

Besides modulation of the phase preference of lipids by interactions with proteins, there is another area of lipid-protein interactions in relation to lipid polymorphism relevant to the barrier function of membranes. It is a general finding that incorporation of integral membrane proteins into lipid vesicles composed of a single liquid-crystalline species, such as $18{:}1_c/18{:}1_c$-PC, leads to a large increase in passive bilayer permeability for a number of charged and neutral solutes (Van der Steen et al., 1982; Van Hoogevest et al., 1983). When these proteins are reconstituted into a heterogenous lipid mixture, for instance, in the case of glycophorin in the total lipid extract of the erythrocyte membrane (Van der Steen et al., 1982; Van Hoogevest et al., 1984), the passive permeability of the protein-containing bilayer is much lower. These observations suggest that a mixture of different lipids is needed to properly seal the protein-lipid interface, or alternatively suggests that the way the protein is incorporated in the bilayer is lipid dependent. Israelachvili (1977) has put forward the original suggestion that there has to be a match of shapes of the irregular outer contour of the intrinsic part of membrane proteins and the surrounding lipids to properly accommodate the protein in the bilayer. Since the shape of a lipid molecule is important in determining the phase preference of a lipid it can be expected that an understanding of the influence of proteins on lipid polymorphism can give insight in the barrier properties of the protein-containing bilayer. This can be exemplified with glycophorin. The presence of this protein into $18{:}1_c/18{:}1_c$-PC bilayer induces the formation of a permeation pathway allowing molecules with a molecular weight of up to approximately 900 to pass the bilayer (Van Hoogevest et al., 1983). Incorporation of increasing amounts of $18{:}1_c/18{:}1_c$-PE strongly reduces the bilayer permeability in this system (Van Hoogevest et al., 1984). As described in Section III.2 glycophorin stabilizes the bilayer structure of this H_{II} phase-preferring lipid. This, then, suggests that the permeability defect of the $18{:}1_c/18{:}1_c$-PC/glycophorin bilayer is related to the existence of an inverted cone shape of part of the hydrophobic domain of glycophorin. The cone-shaped PE molecules could better match this shape than the more cylindrical PC molecules. In addition, the aggregation state of the protein in the bilayer could be influenced by a mismatch in shapes of the proteins and lipids (Israelachvili, 1977). This might also be relevant to the glycophorin-containing bilayer as both freeze-fracture electron microscopy and time-resolved phosphorescence decay studies (Van Hoogevest et al., 1984) indicate that in PE systems the protein is much more strongly aggregated than in PC bilayers.

3. PROTEIN INSERTION AND TRANSLOCATION

At least three different mechanisms are known by which proteins can pass the barrier of a biological membrane. Receptor-mediated endocytosis forms a common pathway by which ligands and receptors, both of which could be (glyco)proteins, are taken up by the cell in an endocytotic event which often involves specialized plasma membrane domains such as coated pits (Simons et al., 1982). For the export of secretory proteins, transport vesicles are used which fuse with the inside of the plasma membrane, thereby releasing the contents of the vesicle at the outside of the cell. Transport vesicles also appear to act as vehicles for transport of proteins between the endoplasmic reticulum and the Golgi apparatus (Rothman, 1981). All these transport processes involve membrane fusion and, thus, could possibly involve non-bilayer lipid structures as indicated in Section IV.1.

The third and most poorly understood mechanism, is that of molecular insertion and translocation of proteins across membranes. This process occurs, in general, in conjunction with protein biosynthesis. The majority of the cellular secretory and membrane proteins are synthesized co-translationally on the rough endoplasmic reticulum membrane (Blobel and Dobberstein, 1975). In short, in eukaryotic cells, this mechanism involves the following steps. The proteins are synthesized with a 10–30 amino acid-long N-terminal extension of the polypeptide chain, which is called the signal peptide, and which is characterized by a high hydrophobicity and the presence of some positively charged amino acids at the N-terminal end of the signal peptide. When, during synthesis which starts on free ribosomes, the signal peptide emerges from the ribosome-RNA complex it becomes recognized by the signal recognition particle (SRP) (Walter et al., 1981), which then stops translation and moves the SRP-ribosome complex to the endoplasmic reticulum membrane where it is recognized by a receptor (Gilmar et al., 1982). Protein synthesis is now continued and coupled to insertion and translocation (of part) of the polypeptide chain across the membrane. During this event, the signal peptide is removed by a signal peptidase.

Another route for molecular insertion and translocation of proteins is that of the post-translational transport of proteins across membranes. This route is followed by the majority of the mitochondrial and chloroplast proteins which are synthesized under control of the nuclear genetic code (Schatz and Mason, 1974). In these cases the proteins are synthesized on free ribosomes in the form of (often quite large) precursors with an N-terminal extension of the polypeptide chain, and are subsequently imported into the organelle, probably involving a receptor-mediated recognition. Depending on the type of protein, either insertion or translocation across one or two membranes has to occur. During this process the precursor is converted into the mature form. A similar mechanism is followed for the insertion into the host cell membrane of some newly synthesized phage membrane proteins (Wickner, 1983).

The molecular details of the actual insertion and translocation steps are obscure. Conceptually it is difficult to imagine how this process can proceed through the protein type of channels. Since proteins are relatively rigid and irregularly shaped, such a transport process would probably lead to loss of vital barrier function. Since lipids are much more flexible, it is easier to envisage how a polypeptide chain can be inserted into, or translocated across, a lipid layer in a way such that the barrier function of the system is maximally preserved. Besides this general consideration, lipid polymorphism offers other arguments for an involvement of lipids in such transport processes.

Short segments of hexagonal H_{II} tubes oriented perpendicular to the membrane surface have been hypothesized as transport routes for bacterial secretory proteins (Nesmeyanova, 1982). A variant of such a mechanism in which some features of the ideas of Fromherz (1983) on the structure of inverted micelles are incorporated is shown in Fig. 31. In this model the basic amino acids preceding a hydrophobic stretch of polypeptide interact with negatively charged lipids,

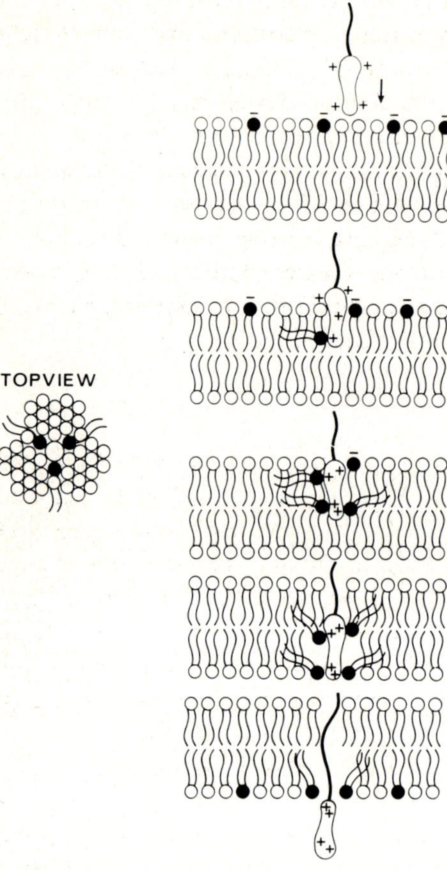

Fig. 31. Model of translocation of the positively charged signal peptide of a protein across a membrane. Upon electrostatic interaction between the signal peptide and negatively charged lipids, non-lamellar lipid structures are formed which enable the signal peptide to cross the membrane. The translocation might be driven, among other factors, by protein synthesis or the membrane potential. The dark line represents a hydrophobic stretch of the polypeptide chain. In the top view it is shown that the acyl chains of the phospholipids adopting this non-lamellar configuration are shielded from the aqueous medium by the headgroups of the lamellar phospholipids.

which then move together with the polypeptide chain across the lipid bilayer in a configuration similar to that found for lipids organized in an H_{II} phase.

Inverted micelles could possibly also act as carriers for proteins. This would, in particular, be relevant for small positively charged hydrophilic proteins, which by interaction with negatively charged lipids could become 'solvated' with the lipids such that the outside of the lipid-protein complex would be hydrophobic. The protein being present inside inverted micelles could then pass a hydrophobic barrier. The formation of the hexagonal H_{II} phase by cytochrome *c* in CL-containing model membranes (De Kruijff and Cullis, 1980b), together with the observation that this protein can form an *iso*-octane soluble complex with negatively charged lipids (Das and Crane, 1964), supports such a mechanism.

As discussed in Section III.1, apocytochrome *c*, the precursor of cytochrome *c*, can greatly affect the molecular organization of PS model membranes. Most interestingly, the apocytochrome *c*-PS interaction is followed by a (partial) translocation of the protein across the bilayer (Rietveld and De Kruijff, 1984).

Progress in the understanding of the mechanism of the molecular transport of proteins across membranes requires reconstitution of the insertion and translocation steps in model membranes. Studying the lipid dependency of the transport, in combination with a characterization of the physical state of the membrane lipids in the recombinant, will give answers concerning the possible involvement of non-bilayer lipid structure in protein translocation.

4. DOMAINS

The fluid mosaic model of biological membranes suggests that membrane proteins and lipids can freely diffuse within the plane of the membrane. However, cell biology studies have demonstrated that, within one membrane, domains can exist with specific protein composition and function. Well-known examples of such domains are the purple membrane of *H. halobium*, coated pits in plasma membranes of eukaryotic cells, and the basolateral and apical parts of the plasma membrane of polarized cells. Protein-protein interaction with the membrane, or between the membrane and cytoskeletal components have been shown to play an important rôle in the formation and maintenance of such domains.

The existence of lipid domains in membranes is less well documented. In the case of some bacteria, phase separation between gel state and liquid-crystalline lipid domains can occur in the membrane due to the high gel/liquid-crystalline phase transition temperature of some of the lipids. Most integral membrane proteins partition in the liquid-crystalline domains.

Although the subject of frequent speculation, there is no clear evidence for domains or (micro) cluster formation of lipids in a continuous liquid-crystalline membrane. An exception is formed by the plasma membrane of polarized cells. Not only the protein (Boulain and Sabatini, 1978), but to a limited extent also the lipid composition (Van Meer and Simons, 1982) of the apical and basolateral part of the membrane, is different. The tight junction which separates these two halves of the plasma membrane thus seems to act as a diffusion barrier. Electron microscopy in combination with selective extraction procedures has shown that the tight junction is most likely formed by intrabilayer tubes of lipids similar to those found in the hexagonal H_{II} phase (Kachar and Reese, 1982; Pinto da Silva and Kachar, 1982). Some possible modes of organization of such isolated lipid cylinders in a bilayer are shown in Fig. 32. The situation of the H_{II} tube lying at the nexus of two intersecting bilayers is probably the most relevant configuration for tight junction structure. It is easy to envisage how these tubes can act as diffusion barriers for intrinsic membrane proteins. This model of tight junction structure predicts a continuous lipid domain of the inner monolayer of the

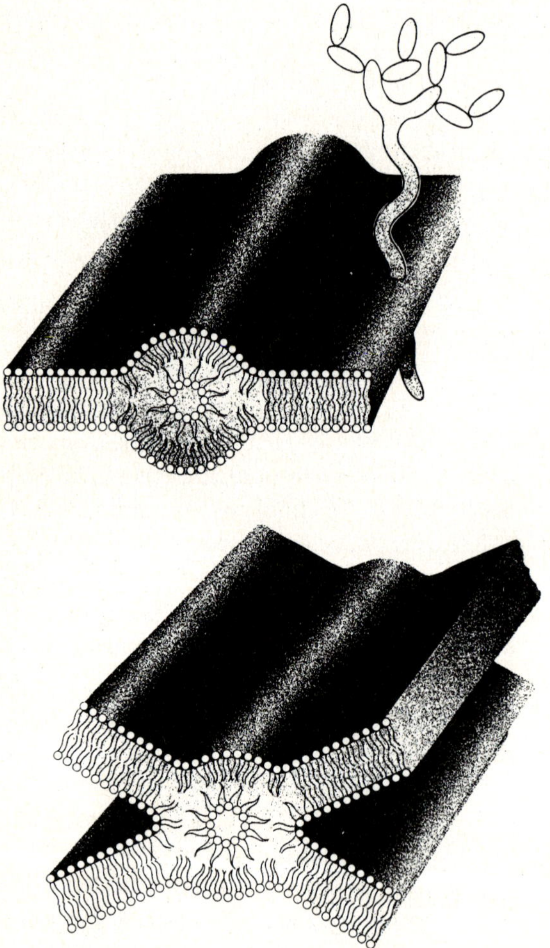

Fig. 32. Inverted lipid cylinders can act as diffusion barriers for membrane protein (upper panel) and membrane lipids.

basolateral and apical part of the membrane, whereas the outer monolayers of both plasma membrane domains would be separated by the tight junction. The tight junction is most probably anchored at its specific location in the cell via protein-protein interaction involving the cytoskeleton. Whether the H_{II} type of intramembraneous lipid cylinders can act as diffusion barriers in other biological membranes is not known.

V. Concluding remarks

The observation of the formation of non-bilayer structures by isolated hydrated membrane lipids under physiological conditions of pH, ionic strength and temperature has widened our perspective on the rôle of lipids in membrane structure and function. The model membrane studies summarized in this review, which showed that (membrane) proteins can modulate lipid polymorphism in a protein- and lipid-specific manner, form a firm indication that local formation of non-bilayer lipid structures in biological membranes can occur and plays significant functional rôles.

It can be expected that the application of the research strategies outlined in this review together with the increasingly sophisticated methods to monitor lipid structure will lead in the coming years to a detailed understanding of the importance of lipid-protein interactions for regulation of lipid structure in membranes.

References

Beyer, K. (1983) A ^{31}P and ^{2}H NMR study on PC's in liquid-crystalline polyoxyethylene detergents. Chem. Phys. Lipids 34, 65–80.

Birrell, G.B. and Griffith, O.H. (1976) Cytochrome *c* induced lateral phase separation in a diphosphatidylglycerol-steroid spin-label model membrane. Biochemistry 15, 2925–2929.

Blaurock, A.E. (1982) Evidence of bilayer structure and of membrane interactions from X-ray diffraction analysis. Biochim. Biophys. Acta 650, 167-207.

Blobel, S. and Dobberstein, B. (1975) Transfer of proteins across membranes. I. Presence of proteolytically processed and unprocessed nascent immunoglobulin light chains on membrane-bound ribosomes of murine myeloma. J. Cell Biol. 835-851.

Boggs, J.M., Stamp, D., Hughes, D.W. and Deber, C.M.. (1981) Influence of ether linkage on the lamellar to hexagonal phase transition of ethanolamine phospholipids. Biochemistry 20, 5728–5735.

Boulain, E. and Sabatini, D.D. (1978) Asymmetric budding of viruses in epithelial monolayers: A model system for study of epithelial polarity. Proc. Natl. Acad. Sci. U.S.A. 75, 5071–5075.

Brown, L.R. and Wüthrich, K. (1977) NMR and ESR studies of the interactions of cytochrome *c* with mixed cardiolipin-phosphatidylcholine vesicles. Biochim. Biophys. Acta 468, 389–410.

Büldt, G., Gally, H.U., Seelig, A., Seelig, J. and Zaccai, G. (1978) Neutron diffraction studies on selectively deuterated phospholipid bilayers. Nature 271, 182–184.

Burnell, E.E., Cullis, P.R. and De Kruijff, B. (1980) Effects of tumbling and lateral diffusion on

phosphatidylcholine model membrane ^{31}P NMR line-shapes. Biochim. Biophys. Acta 603, 63–69.
Chapman, D., Cornell, B.A., Eliasz, A.W. and Perry, A. (1977) Interactions of helical polypeptide segments which span the hydrocarbon region of lipid bilayers. Studies of the gramicidin A lipid-water system. J. Mol. Biol. 113, 517–538.
Chi, E.Y. and Lagunoff, D. (1978) Linear arrays of intramembranous particles in pulmonary tubular myelin. Proc. Natl. Acad. Sci. U.S.A. 75, 6225–6229.
Corless, J.M. and Costello, M.J. (1981) Paracrystalline inclusions associated with the disk membranes of frog retinal rod outer segments. Exp. Eye Res. 32, 217–228.
Cortijo, M. and Chapman, D. (1981) A comparison of the interactions of cholesterol and gramicidin A with lipid bilayers using an infrared data station. FEBS Lett. 131, 245–248.
Crowe, L.M. and Crowe, J.H. (1982) Hydration dependent hexagonal phase lipid in a biological membrane. Arch. Biochem. Biophys. 217, 582–587.
Cullis, P.R. and De Kruijff, B. (1978a) The polymorphic phase behavior of phosphatidylethanolamines of natural and synthetic origin. A ^{31}P NMR study. Biochim. Biophys. Acta 513, 31–42.
Cullis, P.R. and De Kruijff, B. (1978b) Polymorphic phase behaviour of lipid mixtures as detected by ^{31}P-NMR. Evidence that cholesterol may destabilize bilayer structure in membrane systems containing phosphatidylethanolamine. Biochim. Biophys. Acta 507, 207–218.
Cullis, P.R. and Hope, M.J. (1978) Effects of fusogenic agents on membrane structure of erythrocyte ghosts and the mechanism of membrane fusion. Nature 271, 672–675.
Cullis, P.R. and De Kruijff, B. (1979) Lipid polymorphism and the functional role of lipids in biological membranes. Biochim. Biophys. Acta 559, 399–420.
Cullis, P.R. and Hope, M.J. (1980) The bilayer stabilizing role of sphingomyelin in the presence of cholesterol. A ^{31}P-NMR study. Biochim. Biophys. Acta 597, 533–542.
Cullis, P.R., Van Dijck, P.W.M., De Kruijff, B. and De Gier, J. (1978a) Effects of cholesterol on the properties of equimolar mixtures of synthetic phosphatidylethanolamine and phosphatidylcholine. A ^{31}P-NMR and differential scanning calorimetry study. Biochim. Biophys. Acta 513, 21–30.
Cullis, P.R., Verkleij, A.J. and Ververgaert, P.H.J.Th. (1978b) Polymorphic phase behavior of cardiolipin as detected by ^{31}P-NMR and freeze-fracture techniques. Effects of calcium, dibucaine and chlopromazine. Biochim. Biophys. Acta 513, 11–20.
Cullis, P.R., De Kruijff, B., Hope, M.J., Nayar, R. and Schmid, S.L. (1980a) Phospholipids and membrane transport. Can. J. Biochem. 58, 1091–1100.
Cullis, P.R., De Kruijff, B., Hope, M.J., Nayar, R., Rietveld, A. and Verkleij, A.J. (1980b) Structural properties of phospholipids in the rat liver inner mitochondrial membrane. A ^{31} P-NMR study. Biochim. Biophys, Acta 600, 625–635.
Cullis, P.R., De Kruijff, B., Hope, M.J., Verkleij, A.J., Nayar, R., Farren, S.B., Tilcock, C., Madden, T.D. and Bally, M.B. (1983) Structural properties of lipids and their functional roles in biological membranes. In: *Membrane Fluidity*, Vol. 2, pp. 40-79. Editor: R.C. Alioa. Academic Press, New York.
Das, M.L. and Crane, F.L. (1964) Proteolipids. I. Formation of phospholipid-cytochrome c complexes. Biochemistry 3, 696–704.
Deamer, D.W., Leonard R., Tardieu, A. and Branton, D. (1970) Lamellar and hexagonal lipid phases visualized by freeze etching. Biochim. Biophys. Acta 218, 47–60.
Dekker, C.J., Geurts van Kessel, W.S.M., Klomp, J.P.G., Pieters, J. and De Kruijff, B. (1983) Synthesis and polymorphic phase behaviour of polyunsaturated phosphatidylcholines and phosphatidylethanolamines. Chem. Phys. Lipids 33, 93–106.
De Kruijff, B. and Cullis, P.R. (1980a) The influence of poly(L-lysine) on phospholipid polymorphism. Evidence that electrostatic polypeptide-phospholipid interactions can modulate bilayer-nonbilayer transitions. Biochim. Biophys. Acta 601, 235–240.

De Kruijff, B. and Cullis, P.R. (1980b) Cytochrome *c* specifically induces non-bilayer structures in cardiolipin-containing model membranes. Biochim. Biophys. Acta 602, 477–490.

De Kruijff, B., Van Zoelen, E.J.J. and Van Deenen, L.L.M. (1978) Glycophorin facilitates the transbilayer movement of phosphatidylcholine in vesicles. Biochim. Biophys. Acta 509, 537–542.

De Kruijff, B., Verkleij, A.J., Van Echteld, C.J.A., Gerritsen W.J., Mombers, C., Noordam, P.C. and De Gier, J. (1979) The occurrence of lipidic particles in lipid bilayers as seen by ^{31}P-NMR and freeze-fracture electron microscopy. Biochim. Biophys. Acta 555, 200–209.

De Kruijff, B., Rietveld, A. and Cullis, P.R. (1980a) ^{31}P-NMR studies on membrane phospholipids in microsomes, rat liver slices and intact perfused rat liver. Biochim. Biophys. Acta 600, 343–357.

De Kruijff, B., Cullis, P.R. and Verkleij, A.J. (1980b) Non-bilayer lipid structures in model and biological membranes. Trends Biochem. Sci. 5, 79–81.

De Kruijff, B., Verkleij, A.J., Van Echteld, C.J.A., Gerritsen, W.J., Noordam, P.C., Mombers, C., Rietveld, A., De Gier, J., Cullis, P.R., Hope, M.J. and Nayar, R. (1981) Non-bilayer lipids and the inner mitochondrial membrane. In: *International Cell Biology 1980–1981*, pp. 559–571. Editor: H.G. Schweiger. Springer, Berlin.

De Kruijff, B., Verkleij, A.J., Leunissen-Bijvelt, J., Van Echteld, C.J.A., Hille, J. and Rijnbout, H. (1982) Further aspects of the Ca^{2+}-dependent polymorphism of bovine heart cardiolipin. Biochim. Biophys. Acta 693, 1–12.

De Kruijff, B., Cullis, P.R., Verkleij, A.J., Hope, M.J., Van Echteld, C.J.A. and Taraschi, T.F. (1984) In: *The Enzymes of Biological Membranes*. Editor: A. Martinosi. Plenum Press, New York, in press.

Dufourq, C.J., Faucon, J.F., Bernard, E., Perolet, M., Tessier, M., Bougis, P., Van Rietschoten, J., Delori, P. and Rochat, H. (1982) Structure-function relationships for cardiotoxins interacting with phospholipids. Toxicon 20, 165–174.

Eytan, G.D. and Almary, T. (1983) Melittin-induced fusion of acidic liposomes. FEBS Lett. 156, 29–32.

Farren, S,B., Hope, M.J. and Cullis, P.R. (1983) Polymorphic phase preferences of phosphatidic acid: A ^{31}P and ^{2}H-NMR study. Biochem. Biophys. Res. Commun. 111, 675–683.

Fisher, W.R., Tamiuchi, H. and Anfinsin, C.B. (1973) On the role of heme in the formation of the structure of cytochrome *c*. J. Biol. Chem. 248, 3188-3195.

Fromherz, P. (1983) The assembly of lipids and surfactants: Molecular and phenomenological concepts. In: *Biological and Technological Relevance of Reversed Micelles and Other Amphiphilic Structures in a Polar Media*, pp. 10–22. Editor: P.L. Luisi, Plenum Press, New York.

Gally, H.U., Pluschke, G., Overath, P. and Seelig, J. (1980) Structure of *Escherichia coli* membranes. Fatty acyls chain order parameters of inner and outher membranes and derived liposomes. Biochemistry 19, 1638–1643.

Gerritsen, W.J., De Kruijff, B., Verkley, A.J., De Gier, J. and Van Deenen, L.L.M. (1980) Ca^{2+}-induced isotropic motion and phophatidylcholine flip-flop in phosphatidylcholine/cardiolipin bilayers. Biochim. Biophys. Acta 598, 554–560.

Ghosh, R. and Seelig, J. (1982) The interaction of cholesterol with bilayers of phosphatidylethanolamine. Biochim. Biophys. Acta 691, 151–160.

Gilmore, R., Blobel, G. and Walter, P. (1982) Protein translocation across the endoplasmic reticulum. I. Detection in the microsomal membrane of a receptor for the signal recognization particle. J. Cell Biol. 95, 463–469.

Gulik-Krzywicky, T., Shechter, E., Luzatti, V. and Foure, M. (1969) Interactions of proteins and lipids: Structure and polymorphism of protein-lipid-water phases. Nature 223, 1116–1117.

Gulik-Krzywicky, T., Balerna, M., Vincent, J.-P. and Lazdunski, M. (1981) Freeze-fracture study of cardiotoxin action on axonal membrane and axonal membrane lipid vesicles. Biochim. Biophys. Acta 643, 101–114.

Habermann, E. (1980) Mellitin structure and activity. In: *Natural Toxins*, pp. 173–181. Editors:K.

K. Eaher and D. Wadstrom. Pergamon Press, New York.

Hardman, P.D. (1982) Spin-label characterization of the lamellar-to-hexagonal (H_{II}) phase transition in egg phosphatidyl-ethanolamine aqueous dispersions. Eur. J. Biochem. 124, 95–101.

Harlos, K. and Ebil, H. (1981) Hexagonal phases in phospholipids with saturated chains: Phosphatidylethanolamines and phosphatidic acids. Biochemistry 20, 2888–2892.

Hartman, W.and Galla, H.J. (1978) Binding of poly-lysine to charged bilayer membranes: molecular organization of a lipid-peptide complex. Biochim, Biophys. Acta 509, 474–490.

Hennig, B. and Neupert, W. (1981) Assembly of cytochrome *c*. Apocytochrome *c* is bound to specific sites on mitochindria before its conversion to holocytochrome *c*. Eur. J. Biochem. 121, 203–212.

Hoekstra, D. and Martin. O.C. (1982) Transbilayer redistribution of phosphatidylethanolamine during fusion of phospholipid vesicle. Dependence on fusion rate, lipid phase separation and formation of non-bilayer structures. Biochemistry 21, 6097–6103.

Holmes, R.P. and Voss, N.L. (1983) Failure of phosphatidic acid to translate Ca^{2+} across phosphatidylcholine membranes. Nature 305, 637–638.

Hope, M.J.and Cullis, P.R. (1981) The role of non-bilayer lipid structures in the fusion of human erythrocytes induced by lipid fusogens. Biochim. Biophys. Acta 640, 82–90.

Israelachvili, J.N. (1977) Refinement of the fluid-mosaic model of membrane structure. Biochim. Biophys. Acta 469, 221–225.

Ivanov, V. (1982) Gramicidin A. From structure towards the molecular mechanism of action. In: *Peptides 1982*, 17th European Peptide symposium, Prague, Czechoslavakia, August 29–September 3, 1982, pp. 73–90. Editors: K. Blaha, and P. Malon. Walter de Gruyter, Hawthorne.

Kachar, B. and Reese, T.S. (1982) Evidence for the lipidic nature of tight junction strands. Nature 296, 464–466.

Killian, J.A., De Kruyff, B., Van Echteld, C.J.A., Verkleij, A.J., Leunissen-Bijvelt, J. and De Gier, J. (1983) Mixtures of gramicidin and lysophosphatidylcholine form lamellar structures. Biochim. Biophys. Acta 728, 141–144.

Killian, J.A., Leunissen-Bijvelt, J.T. Verkleij, A.J. and De Kruyff, B. (1985) External addition of gramicidin to dioleylphosphatidylcholine model membranes results in fusion and H_{II} phase formation. Biochim. Biophys. Acta, in press.

Kimelberg, H.K. and Papahadjopoulos, D. (1971) Interactions of basic proteins with phospholipid membranes. Binding and changes in the sodium permeability of phosphatidylserine vesicles. J. Biol. Chem 246, 1142–1148.

Lambers, J.W.J., Verkleij, A.J. and Terpstra, W. (1984) Reconstitution of chlorophyllase with mixed plant lipids, in the presence and absence of Mg^{2+}; influence of single and mixed plant lipids on enzyme stability. Biochim. Biophys. Acta 786, 1–9.

Larsson, K., Fontell, K. and Krog, N. (1980) Structural relationships between lamellar, cubic and hexagonal phases in monoglyceride-water systems. Possibility of cubic structures in biological systems. Chem. Phys. Lipids 27, 321–328.

Lee, D.C., Durrani, A.A. and Chapman, D. (1984) A difference infrared spectroscopic study of gramicidin A, alamethicin and bacteriorhodopsin in perdeuterated dimyristoyl phosphatidylcholine. Biochim. Biophys. Acta 769, 49–56.

Lucy, J.A. (1984) Do hydrophobic sequences cleaved from cellular polypeptides induce membrane fusion reactions in vivo? FEBS Lett. 166, 223–231.

Luzatti, V (1968) X-ray diffraction studies of lipid-water systems. In: *Biological Membranes*, pp. 71–123. Editor: D. Chapman. Academic Press, New York.

Luzatti, V., Gulik-Krzywicki, T. and Tardieu, A. (1968) Polymorphism of lecithins. Nature 218, 1031–1034.

Madden, T.D. and Cullis, P.R. (1982) Stabilization of bilayer structure of unsaturated phosphatidylethanolamines by detergents. Biochim. Biophys. Acta 684, 149–153.

Mantsch, H.H., Martin, A. and Cameron, D.G. (1981) Characterization by infrared spectroscopy of the bilayer to non-bilayer phase transition of phosphatidylethanolamines. Biochemistry 20, 3138–3145.

Marsh, D. and Seddon, J.M. (1982) Gel-to-inverted H_{II} (L_β-H_{II}) phase transition in phosphatidylethanolamines and fatty acidphosphatidylcholine mixtures, demonstrated by ^{31}P-NMR spectroscopy and X-ray diffraction. Biochim. Biophys. Acta 690, 117–123.

McDonald, R.J. and McDonald R.C. (1975) Assembly of phospholipid vesicles bearing sialoglycoprotein from erythocyte membrane. J. Biol. Chem. 250, 9206–9214.

Navarro, J., Kinnucan, M.T. and Racker, E. (1984) Effect of lipid composition on the calcium/adenosine 5′-triphosphate coupling ratio of the Ca^{2+}-ATPase of sarcoplasmic reticulum. Biochemistry 23, 130–135.

Nesmeyanova, M.A. (1982) On the possible participation of acid phospholipids in the translocation of secreted proteins through the bacterial cytoplasmic membrane. FEBS Lett. 142, 189–193.

Nicholls, P. (1974) Cytochrome *c* binding to enzymes and membranes. Biochim. Biophys. Acta 346, 261–310.

Noordam, P.C., Van Echteld, C.J.A., De Kruijff, B. and De Gier, J. (1981) Rapid transbilayer movement of phosphatidylcholine in unsaturated phosphatidylethanolamine containing model systems. Biochim. Biophys. Acta 646, 483–487.

Ohno-Iwashita, Y. and Wickner, W. (1983) Reconstitution of rapid and asymmetric assembly of M13 procoat protein into liposomes which have bacterial leader peptidase. J. Biol. Chem. 258, 1895–1900.

Op den Kamp, J.A.F. (1979) Lipid asymmetry in membranes. Ann. Rev. Biochem. 48, 47–71.

Paphadjopoulos, D., Moscarello, M. Eylan, E.H. and Isac, T. (1975) Effects of proteins on thermotropic phase transitions of phospholipid membranes. Biochim. Biophys. Acta 401, 317–335.

Pasquali-Rouchetti, I. Spisni, A., Casali, E., Masotti, L. and Urry, D.W. (1983) Gramicidin A induces lysolecithin to form bilayers. Biosci. Rep. 3, 127–133.

Pinto da Silva, P. and Kachar, B. (1982) On tight-junction structure. Cell 28, 441–450.

Putney, J.W., Weiss, S.J., Van Der Walle, C.M. and Haddas, R.A. (1980) Is phosphatidic acid a calcium ionophore under neurohumoral control? Nature 284, 345–347.

Rand, R.P. and Sengupta, S. (1972) Cardiolipin forms hexagonal structures with divalent cations. Biochim. Biophys. Acta 255, 484–492.

Rice, D. and Oldfield, E. (1979) Deuterium nuclear magnetic resonance studies of the interaction between dimyristoylphosphatidylcholine and gramicidin A′. Biochemistry 18, 3272–3279.

Rietveld, A. and De Kruijff, B. (1984) Is the mitochondrial precursor protein apocytochrome *c* able to pass a lipid bilayer? J. Biol. Chem. 259, 6704–6707.

Rietveld, A., Syens, P., Verkleij, A.J. and De Kruijff, B. (1983) Interaction of cytochrome *c* and its precursor apocytochrome *c* with various lipids. EMBO J. 2, 907–913.

Rothman, J.E. (1981) The Golgi apparatus: Two organelles in tandem. Science 213, 1212–1219.

Ruppel, D., Kapitzka, H.G., Galla, H.J., Sixl, F. and Sackmann, E. (1982) On the microstructure and phase diagram of dimyristoylphosphatidylcholine-glycophorin bilayers. The role of defects and the hydrophilic lipid-protein interactions. Biochim. Biophys. Acta 692, 1–17.

Salmon, D.M. and Honeyman, T.W. (1980) Proposed mechanism of cholinergic action in smooth muscle. Nature 284, 344-345.

Schatz, G. and Mason, T.L. (1974) The biosynthesis of mitochondrial proteins. Ann. Rev. Biochem. 43, 51–87.

Seddon, J.M., Cevc, G. and Marsh, D. (1983a) Calorimetric studies of the gel-fluid (L_β-L_α) and Lamellar-inverted hexagonal (L_α-H_{II}) phase transitions in dialkyl- and diacylphosphatidylethanolamines. Biochemistry 22, 1280–1289.

Seddon, J.M., Kaye, R.D. and Marsh, D. (1983b) Induction of lamellar-inverted hexagonal transition in cardiolipin by protons and monovalent cations. Biochim. Biophys. Acta 734, 347–352.

Seelig, J. (1977) Deuterium magnetic resonance: Theory and application to lipid membranes. Q. Rev.

Biophys. 10, 353–418.

Seelig, J. (1978) ^{31}P nuclear magnetic resonance and the head group structure of phospholipids in membranes. Biochim. Biophys Acta 515, 105–140.

Siegel, D. (1984) Inverted micellar structures in bilayer membranes: Formation rates and half-lives. Biophys. J. in press.

Simons, K., Garoff, H. and Helenius. A. (1982) How an animal virus gets into and out of its host cell. Sci. Am. 246, 46–54.

Simpson, D.J. (1978) Freeze-fracture studies on barley plastid membranes. I. Wild-type etioplast. Carlsberg Res. Commun. 43, 145–170.

Taraschi, T.F. and Mendelsohn, R. (1980) Lipid-protein interaction in the glycophorin-dipalmitoylphosphatidylcholine system: Raman spectroscopic investigation. Proc. Natl. Acac. Sci. U.S.A. 77, 2362–2366.

Taraschi, T.F., De Kruijff, B., Verkleij, A.J. and van Echteld, C.J.A. (1982a) Effect of glycophorin on lipid polymorphism. A ^{31}P-NMR study. Biochim. Biophys. Acta 685, 153–161.

Taraschi, T.F., Van der Steen, A.T.M., De Kruijff, B., Tellier, C. and Verkleij, A.J. (1982b) Lectin-receptor interactions in model membranes: Evidence that binding of wheat-germ agglutinin to glycoprotein-phosphatidylethanolamine vesicles induces non-bilayer structures. Biochemistry 21, 5756-5764.

Taraschi, T.F., De Kruijff, B. and Verkleij. A.J. (1983) The effect of an integral membrane protein on lipid polymorphism in the cardiolipin -Ca^{2+} system. Eur. J. Biochem. 129, 621–625.

Thayer, A.M. and Kohler, S.J. (1981) Posphorous-31 nuclear magnetic resonance spectra characteristic of phosphatidylethanolamine in the bilayer phase. Biochemistry 20, 6831–6834.

Tilcock, C.P.S. and Cullis, P.R. (1982) The polymorphic phase behaviour and miscibility properties of synthetic phosphatidylethanolamines. Biochim. Biophys. Acta 684, 212–218.

Tilcock, C.P.S., Bally, M.B., Farren, S.B. and Cullis, P.R. (1982) Influence of cholesterol on the structural preference of dioleoylphosphatidylethanolamine-dioleoylphosphatidylcholine systems: A phosphorus-31 and deuterium nuclear magnetic resonance study. Biochemistry 21, 4596–4601.

Tomita, M. and Marchesi, V.T. (1975) Amino-acid sequence and oligosaccharide attachment sites of human erythrocyte glycophorin. Proc. Natl. Acad. Sci. U.S.A. 72, 2964–2968.

Tyson, C.A., Zande, H.V. and Green, D.E. (1976) Phospholipids as ionophores. J. Biol. Chem. 251, 1326–1332.

Ulmius, J., Lindblom, G., Wennerstrom, H., Johansson, L.B.A., Fontell, K., Soderman, O. and Arvidson, G. (1982) Molecular organization in the liquid-crystalline phases of lecithin-sodium cholate-water systems studied by nuclear magnetic resonance. Biochemistry 21, 1553–1560.

Urry, D.W., Trapane, T.L. and Prasad, W.U. (1983) Is the gramicidin A channel single-stranded or double-stranded helix a simple unequivocal determination. Science 221, 1064–1067.

Van Dijck, P.W.M., De Kruijff, B., Van Deenen, L.L.M., De Gier, J. and Demel, R.A. (1976) The preference of cholesterol for phosphatidylcholine in mixed phosphatidylcholine-phosphatidylethanolamine bilayers. Biochim. Biophys. Acta 455, 576–587.

Van Echteld, C.J.A., Van Stigt, R., De Kruijff, B., Leunissen-Bijvelt, J., Verkleij, A.J. and De Gier, J. (1981) Gramicidin promotes formation of the hexagonal H_{II} phase in aqueous dispersions of phosphatidylethanolamine and phosphatidylcholine. Biochem. Biophys. Acta 648, 287–291.

Van Echteld, C.J.A., De Kruijff, B., Verkleij, A.J., Leunissen-Bijvelt, J. and De Gier, J. (1982) Gramicidin induces the formation of non-bilayer structures in phosphatidylcholine dispersions in a fatty acid chain length-dependent way. Biochim. Biophys. Acta 692, 126–138.

Van Hoogevest, P., Du Maine, A.P.M. and De Kruijff, B. (1983) Characterization of the permeability increase induced by the incorporation of glycophorin in phosphatidylcholine vesicles. Determination of the size of the non-specific permeation pathway. FEBS Lett. 157, 41–45.

Van Hoogevest, P., Du Maine, A.P.M., De Kruijff, B. and De Gier, J. (1984) The influence of lipid composition on glycophorin-induced bilayer permeability. Biochim. Biophys. Acta 771, 119–126.

Van Hoogevest, P., De Kruijff, B. and Garland, P.B. (1985) The influence of lipid composition and lectin-glycophorin interaction on the rotational diffusion of glycophorin in vesicles as measured by time resolved phosphorescence depolarization. Biochim. Biophys. Acta, 813, 1–9.

Van Meer, G. and Simons, K. (1982) Viruses budding from either the apical or the basolateral membrane domain of MDCK cells have unique phospholipid compositions. EMBO J. 1, 847–852.

Van der Steen, A.T.M., De Kruijff, B. and De Gier, J. (1982) Glycophorin incorporation increases the bilayer permeability of large unilamellar vesicles in a lipid-dependent manner. Biochim. Biophys. Acta 691, 13–23.

Van Venetië, R. and Verkleij, A.J. (1981) Analysis of hexagonal H_{II} phase and its relations to lipidic particles and the lamellar phase. A freeze-fracture study. Biochim. Biophys. Acta 645, 262–269.

Van Venetië, R. and Verkleij, A.J. (1982) Possible role of non bilayer lipids in the structure of mitochondria. Biochim. Biophys Acta 692, 397–405.

Van Venetië, R., Hage, W.J., Bluemink, J.G. and Verkleij, A.J. (1981) Propane jet-freezing: A valid ultra-rapid freezing method for preservation of temperature-dependent lipid phases. J. Microsc. 123, 287–292.

Van Zoelen, E.J.J., Verkleij, A.J., Zwaal, R.F.A. and Van Deenen, L.L.M. (1978a) Incorporation and asymmetric orientation of glycophorin in reconstituted protein-containing vesicles. Eur. J. Biochim. 86, 539–546.

Van Zoelen, E.J.J., De Kruijff, B. and van Deenen, L.L.M. (1978b) Protein-mediated transbilayer movement of lysophosphatidylcholine in glycophorin-containing vesicles. Biochim. Biophys. Acta 508, 97–108.

Van Zoelen, E.J.J. Van Dijck, P.W.M., De Kruijff, B., Verkleij, A.J. and Van Deenen, L.L.M. (1978c) Effect of glycophorin incorporation on the physico-chemical properties of phospholipid bilayers. Biochim. Biophys Acta 514, 9–24.

Vasilenko, I., De Kruijff, B. and Verkleij, A.J. (1982) Polymorphic phase behaviour of cardiolipin from bovine heart and from *Bacillus subtilis* as detected by ^{31}P-NMR and freeze-fracture. Effects of Ca^{2+}, Mg^{2+}, Ba^{2+} and temperature. Biochim. Biophys. Acta 684, 282–286.

Veatch, W.R., Fassel, E.T. and Blout, E.R. (1974) The conformation of gramicidin A. Biochemistry 13, 5249–5256.

Verkleij, A.J. (1984) Lipidic intramembranous particles. Biochim. Biophys. Acta 779, 43–63.

Verkleij, A.J., Mombers C., Gerritsen, W.J., Leunissen Bijvelt, J. and Cullis, P.R. (1979a) Fusion of phospholipid vesicles in association with the appearance of lipidic particles as visualized by freeze fracturing. Biochim. Biophys. Acta 555, 358–362.

Verkleij, A.J., Mombers, C., Leunissen-Bijvelt, J. and Ververgaert, P.H.J.Th. (1979b) Lipidic intramembranous particles. Nature 279, 162–163.

Verkleij, A.J., Van Echteld, C.J.A., Gerritsen, W.J., Cullis, P.R. and De Kruijff, B. (1980) The lipidic particle as an intermediate structure in membrane fusion processes and bilayer to hexagonal H_{II} transitions. Biochim. Biophys. Acta 600, 620–624.

Verkleij, A.J., De Maagd, R., Leunissen-Bijvelt, J. and De Kruijff, B. (1982) Divalent cations and chlorpromazine can induce non-bilayer structures in phosphatidic acid-containing model membranes. Biochim. Biophys. Acta 684, 255–262.

Verpoorte, J.A. (1975) Purification and characterization of glycoprotein from human erythrocyte membranes. Int. J. Biochem. 6, 855–862.

Wallace, B.A., Veatch, W.R. and Blout, E.R. (1981) Conformation of gramicidin A in phospholipid vesicles: Circular dichroism studies of effects of ion binding, chemical modification and lipid structure. Biochemistry 20, 5754–5760.

Walter, P., Ibrakimi, I. and Blobel, G. (1981) Translocation of proteins across the endoplasmic reticulum. I. Signal recognition protein (SRP) binds to in vitro-assembled polysomes synthesizing secretory protein. J. Cell Biol. 91, 545–550.

Wickner, W. (1983) M13 coat protein as a model of membrane assembly. TIBS 8, 90–94.

Weinstein, S., Wallace, B.A., Blout, E.R., Marrow, J.S. and Veath, W. (1979) Conformation of gramicidin A channel in phospholipid vesicles: A ^{13}C and ^{19}F nuclear magnetic resonance study. Proc. Natl. Acad. Sci. U.S.A. 76, 4230–4234.

Watts/De Pont (Eds.)
Progress in Protein-Lipid Interactions

CHAPTER 4

ESR spin label studies of lipid-protein interactions

DEREK MARSH

Max Planck Institut für biophysikalische Chemie, Abt. Spektroskopie, D-3400 Göttingen, F.R.G.

I. Introduction

Electron spin resonance (ESR) spectroscopy of spin-labelled lipids, such as those illustrated in Fig. 1, has proved to be extremely useful in the study of lipid interactions with integral membrane proteins (for a previous review see Marsh and Watts, 1982). The spin label spectrum is sensitive to molecular motion on a timescale determined by the ^{14}N hyperfine splitting anisotropy of the nitroxide free radical group: $\tau_{motion} \gtrsim h/(A_{zz} - A_{xx}) \sim 3.10^{-8}$ s. If lipid exchange at the protein surface is appreciably slower than this, the spin-labelled lipids at the protein interface will therefore be distinguishable from those in the bulk lipid regions of the membrane, provided that there is a difference in lipid molecular mobility between these two states. This has proved to be the case in many lipid-protein systems, as first demonstrated by Jost et al. (1973a) for cytochrome *c*

ESR, electron spin resonance; NMR, nuclear magnetic resonance; M.W., molecular weight; DMPC, 1,2-dimyristoyl-*sn*-glycero(3)phosphocholine; DOPC, 1,2-dioleoyl-*sn*-glycero(3)phosphocholine; n-PCSL, -PESL, -PGSL, -PSSL, -PASL, 1-acyl-2-[*n*-(4′,4′-dimethyl-oxazolidine-*N*-oxyl)stearoyl]-*sn*-glycero(3)phosphocholine, -phosphoethanolamine, -phosphoglycerol, -phosphoserine, -phosphoric acid; *n*-CLSL, 1-(3-*sn*-phosphatidyl)-3-[1-acyl-2-(*n*-(4′,4′-dimethyl-oxazolidine-*N*-oxyl)stearoyl)-glycero(3)phospho]-*sn*-glycerol; *n*-SASL, *n*-(4′,4′-dimethyl-oxazolidine-*N*-oxyl)stearic acid; ASL, 17β-hydroxy-4′,4′-dimethylspiro[15α-androstane-3,2′-oxazolidin]-3′-yloxy; CL, cardiolipin (diphosphatidylglycerol); PA, phosphatidic acid; PS, phosphatidylserine; PG, phosphatidylglycerol; PC, phosphatidylcholine; PE, phosphatidylethanolamine; SM, sphingomyelin; SA, stearic acid.

oxidase. The lipid component at the protein interface has a reduced mobility and is identified either by the lipid-protein ratio dependence of the spectra, or by comparing the spectra from membranes with those from the extracted lipids. To a first approximation the two components may be separated by digital subtraction, which enables study of the structural, dynamic and thermodynamic characteristics of the lipid-protein interaction. Simulation of the entire line-shape is difficult but yields more accurate data, and in addition exact details of the lipid-protein exchange.

In this chapter the following aspects of lipid-protein interactions as studied by spin label spectroscopy are reviewed: stoichiometry, specificity, molecular conformation and mobility, and lipid exchange. A critical assessment is made of the

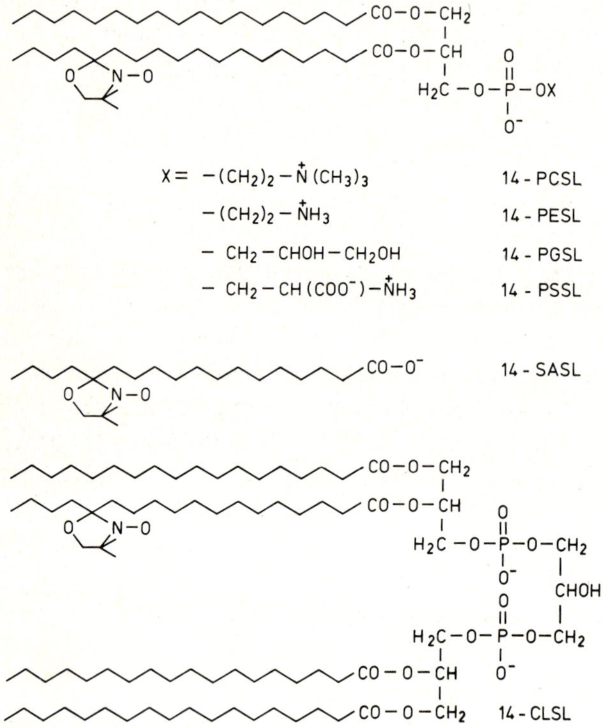

X = $-(CH_2)_2-\overset{+}{N}(CH_3)_3$ 14-PCSL

$-(CH_2)_2-\overset{+}{N}H_3$ 14-PESL

$-CH_2-CHOH-CH_2OH$ 14-PGSL

$-CH_2-CH(COO^-)-\overset{+}{N}H_3$ 14-PSSL

$CO-O^-$ 14-SASL

14-CLSL

Fig. 1. Spin-labelled phospholipids commonly used in the study of lipid-protein interactions in membranes. The nitroxide group is situated at the C-14 position of the *sn*-2 chain of the phospholipid which gives rise to a high degree of motional averaging of the spectral anisotropy in fluid lipid membranes. 14-PCSL, -PESL, -PGSL, -PSSL, -PASL, 1-acyl-2-[14-(4′,4′-dimethyl-oxazolidine-*N*-oxyl)stearoyl]-*sn*-glycero(3)phosphocholine, -phosphoethanolamine, -phosphoglycerol, -phosphoserine, -phosphoric acid; 14-CLSL, 1-(3-*sn*-phosphatidyl)-3-[1-acyl-2-(14-(4′,4′-dimethyl-oxazolidine-*N*-oxyl)stearoyl)glycero(3)phospho]-*sn*-glycerol; 14-SASL, 14-(4′,4′-dimethyl-oxazolidine-*N*-oxyl)stearic acid.

two-component nature of the lipid-protein spectra, since this constitutes the fundamental basis of the method. Particular attention is also drawn to defining the extent to which the conclusions are likely to be sensitive to potential perturbing effects of the spin label reporter group. Finally, attention is drawn to the resolution of several points of controversy in the literature.

An introduction to the spectroscopic aspects of the work discussed in this chapter can be found in Knowles et al. (1976) and Marsh (1981, 1982).

II. Two-component spectra from lipid-protein systems

A typical ESR spectrum of a lipid spin label in a reconstituted lipid-protein complex of moderately low lipid/protein ratio is given in Fig. 2a. This spectrum apparently consists of two components, as seen by comparison with the spectra of the same spin label in bilayers of the lipid alone (Fig. 2c) and bound to the fully delipidated protein (Fig. 2b). Choice of a lipid probe spin labelled close to

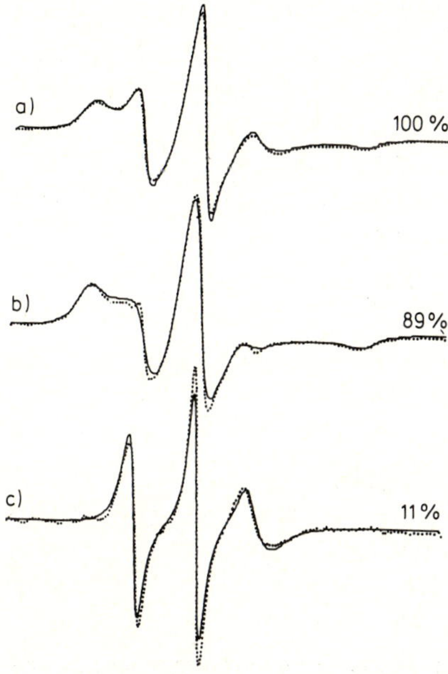

Fig. 2. Spectral subtraction and addition with the 14-PASL spin label in a lipid-protein complex. Full lines represent original ESR spectra: (a) myelin proteolipid apoprotein/dimyristoyl phosphatidylcholine complex of lipid/protein ratio 12:1 at 30°C; (b) proteolipid alone at 32°C; (c) dimyristoyl phosphatidylcholine alone at 25°C. Dotted lines represent summed spectra and difference spectra: (a) 11% lipid-alone spectrum plus 89% protein-alone spectrum; (b) recombinant spectrum minus 11% lipid-alone spectrum; (c) recombinant spectrum minus 89% protein-alone spectrum (Brophy et al., 1984).

the terminal methyl end of the chain (in the C-14 position, c.f., Fig. 1) ensures that a narrow, extensively motional-averaged spectrum is obtained in fluid bilayers (c.f., Fig. 2c). Thus, a large range of averaged spectral anisotropy is available in the wings of the spectrum to detect any motional restriction of the label induced by the protein, as in Fig. 2b. It should also be noted that the myelin proteolipid protein is somewhat exceptional amongst integral proteins in being readily obtainable as a completely delipidated apoprotein, and thus serves as a particularly instructive example.

The apparent two-component nature of the spectra can be substantiated by spectral subtraction or addition as illustrated in Fig. 2. Subtraction of the spectrum of the lipid alone from that of the lipid-protein complex yields a difference spectrum very similar to that of the protein alone (dotted spectrum in Fig. 2b). Conversely subtraction of the protein-alone spectrum yields a difference spectrum which is very similar to the lipid-alone spectrum (dotted spectrum in Fig. 2c). However, it will be noticed that it is necessary to use a lipid-alone spectrum recorded at a temperature lower than that of the complex, and a protein-alone spectrum recorded at a higher temperature, in order to match these two components in the spectrum of the complex. As will be seen in a later section, one reason for this discrepancy is (slow) exchange between the two lipid components which modifies the spectral line-shape in such a way as to reduce the apparent mobility of the fluid component and to increase that of the motionally-restricted component. Matching with single-component spectra at different temperatures tends to compensate for this exchange effect. The exchange rate between the two components in Fig. 2a turns out to be rather slow (see below), which is one of the reasons why such a good fit to the total line-shape is obtained by summation of the two components (dotted spectrum in Fig. 2a). In situations with somewhat faster exchange, spectral subtraction with optimally matched single components as in Fig. 2, still provides a reasonable first approximation for quantitating the relative amounts of the fluid and motionally restricted lipid populations, and for assessing the mobility of the two components.

Further evidence for the essential two-component nature of the lipid-protein spectra comes from the systematic changes observed with varying lipid/protein ratio, with different spin label phospholipid headgroups, and with pH and salt titration. The features of the lipid-protein interaction revealed by such studies are discussed in the subsequent sections.

III. Stoichiometry of lipid association

Titration of the ratio of the amount of fluid to motionally-restricted spin-labelled lipid as a function of the total lipid/protein ratio in a lipid-protein recombinant gives information on the stoichiometry and on the strength of the lipid association with the protein. The association can be modelled as an exchange equilibrium between labelled and unlabelled lipids, L^* and L, respectively, competing for sites on the protein (Brotherus et al., 1981):

$$B_iL + L^* \rightleftharpoons B_iL^* + L \tag{1}$$

Here it is assumed that there are m classes of different association sites, B_i (i = 1, ... , m), each of which can accommodate n_i lipid molecules. The relative association constant of a labelled lipid for a given site is then:

$$K'_i = \frac{L^*_{b_i} \cdot L_f}{L_{b_i} \cdot L^*_f} \left(\frac{\gamma^*_{b_i} \cdot \gamma_f}{\gamma_{b_i} \cdot \gamma^*_f} \right) \tag{2}$$

where $L^*_{b_i}$ and L_{b_i} are the moles of labelled and unlabelled lipid, respectively, associated with the protein sites B_i. L^*_f and L_f are similarly defined for the free (non-associated) lipid populations in the bilayer and γ_f values are the corresponding activity coefficients (defined in terms of mole fractions). The standard state for the unlabelled lipid in the bilayer is the pure bilayer of unlabelled lipids, and that for the labelled lipids is chosen such that the activity becomes equal to the mole fraction as the latter tends to zero. If activity coefficients are assumed to be unity, which should be the case for low concentrations of protein and labelled lipids (in the absence of specific protein-protein or lipid-lipid interactions), a new effective association constant may be defined solely in terms of moles of lipid:

$$K_i = K'_i \left(\frac{\gamma_{b_i} \cdot \gamma^*_f}{\gamma^*_{b_i} \cdot \gamma_f} \right) = \frac{L^*_{b_i} \cdot L_f}{L_{b_i} \cdot L^*_f} \tag{3}$$

The conservation of the total number of labelled and unlabelled lipids and of protein association sites, leads to the following respective equations:

$$L^*_f + \sum_{i=1}^{m} L^*_{b_i} = L^*_t \tag{4}$$

$$L_f + \sum_{i=1}^{m} L_{b_i} = L_t \tag{5}$$

$$L_{b_i} + L^*_{b_i} = n_i P \tag{6}$$

where L^*_t and L_t are the total numbers of moles of labelled and unlabelled lipids, respectively, and P is the total number of moles of protein. Combination of Eqns.

3 and 4–6, leads to the following equation for the equilibrium association of lipids with the protein (Brotherus et al., 1981):

$$\sum_{i=1}^{m} \frac{n_i K_i (n_f^*/n_b^*)}{[1 + K_i(n_f^*/n_b^*)][1 + n_f^*/n_b^*]^{-1} n_t^* + n_t - N_1} = 1 \tag{7}$$

where

$$N_1 = \sum_{i=1}^{m} n_i$$

is the total number of first shell association sites on the protein, $n_t^* = L_t^*/P$ and $n_t = L_t/P$ are the total numbers of labelled and unlabelled lipids per protein, and n_f^*/n_b^* is the ratio of labelled lipids free in the bilayer to those associated in first shell sites on the protein, i.e., n_f^*/n_b^* is the ratio of fluid to motionally restricted spin labelled lipids obtained by ESR spectroscopy.

In principle, Eqn. 7 is quite general and symmetrical, in that the roles of labelled and unlabelled lipids may be interchanged and there is no restriction on the relative amounts of labelled and unlabelled lipids and protein. The commonly used approximation in interpreting ESR experiments is that appropriate to low concentrations of labelled lipid, i.e.,

$$[1 + K_i\,(n_f^*/n_b^*)][1 + (n_f^*/n_b^*)]^{-1}\, n_t^* \ll n_t - N_1.$$

The equilibrium association equation can then be written:

$$(n_f^*/n_b^*) = n_t/(N_1 \cdot K_r^{av}) - 1/K_r^{av} \tag{8}$$

where the average relative association constant is defined by:

$$K_r^{av} = \sum_{i=1}^{m} n_i K_i / N_1 \tag{9}$$

i.e., K_r^{av} is the mean of all the different classes of association constants weighted by the number of sites in each class. The spin label ESR experiment thus allows the determination of the total number of first shell association sites on the protein, N_1, i.e., the stoichiometry of the lipid-protein interaction, and the average association constant of the labelled lipids relative to the unlabelled lipids, K_r^{av}.

An illustration of the application of the above analysis is given in Fig. 3, which depicts the equilibrium association of various spin-labelled phospholipids with the myelin proteolipid apoprotein in recombinants with unlabelled dimyristoyl phosphatidylcholine of various lipid/protein ratios, n_t. The spin-labelled lipid is

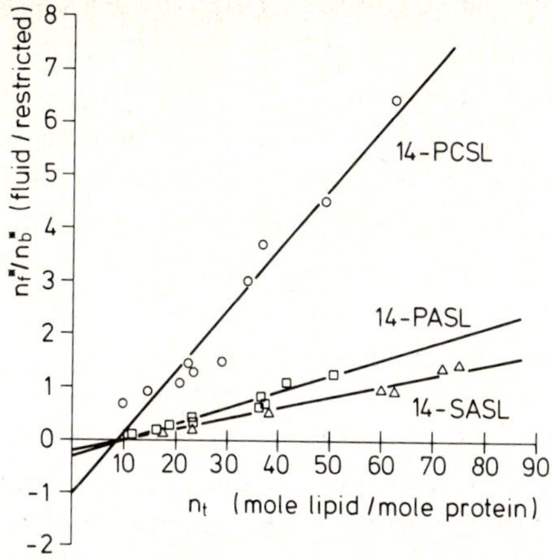

Fig. 3. Lipid/protein titration of myelin proteolipid protein-dimyristoyl phosphatidylcholine recombinants from the ESR difference spectra of the 14-PCSL phosphatidylcholine (O), 14-PASL phosphatidic acid (□), and 14-SASL stearic acid (△) spin labels at 30°C. n_f^*/n_b^* is the ratio of the double-integrated intensity of the fluid and motionally restricted components in the ESR spectra of the recombinants and n_t is the total lipid/protein ratio (Brophy et al., 1984).

present at a concentration of ca. 1 mol% relative to the unlabelled lipid, and thus the approximation used in Eqn. 8 should hold. It is seen, firstly, from Fig. 3 that the data for the phosphatidylcholine spin label, 14-PCSL, follows the linear dependence of Eqn. 8 with an intercept of -1 on the y-axis, indicating that $K_r^{av} = 1$. Therefore, the spin-labelled phosphatidylcholine has the same average association constant as the unlabelled dimyristoyl phosphatidylcholine host lipid, showing that the presence of the spin label group does not have an appreciable perturbing effect on the thermodynamics or stoichiometry of association of phosphatidylcholine with the protein. This result has been confirmed in several other reconstituted lipid-protein systems, including those of cytochrome oxidase (Knowles et al., 1979; Brotherus et al., 1981), Na^+,K^+-ATPase (Brotherus et al., 1981), Ca^{2+}-ATPase (Silvius et al., 1984) and acetylcholine receptor (Ellena et al., 1983). The total number of lipid sites on the protein is given by the intercept on the x-axis in Fig. 3, in this case $N_1 \approx 10$ for phosphatidylcholine.

The data for the phosphatidic acid spin label, 14-PASL, and the stearic acid spin label, 14-SASL, in Fig. 3 also conform to Eqn. 8, but in this case there is clearly a preferential association of these two lipids with respect to the unlabelled dimyristoyl phosphatidylcholine (see further below). Nevertheless the total number of lipid association sites is essentially the same as found for phosphatidylcholine, in accordance with the basic formulation of the model. Thus, the data for phosphatidic and stearic acids further support the concept of a fixed stoichiometry of protein association sites, independent of total lipid/protein ratio. In this sense the motionally restricted lipids observed by ESR spectroscopy can be operationally defined as a first shell of lipids interacting directly with the intramembraneous hydrophobic surface of the protein. The proviso must always be made, of course, that the protein concentration is sufficiently low that direct protein-protein contacts may be neglected. For cytochrome oxidase in dimyristoyl phosphatidylcholine complexes, for instance, it has been observed that the stoichiometry of the motionally restricted lipid falls off at very low lipid/protein ratios, for which the protein also progressively loses activity (Knowles et al., 1979), and that this decrease in stoichiometry can be explained in terms of random protein-protein contacts which exclude lipid from the protein interface (Hoffmann et al., 1981; Marsh et al., 1982).

The stoichiometries of lipid association with a number of different integral membrane proteins have now been determined in reconstituted systems using the method described above. Effective stoichiometries have also been determined in purified membrane systems which naturally contain a high proportion of a single integral protein, e.g., rod outer segment discs, sarcoplasmic reticulum or Na^+,K^+-ATPase membranes. In the latter cases some assumptions must be made since a lipid/protein titration is not possible. Spin-labelled phosphatidylcholine was used and it was assumed that K_r^{av} (PC*) $\simeq 1$, since phosphatidylcholine was the majority background lipid and the one which showed least specificity (if any) in all the membranes studied. The results of these stoichiometry determinations are summarized in Table 1. The first shell stoichiometries, N_1, correlate with the protein molecular weight, in most cases being greater for the larger proteins. With a few exceptions, the stoichiometries scale with the square root of the protein molecular weight, as would be expected for proteins of roughly cylindrical cross-section which protrude to equivalent extents from the membrane surface.

To obtain a more precise structural assignment of the first shell of motionally restricted lipids detected by ESR requires comparison with the protein structure. Cytochrome oxidase is one of the integral proteins whose structure is currently known with some degree of accuracy. The structure determined by a combination of electron microscopy and electron diffraction is given in Fig. 4. The structure of the intramembraneous domains, designated M_1 and M_2 in Fig. 4, has been determined to a resolution of 10–15 Å from specimens embedded in glucose, but only as the total projection on the plane of the membrane. The two domains M_1 and M_2 can be approximated as cylinders of diameter 30 and 20 Å, respectively. The cylinders are centred approximately 35 Å apart, hence a continuous shell of lipid can be accommodated around both cylinders. Assuming a lipid chain diameter of 4.8 Å it can be calculated that approximately 40 phospholipid molecules can be situated around a perimeter 2.4 Å out from the protein cylinder surfaces (allowing for two chains per phospholipid and both halves of the bilayer). A similar value of 45 phospholipid molecules is also obtained by fitting the cross-sectional projection of the lipid chains around the perimeter of the electron density profile of the intramembraneous protein domains as shown in Fig. 4B. This estimated number of lipids which can be accommodated around the protein surface correlates rather well with the 45 ± 4 motionally restricted lipids per 165 000 Da protein observed by ESR spectroscopy. It should be noted that the latter figure has been corrected to a molecular weight of 165 000 from the value of 200 000 assumed in the original publication (Knowles et al., 1979). The molecular weight of 165 000 is that calculated from the amino acid sequence of the cytochrome oxidase subunits, assuming a unit stoichiometry, and conforms better to the volume of the cytochrome oxidase structure in Fig. 4 (Deatheridge et al., 1982). Thus, in spite of the rather complex intramembraneous shape of cytochrome oxidase, it appears that the first shell of motionally restricted lipids may form a continuous boundary around it.

TABLE 1
Stoichiometries of the motionally restricted lipid component in various lipid-protein systems

Protein/Membrane	M.W. $\times 10^{-3}$	N_1^{exp} (mol/mol)	$N_1^{exp}/\sqrt{M.W.}$	N_1^{calc} (mol/mol)	Ref.
Na^+,K^+-ATPase-DOPC	314	63 ± 3	0.112 ± 0.005	(~57–72)	1
Na^+,K^+-ATPase shark rectal gland	265	58 ± 4	0.112 ± 0.008	(~57–72)	2
Cytochrome oxidase-DMPC	165	45 ± 4	0.110 ± 0.011	40–45	3
Acetylcholine receptor-DOPC	250	40 ± 7	0.080 ± 0.014	43	4
Ca^{2+}-ATPase-egg PC	115	22 ± 2	0.065 ± 0.006	(~23–27)[a]	5
Sarcoplasmic reticulum/Ca^{2+}-ATPase	115	24	0.071	(~23–27)[a]	6
Bovine rod outer segment disc/rhodopsin	39	25 ± 3	0.125 ± 0.016	24 (± 2)	7
Frog rod outer segment disc/rhodopsin	39	23 ± 2	0.114 ± 0.010	(24)	2
Myelin proteolipid apoprotein-DMPC	25	10 ± 2	0.063 ± 0.013	(~10–12)[b]	8

a. Calculated assuming a dimer, with monomer radius 20 Å (Ref. 9).
b. Calculated assuming hexamer; DMPC, dimyristoyl phosphatidylcholine; DOPC, dioleoyl phosphatidylcholine; M.W., molecular weight.
References: 1, Brotherus et al. (1981); 2, Marsh et al. (1982); 3, Knowles et al. (1979); 4, Ellena et al. (1983); 5, Silvius et al. (1984); 6, Thomas et al. (1982); 7, Watts et al. (1979); 8, Brophy et al. (1984); 9, Marsh and Watts (1982).

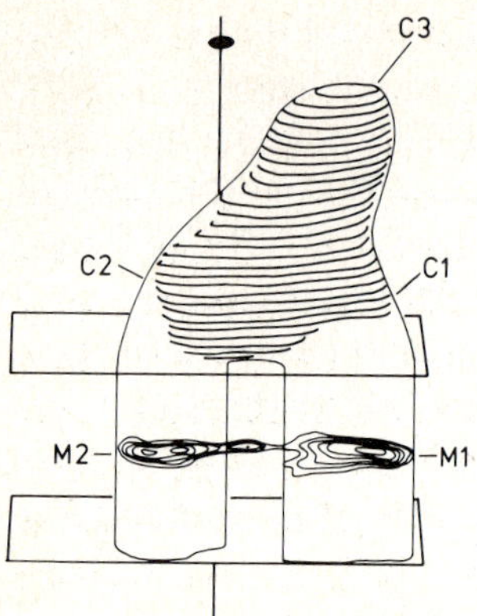

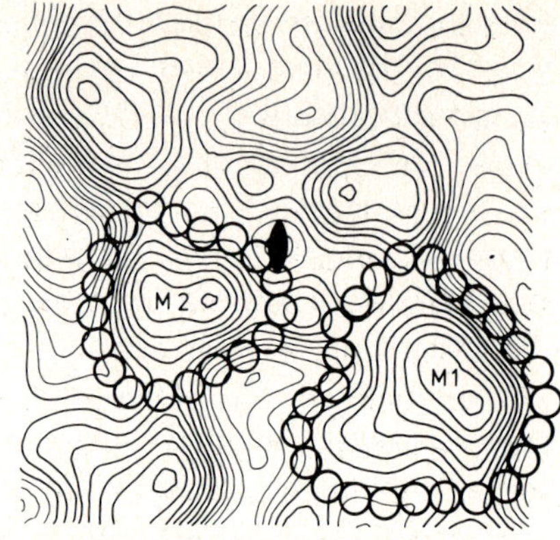

Fig. 4. Map of the cytochrome oxidase structure obtained from electron microscopy of two-dimensional crystals (Deatheridge et al., 1982). (A) The upper and lower surfaces of the bilayer are represented by the planes perpendicular to the two-fold axis of the crystal. The upper side of the bilayer corresponds to the cytoplasmic side of the mitochondrial inner membrane. C_1, C_2 and C_3 are the exposed domains of the protein which contain the cytochrome *c* binding site. (B) The structure of the intramembraneous domains of the protein, M_1 and M_2, has been determined only in two-dimensional projection, as indicated. The enclosing shell of circles is drawn to scale for lipid chains.

The structure of the acetylcholine receptor has also been determined by electron microscopy (Kistler et al., 1982). The intramembraneous portion is approximately cylindrical with a radius of 29 Å. Thus, approximately 43 lipid molecules can be accommodated around its perimeter in agreement with the number of motionally restricted lipids per protein determined from ESR spectroscopy (Ellena et al., 1983). Less is known of the detailed structure of the other proteins listed in Table 1. The Na^+,K^+-ATPase has a maximum intramembraneous diameter of 50 Å for the $\alpha\beta$-monomer, and is probably present in the membrane as the $\alpha_2\beta_2$-dimer of molecular weight 265 000-314 000 (Hebert et al., 1982; Deguchi et al., 1977). Thus, a maximum of 72 lipids can be accommodated around the dimer if there is lipid in the intervening region between the two monomers. This value decreases to approximately 57 lipids per dimer if the region between the monomers is inaccessible to lipid. The values for the numbers of motionally restricted lipids observed by spin label spectroscopy again lie within this range. X-ray scattering measurements have yielded a radius of gyration which indicates an elongated structure for the rhodopsin molecule and, amongst other possible shapes, is consistent with either a cylinder of 31.5 Å diameter or a prolate ellipsoid of minor axis 34 Å (Sardet et al., 1976). The results of the low resolution structure from two-dimensional negatively stained rhodopsin crystals, which reveal a dimer 20–25 Å in width and 70–80 Å in length, are also approximately in accord with these structures (Corless et al., 1982). The number of lipid molecules which can be accommodated around the cylindrical or ellipsoidal structures is 24 per protein monomer in each case. It is also relevant to note that 25 lipid molecules may be accommodated around the bacteriorhodopsin monomer structure (Michel et al., 1980), since the dimer of the latter has almost exactly the same projected area as the rhodopsin dimer (Corless et al., 1982). These estimates are again in accord with the number of motionally restricted lipids per rhodopsin determined by ESR spectroscopy. (Note that the latter figures have been corrected in Table 1 to a molecular weight of 39 000 as determined from the amino acid sequence.) A double-cylinder model has been proposed from X-ray scattering measurements on the Ca^{2+}-ATPase, with the intramembraneous portion being 40 Å in diameter (Le Maire et al., 1981). The intramembraneous cross-sectional area of the monomer has been determined by X-ray diffraction to be 2375 ± 130 Å^2 (corresponding to an effective diameter of 55 ± 2 Å) and there is evidence that the Ca^{2+}-ATPase exists as a dimer in the membrane (Blasie et al., 1982; Napolitano et al., 1983). A dimer of 40 Å diameter cylinders could accommodate approximately 23 lipids per monomer around its perimeter, and a cylindrical dimer of cross-sectional area 2 × 2375 Å^2 could accommodate 27 ± 1 lipids per monomer around its perimeter. The number of motionally restricted lipids determined for the Ca^{2+}-ATPase thus lies within the range estimated for a single lipid shell around the dimer. Assuming a cylindrical shape for the myelin proteolipid protein, the intramembraneous perimeter can be estimated from the protein molecular weight and partial specific volume. The number of first shell lipids varies from 23 to 30 lipids per monomer depending on whether an appreciable portion of the protein protrudes into the aqueous phase (Curatolo et al., 1977). This number is reduced to 10–12 lipids per monomer if it is assumed that the protein is a hexamer in the membrane, as is found in detergents (Smith et al., 1984). The latter value is in good agreement with the number of motionally restricted lipids observed by ESR.

Thus, the number of motionally restricted lipids correlates with estimates of the number of lipids which can be accommodated around the protein perimeter, based on the limited amount of structural information available for the various membrane proteins. It should be stressed, however, that the exact surface contour of the proteins is not known accurately; thus, it cannot be excluded that the motionally restricted lipids are accommodated within surface invaginations, rather than covering the entire surface.

In concluding this section on the stoichiometry of the lipid-protein association the resolution of several controversial points will be discussed. The first, considers the temperature dependence of the fraction of motionally restricted lipid. The stoichiometries in Table 1 were all determined at relatively low temperatures, for which the exchange between the fluid and motionally restricted components is relatively slow. Under these circumstances spectral subtractions reflect the true fraction of lipid associated with the protein, provided that fluid and motionally restricted components which accurately match the linesplittings and linewidths in the composite spectrum are chosen for the subtractions. Failure to do this will lead to erroneous results (Brotherus et al., 1980), as was the case in studies by Davoust et al. (1979), c.f., also Davoust and Devaux (1982). The effects of exchange on the fidelity of spectral subtractions will be discussed further in a subsequent section.

A further controversial point regards the motionally restricted lipid associated with rhodopsin in rod outer segment disc membranes. It has been conclusively demonstrated (Watts et al., 1981) that the motionally restricted lipid component in normal, unbleached membranes which is referred to in Table 1, is quite distinct from the more strongly immobilized lipid component which can be artefactually induced by extensive bleaching or delipidation (Favre et al., 1979). Thus, the claims of the latter publication that motionally restricted lipid is only seen under artefactual conditions can be safely ignored. In addition, the spectra of spin-labelled fatty acids covalently attached to rhodopsin (Davoust et al., 1979, 1980) have subsequently been shown to arise from exchange between the first and second shells of lipid surrounding the protein (Davoust and Devaux, 1982). Thus, the claims of the former publications that the motionally restricted lipid component arose from a temperature-dependent protein aggregation must also be disregarded.

Controversy has also surrounded the original observation of motionally restricted lipids in acetylcholine receptor membranes (Marsh and Barrantes, 1978). The claim that a motionally restricted component was seen only with fatty acids and not with phospholipid spin labels (Rousselet et al., 1979) was successfully refuted by Marsh et al. (1981). The subsequent measurements with reconstituted systems (Ellena et al., 1983) establish beyond doubt the association of motionally restricted lipids with the receptor. This latter work very elegantly vindicates the original observations, but also demonstrates that the numbers of first shell motionally restricted lipids were overestimated (Marsh and Barrantes, 1978) because of the specificity of fatty acids and androstanol for association with the receptor.

IV. Specificity of lipid-protein associations

The results of Fig. 3 have already demonstrated the possibility of a selectivity between the different lipid types for association with the protein. The results of the lipid/protein titration show that the origin of the specificity in this case, and also for the acetylcholine receptor (Ellena et al., 1983), cytochrome oxidase (Knowles et al., 1981) and the Na^+,K^+-ATPase (Brotherus et al., 1981) arises from changes in the average relative association constant K_r^{av} rather than in the total number of first shell association sites, N_1. This is in accordance with the model present in the previous section. Thus, experiments on natural membranes for which the lipid/protein ratio is fixed can also be interpreted according to this model in terms of changes in the relative association constant. From Eqn. 8 with N_1 constant we get:

$$(n_f^*/n_b^*)^{PC}/(n_f^*/n_b^*)^L = K_r^{av}\,(L)/K_r^{av}\,(PC) \tag{10}$$

where K_r^{av} $(L)/K_r^{av}$ (PC) is the ratio of the average relative association constant for a particular spin-labelled lipid, L, to that for spin-labelled phosphatidylcholine (PC). Thus, relative values for the association constants may be obtained from spectral subtractions at a single lipid/protein ratio which yield values for the left-hand side of Eqn. 10.

The order of selectivity of lipids obtained in this way for a variety of different integral membrane proteins is given in Table 2. The myelin proteolipid protein, the Na^+,K^+-ATPase, cytochrome oxidase and the acetylcholine receptor all display a well-defined specificity pattern, with the relative association constant, K_r^{av}, reaching maximum values of 7.0, 3.8, 5.4 and 4.1 relative to phosphatidylcholine for these proteins, respectively. For Ca^{2+}-ATPase the selectivities are considerably smaller with a maximum value of $K_r^{av} \lesssim 2.0$ relative to phosphatidylcholine, and for rhodopsin there is practically no selectivity at all with $K_r^{av} \approx 1.0$ for all lipids. From Table 2 it is clear that there is a different pattern of lipid specificity for the different proteins. This indicates that the selectivity is not a property of the lipid alone but depends, as one might expect, also on the particular amino acid configuration and structure of the different proteins. In particular, it is of interest to note that cardiolipin is not invariably the lipid which displays the highest selectivity, and that for rhodopsin it displays no selectivity at all. It has been argued (Lee, 1983) that there should be an apparent specificity between the different lipids depending on the number of chains per lipid. This was based on a model of the protein interface as a regular linear array of chain binding sites which is clearly unrealistic and makes predictions which are inconsistent with the experimental findings of Table 2. Most crucially, there is no selectivity with rhodopsin, independent of the number of chains, demonstrating that the latter does not play a critical role. It is also interesting to note that there does not

TABLE 2
Order of selectivity of spin-labelled lipids for association with integral membrane proteins

Myelin proteolipid:	SA > PA > CL $\gtrsim$ PS > PG $\approx$ SM $\approx$ PC > PE	Refs. 1, 7
Na^+,K^+-ATPase:	CL > PS $\approx$ SA $\gtrsim$ PA > PG $\approx$ SM $\approx$ PC $\approx$ PE	Refs. 2, 7
Cytochrome oxidase:	CL > PA $\approx$ SA > PS $\approx$ PG $\approx$ SM $\approx$ PC $\approx$ PE	Refs. 3, 7
Acetylcholine receptor:	SA > PA > PS $\approx$ PC $\approx$ PE	Ref. 4
Ca^{2+}-ATPase:	CL > PS $\approx$ SA $\gtrsim$ PA $\gtrsim$ PG $\approx$ SM $\approx$ PC $\approx$ PE	Refs. 5, 7
Rhodopsin:	CL $\approx$ PA $\approx$ SA $\approx$ PS $\approx$ PG $\approx$ SM $\approx$ PC $\approx$ PE	Refs. 6, 7

CL, cardiolipin; PA, phosphatidic acid; PS, phosphatidylserine; PG, phosphatidylglycerol; PC, phosphatidylcholine; PE, phosphatidyletyhanolamine; SM, sphingomyelin; SA, stearic acid.
References: 1, Brophy et al. (1984); 2, Esmann et al. (1985); 3, Knowles et al. (1981); 4, Ellena et al. (1983); 5, Hidalgo and Marsh (1985); 6, Watts et al. (1979); 7, Marsh et al. (1985).

appear to be any significant selectivity of the sphingolipid sphingomyelin relative to phosphatidylcholine (Marsh et al., 1985).

The average relative association constants may be used to yield information regarding the thermodynamics of the lipid selectivity. An effective free energy of association relative to phosphatidylcholine may be written as:

$$\Delta G_L - \Delta G_{PC} = -RT \ln K_r^{av}(L)/K_r^{av}(PC) \tag{11}$$

Due to the nature of the averaging of the association constants this is only a true free energy if all sites are equivalent. The maximum values for the relative association constants thus correspond to free energy differences in the range: $\Delta G_L - \Delta G_{PC} \sim 0.8\text{–}1.2$ kcal/mol, and thus can be appreciable relative to thermal energies. In principle the relative contributions of enthalpic and entropic terms to the free energy of association may be determined from the temperature dependence of the fraction of motionally restricted lipid, using Eqns. 10, 11 and the normal thermodynamic relations: $\Delta H_L - \Delta H_{PC} = -R\partial \ln (K_r^L/K_r^{PC})/\partial(1/T)$ and $\Delta G_L - \Delta G_{PC} = \Delta H_L - \Delta H_{PC} - T(\Delta S_L - \Delta S_{PC})$. However, as discussed above and further in the section on exchange rates below, there are other factors which introduce uncertainties into the apparent temperature dependence. Some data are available for Na^+,K^+-ATPase membranes which indicate that the temperature variation is slight, and hence suggest that the enthalpy/entropy terms are of comparable size to the overall free energy associated with the specificity of interaction (Esmann et al., 1985). Thus, it seems likely that there is no extensive entropy-enthalpy compensation and the free energies are a direct indication of the strength of the interaction specificity.

Whereas the greatest specificity is observed for negatively-charged lipids in Table 2, this does not bear a direct relation to the net charge on the lipid, and furthermore the selectivity pattern amongst the various negatively charged lipids is different for different proteins. This suggests that the selectivity does not have purely a simple electrostatic origin, but depends in detail on the chemical nature of the groups involved. More insight into the origin of this selectivity can be obtained from pH and salt titration as illustrated in Figs. 5 and 6. The fractions, f, of motionally restricted stearic acid and phosphatidic acid spin labels associated with the Na^+,K^+-ATPase titrate with pH according to:

$$f = f_{min} + (f_{max} - f_{min})/(1 + [H^+]/K_a) \tag{12}$$

with $pK_a = -\log_{10} K_a = 8.00$ and 6.59, respectively. Spin-labelled phosphatidylserine, which exhibits a selectivity but has no pK in this pH range, shows no titration indicating that there are no titratable groups on the protein within this range which contribute to the lipid-protein interaction. Most significantly the fraction of motionally restricted stearic and phosphatidic acid labels in the proto-

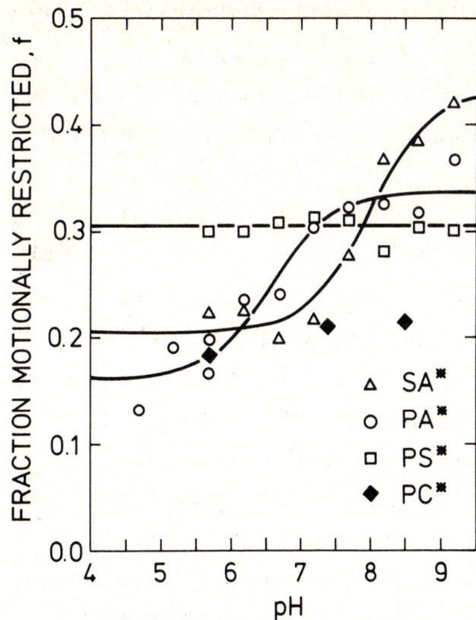

Fig. 5. pH dependence of the fraction of motionally restricted lipid spin labels in Na^+,K^+-ATPase membranes in 10 mM Na_2HPO_4. (—△—) stearic acid spin label, 14-SASL; (—○—) phosphatidic acid spin label, 14-PASL; (—□—) phosphatidylserine spin label, 14-PSSL; (◆)phosphatidylcholine spin label, 14-PCSL (Esmann and Marsh, 1985).

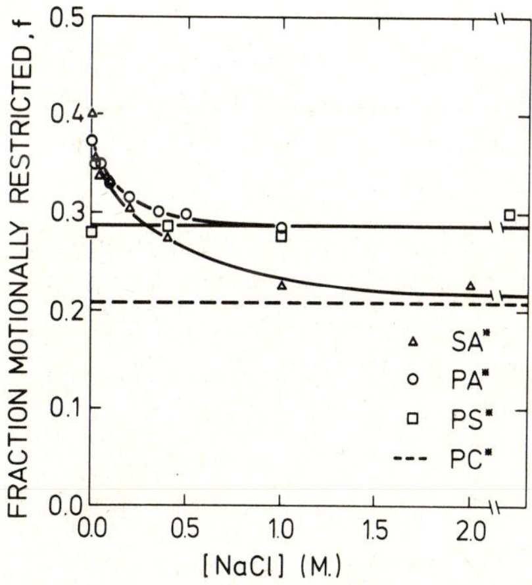

Fig. 6. Salt dependence of the fraction of motionally restricted lipid spin labels in Na^+,K^+-ATPase membranes in 20 mM Tris (pH 9.2) for 14-PASL and 14-SASL, and in 30 mM histidine (pH 7.6) for 14-PSSL. Symbols are as defined in Fig. 5. The highest concentration point for 14-PSSL is 3 M NaCl. The dashed line indicates the fraction of motionally restricted phosphatidylcholine spin label, 14-PCSL, from Fig. 5 (Esmann and Marsh, 1985).

nated form, f_{min}, is quite close to that for the zwitterionic phospholipid phosphatidylcholine. Thus, protonation appears to remove the specificity of interaction with these two lipids. Whereas in the case of stearic acid the negative charge is totally removed, for phosphatidic acid the number of negative charges is merely reduced from two to one.

A similar difference is seen in the salt dependence in Fig. 6. A progressive decrease in the fraction of motionally restricted lipid is observed corresponding to screening of the electrostatic interaction between protein and lipid. This screening arises because the activity coefficients, γ_j, of the charged species are reduced by the counterion cloud. Hence, according to Eqn. 3, the average relative lipid association constant in the presence of ions becomes:

$$K_r = K_r^o(\gamma_{L^*}\gamma_P/\gamma_{L^*P}) \tag{13}$$

where K_r^o is the association constant in the absence of screening, and γ_{L^*}, γ_P and γ_{L^*P} are the activity coefficients of spin label, protein and lipid-protein complex, respectively, in the presence of ions. The salt dependence can thus be understood, at least semi-quantitatively, in terms of Debye-Hückel theory (Esmann and Marsh, 1985), where the activity coefficients are given by:

$$\ln \gamma_j = \frac{-Z_j^2 e^2}{8\pi\varepsilon_o \varepsilon kT} \frac{\varkappa}{1 + \varkappa a_j} \tag{14}$$

with Z_j the charge on species j, a_j the interaction distance of species j with counterions, and the inverse screening length: $\varkappa = (2000\ N_a e^2\ I/\varepsilon_o \varepsilon kT)^{1/2}$ where I is the ionic strength. The fraction of motionally restricted lipid levels off to an approximately constant value for stearic and phosphatidic acids at the higher salt concentrations as predicted by the theory. However, the limiting values are different for the two lipids and for phosphatidylserine there is virtually no salt dependence. Significantly the limiting value of the fraction of motionally restricted stearic acid observed at high salt is only very slightly higher than that found on removing the charge by protonation (c.f., Fig. 5), and is also comparable to that observed for the zwitterionic phosphatidylcholine. Thus, it appears that the selectivity of interaction of stearic acid with Na^+,K^+-ATPase is essentially of simple electrostatic origin, in common with previous findings on other single-tailed lipids (Brotherus et al., 1980). This is not so, however, for phosphatidic acid and phosphatidylserine. In the former case the proportion of the motionally restricted lipid which can be displaced by salt is considerably less than that which is removed on protonating the lipid, and for phosphatidylserine salt causes no displacement. Thus, direct electrostatic effects alone cannot account for the whole of the observed specificity of interaction of these two phospholipids. Lipid hydration and other chemical effects must also be important. Lipid hydration and hydrogen bonding interactions are also presumably involved in determining the difference in specificity of the two zwitterionic lipids phosphatidylcholine and phosphatidylethanolamine with the myelin proteolipid protein (c.f., Table 2).

Spin-labelled steroid molecules also give an interesting pattern of specificity of interaction with the protein. Whereas there is little selectivity between spin-labelled androstanol and phosphatidylcholine in interaction with cytochrome oxidase (Jost et al., 1973c) or with rhodopsin (Watts et al., 1979) or the Na^+,K^+-ATPase (Esmann et al., 1985), there is a positive specificity ($K_r^{av} \simeq 4.3$) in the interaction of androstanol with the acetylcholine receptor (Ellena et al., 1983) and a negative specificity for the interaction with the myelin proteolipid protein ($K_r^{av} \sim 0.34$; Brophy et al., 1984). There is also a negative selectivity ($K_r^{av} \sim 0.65$) in the interaction of a spin-labelled cholesterol analogue with the Ca^{2+}-ATPase from sarcoplasmic reticulum (Silvius et al., 1984). A lipid/protein titration has been performed both in the latter case and for the acetylcholine receptor. In spite of the different selectivities it is found that the steroids yield approximately the same numbers of sites, N_1, as obtained with phosphatidylcholine, in both cases. This gives further support for the concept of a fixed site stoichiometry in the lipid-protein interaction, as discussed in the previous section. Clearly the different selectivities observed for the steroid interactions cannot be explained in simple electrostatic terms. Presumably the specificities must have steric, hydrogen-bonding and possibly hydration interactions as an origin.

Lipid specificities have mostly been interpreted using Eqn. 8 in terms of an average relative association constant, K_r. It is thus not known whether the specificity arises from a uniform increase in association constant of all N_1 sites, or is due to a much higher specificity of just a few sites. In principle it can be decided between these two possibilities by varying the amount of the specific (labelled) lipid according to Eqn. 7. Calculations are given in Fig. 7 for two classes of binding sites for which Eqn. 7 becomes:

$$\frac{n_1K_1(n_f^*/n_b^*)}{[1+K_1(n_f^*/n_b^*)][1+n_f^*/n_b^*]^{-1}n_t^*+n_t-N_1} + \frac{n_2K_2(n_f^*/n_b^*)}{[1+K_2(n_f^*/n_b^*)][1+n_f^*/n_b^*]^{-1}n_t^*+n_t-N_1} = 1 \qquad (15)$$

One of the sets of sites is assumed to be non-selective: $K_2 = 1$ and the other is chosen such that the average relative association constant (c.f., Eqn. 9) is: $K_r^{av} = (n_1K_1 + n_2K_2)/N_1 = 5.4$ with $N_1 = 50$, as for the interaction of cardiolipin with cytochrome oxidase (Powell et al., 1985). The variation of the fraction of motionally restricted cardiolipin, $f = (1 + n_f^*/n_b^*)^{-1}$, is given as a function of total cardiolipin/protein mole ratio, n_t^*, for various numbers of specific sites. It is seen

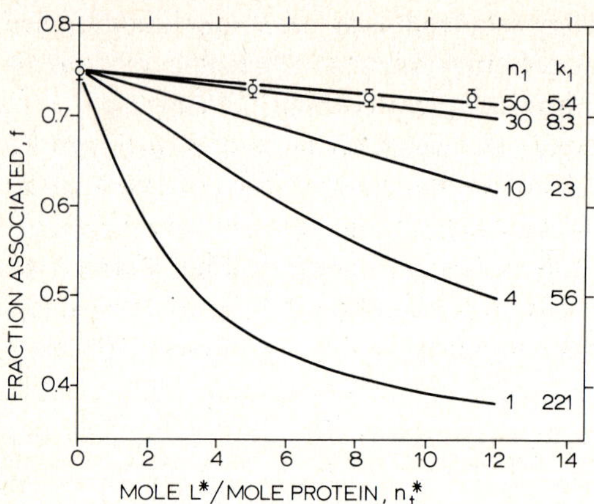

Fig. 7. Fraction, f, of protein-associated specific lipid L^* as a function of total amount of L^* per protein, n_t^*. Calculated according to Eqn. 15 with $n_t = 140$, $N_1 = 50$, $K_r^{av} = 5.4$ and $K_2 = 1$ for various values of n_1 (and K_1). The experimental points for cytochrome oxidase reconstituted at lipid/protein ratio 140:1 with dimyristoyl phosphatidylcholine plus varying amounts of cardiolipin, and spin-labelled with a fixed amount of the cardiolipin spin label, 14-CLSL, are given by the circles (Powell et al., 1984).

that the experimental results vary only a little with total cardiolipin content, and therefore can only be explained in terms of a relatively large number of relatively low specificity sites. The situation approximates a uniform increase in relative association constant for nearly all sites. Clearly there may be other systems for which this is not the case, but the amount of experimental evidence is currently rather limited.

V. Conformational order of the lipids at the protein interface

Evaluation of the spectroscopic properties of the motionally restricted lipid component, such as in Fig. 2c, should give information on the structure and dynamics of the lipid chains in contact with the protein surface. The degree of ordering of the lipid chains relative to the membrane normal is of interest because it gives a direct measure of the chain configuration. Since the spectra of the motionally restricted component lie close to the limits of motional sensitivity of conventional ESR spectroscopy, information about chain ordering can only be obtained from oriented specimens. The configurational disorder of the chains can be depicted by an orientational distribution function, e.g., of the Gaussian type:

$$\varrho(\theta_i) \cdot sin\, \theta_i = \exp[-(\theta_i - \overline{\theta})^2/2\theta_o^2] \cdot \sin \theta_i \qquad (16)$$

where θ_i is the angle between the ith chain segment axis (or nitroxide z-axis) and the normal to the membrane. $\bar{\theta}$ is the angle about which the Gaussian distribution is centred and θ_o represents the width of the distribution. For simplicity axial symmetry has been assumed. In principle the distribution function $\varrho(\theta)$ can be mapped out from the angular dependence of the spectra from oriented membranes, provided that these spectra are in the slow-motion regime. This is in contrast to the situation in the fast-motional regime found in fluid bilayers. In the latter case the rapid motion gives rise to an averaging of the spectral anisotropy to an extent specified by the order parameter (see, e.g., Marsh, 1981):

$$S = \frac{\int_0^{\pi} \tfrac{1}{2}(3\cos^2\theta - 1)\varrho(\theta)\cdot\sin\theta\cdot d\theta}{\int_0^{\pi} \varrho(\theta)\cdot\sin\theta\cdot d\theta} \tag{17}$$

Clearly, for a very narrow distribution ($\theta_o \to 0$), the order parameter is $S \approx \tfrac{1}{2}(3\cos^2\bar{\theta} - 1)$ and varies from 1 to $-\tfrac{1}{2}$ depending on $\bar{\theta}$, and for a very broad distribution ($\theta_o \to \infty$) the order parameter goes to zero, as is the case for isotropic motion. This connection between distribution function and order parameter gives a useful method of comparison with the results of deuterium NMR. For the latter it is thought that the lipid environments at the surface and away from the protein are averaged by an exchange of lipid molecules that is fast on the ^{2}H-NMR timescale.

Relatively few ESR studies have been reported on oriented lipid-protein systems. Jost et al. (1973b) prepared partially oriented samples of cytochrome oxidase with varying lipid/protein ratio. Very little angular dependence was observed for the spectra at lower lipid/protein ratios, although two-component spectra were not resolved because the label was situated on the 5-C atom of the fatty acid chain. The spectra were interpreted as indicating a wide distribution of angular orientations for the motionally restricted lipid, although macroscopic disorder of the sample could not be excluded. Samples of myelin proteolipid-dimyristoyl phosphatidylcholine prepared on a flat substrate yielded a second disordered spectral component from a spin-labelled steroid, which was induced by the protein (Post and Dijkema, 1983). However, the disordered component did not appear to be significantly motionally restricted, and thus most probably corresponded to macroscopic or long-range disordering of the fluid lipids by the protein.

More detailed studies have been performed on partially oriented rod outer segment disc membranes (Pates and Marsh, 1985). Because the orientation was not perfect, the unoriented component has been digitally subtracted from the spectra in Fig. 8 to yield only the oriented component. Experiments were performed with stearic acids labelled either on the 5-C atom (5-SASL) or on the

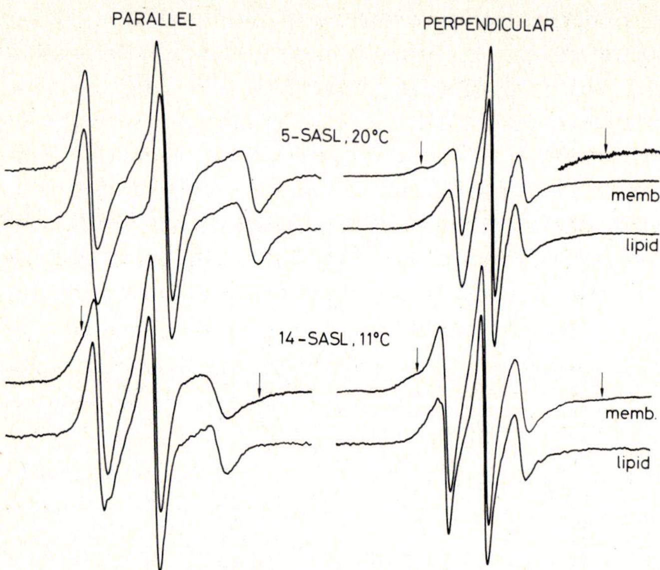

Fig. 8. ESR spectra of oriented rod outer segment disc membranes and their extracted lipids. Left-hand side is with the magnetic field oriented parallel to the membrane normal; right-hand side is with the magnetic field oriented perpendicular to the membrane normal. Upper pairs of spectra are from the 5-C atom labelled stearic acid, 5-SASL, recorded at 20°C, and the lower pairs of spectra are from the 14-C atom labelled stearic acid, 14-SASL, at 11°C. The upper spectrum of each pair is from the membranes and the lower spectrum of each pair is from the extracted lipids. Arrows indicate the motionally restricted component. Total scan range = 100 gauss (Pates and Marsh, 1985).

14-C atom (14-SASL) of the chain. Comparison of the spectra taken with the spectrometer field parallel ($\|$) and perpendicular ($\perp$) to the normal to the plane of the membrane demonstrate the anisotropy in the hyperfine splittings of the fluid lipid component for both labels, but more obviously for 5-SASL which has the larger anisotropy. The motionally restricted component (indicated by arrows) is seen in the wings of the spectra from 14-SASL for both orientations of the magnetic field. Since the maximum splitting of the peaks from this component is approximately the same (~ 58 gauss) for both field orientations and similar to that found in randomly oriented samples it may be concluded that the motionally restricted component must have a wide orientational distribution. For 5-SASL the motionally restricted lipid component is resolved only with the magnetic field in the perpendicular direction. However, since the hyperfine splitting is rather large (~ 58 gauss) this immediately indicates a broad orientational distribution width also at this label position. As with the 14-SASL, subtraction of the expected amount of a randomly oriented motionally restricted component yields spectra similar to those of the lipids alone for both field orientations. This further suggests that the motionally restricted component has a broad distribution of orientations.

Thus, the available evidence suggests that the spin labelled lipid chains are disordered on the surface of the protein. This has been established with reasonable certainty for both the 5-C atom and the 14-C atom segments of stearic acid at the lipid-protein interface of rhodopsin. Although, unfortunately, the data was not extensive enough to determine the complete distribution function. At this point it should be remembered, however, that the data refer to the labelled lipid chains. It is quite possible that the disordering effect may be due to the steric interactions of the spin-label group itself with the protein, and may not be representative of the behaviour of the unlabelled chain segment. The agreement of the data from two well-separated segments of the chain, however, make this seem less likely. A recent detailed analysis of the behaviour of such labels in fluid lipid bilayers has demonstrated that the spin label spectra overestimate the degree of *gauche* chain disorder but faithfully record the degree of order of the chain long axis (Lange et al., 1985). Given these reservations, the results are consistent with current interpretations of deuterium and phosphorus NMR order parameter measurements on lipid-protein systems. These latter suggest that the degree of lipid order is decreased somewhat by the presence of the protein (Seelig et al., 1982), which is in agreement with the wide orientational distribution observed for the motionally restricted lipid spin label population. Thus, there is no inconsistency between the ESR and NMR results although the spectral appearances are quite different, since the lipid exchange at the protein surface is slow on the timescale of ESR but fast on that of NMR (see also Chapters 1, 2 and 3). A further discussion of this comparison and of ESR on oriented lipid-protein systems is given in Marsh (1983).

VI. Exchange of the lipids at the protein interface

In the foregoing sections the spectral analysis has assumed two independent lipid components: the fluid and the motionally restricted. As mentioned in the discussion of the two-component nature of the spectra, these two lipid components are coupled by exchange of lipids on and off the surface of the protein. It is the purpose of this section to analyse the spectral effects of this exchange, with the two-fold aim of measuring the exchange rate and of testing the reliability of the subtraction method for determining the fraction of motionally restricted lipid.

The spectra may be simulated by the two-site exchange model normally used in magnetic resonance spectroscopy (McConnell, 1958). The rate equation for the spin magnetization associated with the motionally restricted component, M_b, is given by:

$$\frac{dM_b}{dt} = -\tau_b^{-1} M_b + \tau_f^{-1} M_f \tag{18}$$

where τ_b^{-1} and τ_f^{-1} are the probabilities per unit time of transfer from the motionally restricted to the fluid component and vice-versa, respectively. A similar rate equation holds for the spin magnetization associated with the fluid lipid component, M_f. These equations assume rapid mixing of the fluid pool by lateral diffusion, such that all fluid lipids sample the lipid-protein interface within the lifetime, τ_b. At exchange equilibrium $dM_b/dt = dM_f/dt = 0$ and hence from equation (18):

$$\tau_b^{-1}/\tau_f^{-1} = M_f/M_b = (1 - f)/f \tag{19}$$

where f is the fraction of motionally restricted lipid. Incorporation of Eqn. 18 (and its equivalent for dM_f/dt) into the Bloch equations for the time evolution of the spin magnetization in the magnetic field leads ultimately to an expression for the line-shape (see, e.g., Atherton, 1973). The general result is rather complicated, but for slow exchange, which holds at least approximately for several lipid-protein systems, a simple physical picture emerges. In this latter case, $1/\tau_b$, $1/\tau_f \ll (\omega_b - \omega_f)$, where ω_b and ω_f are the angular resonance frequencies in the motionally restricted and fluid components, respectively. The out-of-phase component of the magnetization, which corresponds to the resonance absorption, is then given by:

$$v(\omega) \propto \frac{f(T_{2,b}^{-1} + \tau_b^{-1})}{(T_{2,b}^{-1} + \tau_b^{-1})^2 + (\omega_b - \omega)^2} + \frac{(1 - f)(T_{2,f}^{-1} + \tau_f^{-1})}{(T_{2,f}^{-1} + \tau_f^{-1})^2 + (\omega_f - \omega)^2} \tag{20}$$

where $T_{2,b}$ and $T_{2,f}$ are the transverse relaxation times which specify the intrinsic linewidths of the motionally restricted and fluid components, respectively. Clearly, in this case the line-shape, written here in terms of the angular frequency, ω, of the microwave radiation, is a weighted sum of two Lorentizians whose linewidths are increased due to lifetime broadening. The amount of the increase in linewidth is given by the reciprocal of the exchange lifetime, i.e.

$$T_2^{-1} = T_{2,f}^{-1} + \tau_f^{-1} \tag{21}$$

and a similar expression for the motionally restricted component. Equation 21 in principle gives a method for determining the exchange lifetime from the broadening of the fluid lipid component relative to the spectra of the fluid lipids alone (Marsh et al., 1982).

To simulate the entire line-shape a summation must be made over all angular orientations relative to the magnetic field, both with respect to the component to which the transfer is taking place and that from which the transfer originates (Davoust and Devaux, 1982). Comparison of various simulated spectra with the experimental spectrum for a myelin proteolipid protein/dimyristoyl phosphatidylcholine recombinant labelled with the 14-PASL phosphatidic acid spin label is given in Fig. 9. In contrast to the method employed for the spectral subtractions in Fig. 2, the pure lipid and the 'protein-alone' components were taken to be those recorded at the same temperature (30°C) as that for the recombinant. From Fig. 9 it can be seen that a reasonably good fit to the experimental spectrum can be obtained with an exchange rate in the region of 4×10^6 s^{-1}. The fit is not perfect in the high-field region of the fluid component, suggesting that there may also be some direct static perturbation by the protein causing an increase in the anisotropy of motion of the fluid lipids. However, it should be noted that a simplified method which does not take adequate account of motional anisotropy had to be used to simulate the individual components. Nonetheless, the agreement with the experimental spectrum is rather good in the low-field and central regions of the fluid component, that is, in those regions which might be expected to be the most sensitive to exchange.

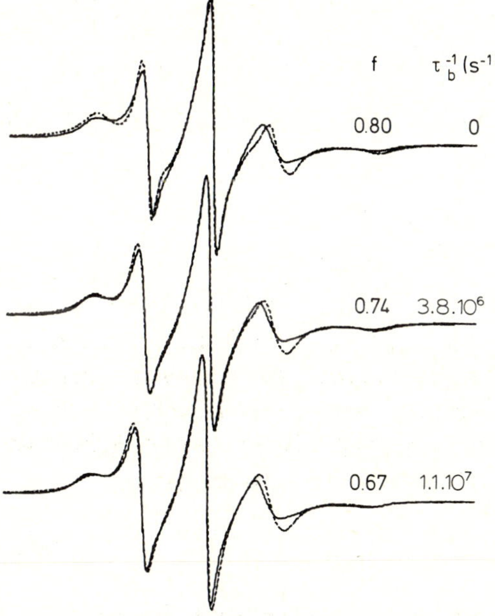

Fig. 9. Comparison of experimental and simulated ESR spectra at various exchange rates. Full lines in all three traces: spectra of 14-PASL phosphatidic acid spin label in myelin proteolipid apoprotein/dimyristoyl phosphatidylcholine recombinants of lipid/protein ratio 24:1 mol/mol at 30°C. Broken lines: simulations for various values of exchange probability, τ_b^{-1}, with the values of the fraction of motionally restricted lipid, f, chosen to optimize the fit in each case. (a) $\tau_b^{-1} = 0$ s^{-1}, $f = 0.80$; (b) $\tau_b^{-1} = 3.8 \times 10^6$ s^{-1}, $f = 0.74$; (c) $\tau_b^{-1} = 1.1 \times 10^7$ s^{-1}, $f = 0.67$ (Horvath et al., 1985).

TABLE 3
Best fit parameters from two-component exchange simulations of the spectra from myelin proteolipid apoprotein/dimyristoyl phosphatidylcholine recombinants at 30°C (Horvath et al., 1985)

Spin probe	lipid/protein (mol/mol)	f	τ_b^{-1} (s^{-1})
14-PASL	27	0.66	5.3×10^6
	24	0.72	4.2×10^6
	23	0.76	4.5×10^6
	17	0.79	4.9×10^6
	12	0.84	3.4×10^6
			Mean 4.5×10^6
14-PCSL	34	0.34	1.9×10^7
	29	0.47	1.6×10^7
	24	0.49	1.4×10^7
	10	0.49	1.5×10^7
	12	0.59	1.4×10^7
			Mean 1.6×10^7

The results of a systematic study of the exchange properties as a function of lipid/protein ratio for myelin proteolipid protein and two labelled lipids of different specificities are given in Table 3. As seen from the Table, the off-rate constant, τ_b^{-1}, is approximately constant independent of the total lipid/protein ratio for both labels. (Although increases in exchange rate are observed at higher lipid/protein ratios.) In contrast the off-rate constants for the phosphatidic acid spin label, 14-PASL, are consistently lower than those for the phosphatidylcholine label, 14-PCSL. This can be readily understood in terms of the relative specificities of the two lipids. The right-hand side of Eqn. 19 may be identified with the quantity (n_f^*/n_b^*) in the equilibrium association Eqn. 8. Hence, the ratio of rate constants can be expressed in terms of the average relative association constant according to:

$$\tau_b^{-1}/\tau_f^{-1} = n_t/(N_1 K_r^{av}) - 1/K_r^{av} \qquad (22)$$

If it is assumed that the on-rate constant, τ_f^{-1}, is the same for all lipids, i.e., that it is diffusion controlled, then for different spin labels, say L and PC, the ratio of off-rate constants is:

$$\tau_b^{-1}\,(\mathrm{PC})/\tau_b^{-1}\,(L) = K_r^{av}\,(L)/K_r^{av}\,(\mathrm{PC}) \qquad (23)$$

From Table 3, the ratio of the rate constants is: τ_b^{-1} (PC)/τ_b^{-1} (PA) $\approx$ 3.5, compared with a value of K_r^{av} (PA)/K_r^{av} (PC) = 2.5, deduced from titration of the values for the fraction of motionally restricted lipid in Table 3, according to Eqn. 8.

Thus, the experimental values of the exchange rate constants for the myelin proteolipid display the expected dependence on lipid-protein ratio and lipid selectivity. This gives some confidence in the reliability of the simulations and strongly suggests that the main source of spectral broadening comes from exchange rather than more direct lipid-protein interactions. The values of the exchange constants for phosphatidic acid are significantly smaller than those for the exchange of lipids due to lateral diffusion in fluid lipid bilayers: $\tau_{diff}^{-1} = 4\, D_{lat}/\langle x^2 \rangle \sim 10^7\ s^{-1}$ (Träuble and Sackmann, 1972; Devaux et al., 1973). Thus, there is a significant binding of the lipid in the case of the phosphatidic acid label. Qualitative comparison with the spectra from other lipid/protein systems suggests that the myelin proteolipid apoprotein may have one of the slowest rates of lipid exchange. For rod outer segment disc membranes the increase in first derivative peak-to-peak linewidth of the fluid lipid component in membranes relative to the extracted lipids is δH_{pp} = 0.8 gauss (Marsh et al., 1982). From Eqn. 21 the on-rate constant is: $\tau_f^{-1} = (\sqrt{3}/2)\ (g\beta/\hbar) \cdot \delta H_{pp} = 1.2 \times 10^7\ s^{-1}$, and from Eqn. 19 the off-rate constant is: $\tau_b^{-1} = \tau_f^{-1}\ (1 - f)/f = 4.0 \times 10^7\ s^{-1}$. These values are very close to the diffusion controlled limit. Double labelling experiments involving collisions between ^{14}N and ^{15}N spin labels have also suggested that the lipid exchange rates at the protein boundary are of the same order of magnitude as those in the bulk lipid phase (Davoust et al., 1983). Thus, it seems that for rhodopsin at least the lipid exchange rates are considerably faster than found for the myelin proteolipid protein.

A final point regards the comparison of the fraction of motionally restricted lipid deduced from the subtraction method, as in Fig. 2, with that deduced by simulation allowing for line-shape modifications due to exchange, as in Fig. 9 and Table 3. The two sets of values for f deduced from myelin proteolipid/dimyristoyl phosphatidylcholine recombinants of various lipid/protein ratios and with different spin-labelled lipids are plotted against each other in Fig. 10. It is seen that there is quite good agreement between the two methods for the data points from 14-PASL ($f \gtrsim 0.6$). For 14-PCSL for which the factors are lower ($f \lesssim 0.6$) and the exchange rates faster, the subtraction method yields lower values than obtained by simulation. However, the maximum discrepancy is of the order of 30%, which is probably not so much greater than the combined uncertainty in the subtraction endpoint and the fitting of the simulated spectrum. It thus appears that the empirical subtraction strategy of choosing single components at different temperatures, such that they best correspond to the apparent components in the composite spectrum is reasonably effective in allowing for the effects of slow exchange between the two components. For systems with faster exchange rates

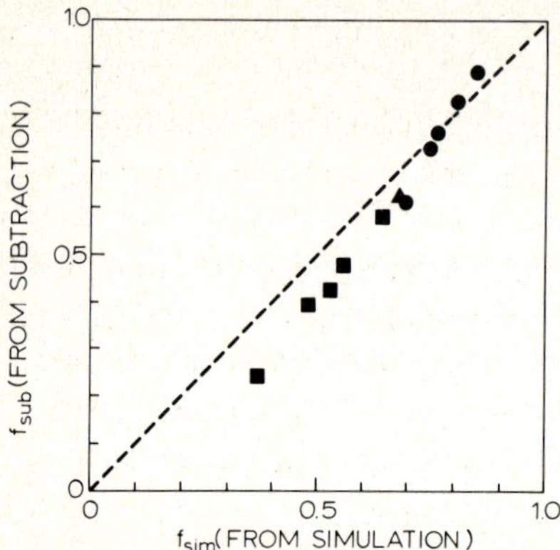

Fig. 10. Comparison of the fractions of motionally restricted lipid in myelin proteolipid apoprotein/dimyristoyl phosphatidylcholine recombinants obtained by spectral subtraction, f_{sub}, and by spectral simulation including exchange, f_{sim}. Spectra from recombinants at various lipid/protein ratios were recorded at 30°C: (●) 14-PASL phosphatidic acid spin label; (■) 14-PCSL phosphatidylcholine spin label. Best subtractions were obtained choosing a pure lipid spectrum recorded at 27°C and a 'protein-alone' spectrum recorded at 32°C, in order to allow for line-broadening resulting from exchange (Horvath et al., 1985).

this strategy might be less effective, and in these cases it would be advisable, and is often the practice, to perform subtractions on spectra recorded at lower temperatures in order to reduce the exchange rate. Davoust and Devaux (1982) have estimated by simulation that the effective fraction of motionally restricted lipid decreases drastically as the exchange frequency is increased from 10^5 to 5×10^6 s^{-1}. However, these subtractions were all performed with the same motionally restricted component and do not correspond to the most reliable subtraction method. The sensitivity to exchange frequency is therefore considerably overestimated at the higher exchange frequencies, as noted previously (Brotherus et al., 1980).

VII. Conclusions

In the above discussion it has been attempted to present in a critical manner the comparative aspects of those spin label measurements that now constitute the classical approach to the study of lipid-protein interactions, namely that of 'boundary counts' in determining the stoichiometry and specificity of interaction. The remainder of the discussion has centred on more recent approaches, either to defining the origin of the specificity of the interactions, or to investigating further structural and dynamic properties: the degree of chain ordering and the rate of lipid exchange. The data on these latter topics is not nearly so extensive and, although it is not possible to generalize at the moment, some important trends are emerging.

With the above reservations in mind, the present state of knowledge of lipid-protein interactions determined by spin label spectroscopy may be summarized as follows.

(1) The number of motionally restricted lipids bears an approximately fixed stoichiometry with respect to protein, except at low lipid contents where protein-protein aggregation becomes appreciable. The number of motionally restricted lipids per protein correlates roughly with the intramembraneous perimeter of the protein, but for most proteins neither the structure nor the degree of aggregation in the membrane is known with any high degree of accuracy.

(2) The perturbation of the lipids by the protein is of short range. Any perturbation beyond the first shell of motionally restricted lipids is small and may partly be the effect of exchange.

(3) There is a selectivity of the protein for different phospholipid headgroups and the selectivity pattern is different for different proteins. No single lipid has the highest specificity for all proteins tested. The specificity is highest for certain, but not all, negatively charged phospholipids. It is sensitive to salt and pH titration, but is not exclusively due to electrostatic interactions.

(4) The spin-labelled lipid chains are disordered on the surface of the protein, at least for rhodopsin and probably for cytochrome oxidase.

(5) The protein-interacting lipids exchange rapidly with the other lipids in the membrane. The rate of exchange can be either comparable to (rhodopsin) or appreciably slower (myelin proteolipid protein phosphatidic acid) than the diffusive exchange in fluid lipid bilayers.

All measurements are made with labelled lipid molecules, but only the measurements of the chain configurational ordering are likely to be appreciably affected by the perturbing effect of the spin label group. For determination of the stoichiometry, lipid/protein titrations have demonstrated that the spin label group does not appreciably perturb the thermodynamics of the interaction: K_r (PC*: PC) $\approx$ 1. Also, the determinations of the relative lipid specificities are unaffected by possible perturbing effects of the spin label group, since this part of the molecule is maintained constant in these experiments. For the above reasons it also seems unlikely that the lipid exchange rates will be appreciably affected by the presence of the spin label group, particularly in the comparison with spin label measurements of lateral diffusion rates in fluid lipid bilayers.

In conclusion, it can be said that the characteristic timescale of spin label ESR is optimally suited to detecting motional restriction of the fluid membrane lipids by integral proteins. The resulting two-component spectra provide an extremely sensitive method of studying the structural, dynamic and thermodynamic features of the lipid-protein interactions. Such studies can also provide the basis for interpreting the lipid dependence of the functional characteristics of integral membrane enzymes (see, e.g., Sandermann, 1978, 1982).

Acknowledgement

I would like to thank Rob Pates, Laszlo Horvāth, Gary Powell, Cecilia Hidalgo and Mikael Esmann for the opportunity to quote from our collaborative work.

References

Atherton, N.M. (1973) *Electron Spin Resonance*. Ellis Horwood, Chichester.

Blasie, J.K., Herbette, L., Pierce, D., Pascolini, D., Scarpa, A. and Fleischer, S. (1982) Static and time-resolved structural studies of the Ca^{2+}-ATPase of isolated sarcoplasmic reticulum. Ann. N.Y. Acad. Sci. 402, 485–514.

Brophy, P.J., Horváth, L.I. and Marsh, D. (1984) Stoichiometry and specificity of lipid-protein interaction with myelin proteolipid protein studied by spin-label electron spin resonance. Biochemistry 23, 860–865.

Brotherus, J.R., Jost, P.C., Griffith, O.H., Keana, J.F.W. and Hokin, L.E. (1980) Charge selectivity at the lipid-protein interface of membranous Na,K-ATPase. Proc. Natl. Acad. Sci. U.S.A. 77, 272–276.

Brotherus, J.R., Griffith, O.H., Brotherus, M.O., Jost, P.C., Silvius, J.R. and Hokin, L.E. (1981) Lipid-protein multiple binding equilibria in membranes. Biochemistry 20, 5261–5267.

Corless, J.M., McCaslin, D.R. and Scott, B.L. (1982) Two-dimensional rhodopsin crystals from disk membranes of frog retinal rods. Proc. Natl. Acad. Sci. U.S.A. 79, 1116–1120.

Curatolo, W., Sakura, J.D., Small, D.M. and Shipley, G.G. (1977) Protein-lipid interactions: recombinants of the proteolipid apoprotein of myelin with dimyristoyllecithin. Biochemistry 11, 2313–2319.

Davoust, J., Schoot, B.M. and Devaux, P.F. (1979) Physical modifications of rhodopsin boundary lipids in lecithin-rhodopsin complexes: a spin label study. Proc. Natl. Acad. Sci. U.S.A. 76, 2755–2759.

Davoust, J., Bienvenüe, A., Fellman, P. and Devaux, P.F. (1980) Boundary lipids and protein mobility in rhodopsin-phosphatidylcholine vesicles. Effects of lipid phase transitions. Biochim. Biophys. Acta 596, 28–42.

Davoust, J. and Devaux, P.F. (1982) Simulation of electron spin resonance spectra of spin-labelled fatty acids covalently attached to the boundary of an intrinsic membrane protein. A chemical exchange model. J. Magn. Reson. 48, 475–494.

Davoust, J., Seigneuret, M., Hervé, P. and Devaux, P.F. (1983) Collisions between nitrogen-14 and nitrogen-15 spin labels. 2. Investigations on the specificity of the lipid environment of rhodopsin. Biochemistry 22, 3146–3151.

Deatheridge, J.F., Henderson, R. and Capaldi, R.A. (1982) Relationship between membrane and cytoplasmic domains in cytochrome *c* oxidase by electron microscopy in media of different density. J. Mol. Biol. 158, 501–514.

Deguchi, N., Jørgensen, P.L. and Maunsbach, A.B. (1977) Ultrastructure of the sodium pump. Comparison of thin sectioning, negative staining, and freeze-fracture of purified membrane-bound (Na^+,K^+)-ATPase. J. Cell. Biol. 75, 619–634.

Devaux, P., Scandella, C.J. and McConnell, H.M. (1973) Spin-spin interactions between spin-labelled phospholipids incorporated into membranes. J. Magn. Reson. 9, 474–485.

Ellena, J.F., Blazing, M.A. and McNamee, M.G. (1983) Lipid-protein interactions in reconstituted membranes containing acetylcholine receptor. Biochemistry 22, 5523–5535.

Esmann, M. and Marsh, D. (1985) Spin label studies on the origin of the specificity of lipid-protein interactions in Na^+,K^+-ATPase membranes from *Squalus acanthias*. to be published.

Esmann, M., Watts, A. and Marsh, D. (1985) Spin label studies of lipid-protein interactions in (Na^+,K^+)-ATPase membranes from rectal glands of *Squalus acanthias*. Biochemistry, in press.

Favre, E., Baroin, A., Bienvenüe, A. and Devaux, P.F. (1979) Spin-label studies of lipid-protein interactions in retinal rod outer segment membranes. Fluidity of the boundary layer. Biochemistry 18, 1156–1162.

Hebert, H., Jørgensen, P.L., Skriver, E. and Maunsbach, A.B. (1982) Crystallization patterns of membrane-bound (Na^+,K^+)-ATPase. Biochim. Biophys. Acta 689, 571–574.

Hidalgo, C. and Marsh, D. (1985) Spin label ESR studies of the stoichiometry and specificity of lipid-protein interactions with sarcoplasmic reticulum Ca^{2+}-ATPase. to be published.

Hoffmann, W., Pink, D.A., Restall, C. and Chapman, D. (1981) Intrinsic molecules in phospholipid bilayers. Fluorescence probe studies. Eur. J. Biochem. 114, 585–589.

Horváth, L.I., Brophy, P.J. and Marsh, D. (1984) Exchange rate and specificity of the motionally restricted lipids interacting with the myelin proteolipid protein in dimyristoylphosphatidylcholine recombinants. A spin label study. to be published.

Jost, P.C., Griffith, O.H., Capaldi, R.A. and Vanderkooi, G.A. (1973a) Evidence for boundary lipid in membranes. Proc. Natl. Acad. Sci. U.S.A. 70, 4756–4763.

Jost, P.C., Griffith, O.H., Capaldi, R.A. and Vanderkooi, G.A. (1973b) Identification and extent of fluid bilayer regions in membranous cytochrome oxidase. Biochim. Biophys. Acta 311, 141–152.

Jost, P.C., Capaldi, R.A., Vanderkooi, G. and Griffith, O.H. (1973c) Lipid-protein and lipid-lipid interactions in cytochrome oxidase model membranes. J. Supramol. Struct. 1, 269–280.

Kistler, J., Stroud, R.M., Klymkowsky, M.W., Lalancette, R.A. and Fairclough, R.H. (1982) Structure and function of an acetylcholine receptor. Biophys. J. 37, 371–383.

Knowles, P.F., Marsh, D. and Rattle, H.W.E. (1976) *Magnetic Resonance of Biomolecules*. Wiley, London, New York.

Knowles, P.F., Watts, A. and Marsh, D. (1979) Lipid immobilization in dimyristoyl phosphatidylcholine-substituted cytochrome oxidase. Biochemistry 18, 4480–4487.

Knowles, P.F., Watts, A. and Marsh, D. (1981) Spin label studies of head-group specificity in the interaction of phospholipids with yeast cytochrome oxidase. Biochemistry 20, 5888–5894.

Lange, A., Marsh, D., Waßmer, K.-H., Meier, P. and Kothe, G. (1985) Electron spin resonance study of phospholipid membranes employing a comprehensive lineshape model. to be published.

Lee, A.G. (1983) An overlapping site model for the lipid annulae of membrane proteins. FEBS Lett. 151, 297–302.

Le Maire, M., Møller, J.V. and Tardieu, A. (1981) Shape and thermodynamic parameters of a Ca^{2+}-dependent ATPase. A solution X-ray scattering and sedimentation equilibrium study. J. Mol. Biol. 150, 273–296.

Marsh, D. (1981) Electron spin resonance: spin labels. In: *Membrane Spectroscopy*, pp. 51–142. Editor: E. Grell. Springer, Berlin, Heidelberg, New York.

Marsh, D. (1982) Electron spin resonance: spin label probes. In: *Techniques in Lipid and Membrane Biochemistry*, Vol. B4/II, pp. B426/1–B426/44. Editors: J.C. Metcalfe and T.R. Hesketh. Elsevier, Ireland.

Marsh, D. (1983) Spin label answers to lipid-protein interactions. Trends Biochem. Sci. 8, 330–333.

Marsh, D. and Barrantes, F.J. (1978) Immobilized lipid in acetylcholine receptor-rich membranes from *Torpedo marmorata*. Proc. Natl. Acad. Sci. U.S.A. 75, 4329–4333.

Marsh, D., Watts, A. and Barrantes, F.J. (1981) Phospholipid chain immobilization and steroid rotational immobilization in acetylcholine receptor-rich membranes from *Torpedo marmorata*. Biochim. Biophys. Acta 645, 97–101.

Marsh, D. and Watts, A. (1982) Spin labeling and lipid-protein interactions in membranes. In: *Lipid-Protein Interactions*, Vol. 2, pp. 53–126. Editors: P.C. Jost and O.H. Griffith. Wiley, New York.

Marsh, D., Watts, A., Pates, R.D., Uhl, R., Knowles, P.F. and Esmann, M. (1982) ESR spin-label studies of lipid-protein interactions in membranes. Biophys. J. 37, 265–274.

Marsh, D., Brophy, P.J., Esmann, M., Hidalgo, C., Horváth, L.I., Knowles, P.F., Pates, R.D., Hoffmann-Bleihauer, P. and Sandhoff, K. (1985) Sphingolipid-protein interactions studied by spin label ESR spectroscopy. to be published.

McConnell, H.M. (1958) Reaction rates by nuclear magnetic resonance. J. Chem. Phys. 28, 430–431.

Michel, H., Oesterhelt, D.S. and Henderson, R. (1980) Orthorhombic two-dimensional crystal form of purple membrane. Proc. Natl. Acad. Sci. U.S.A. 77, 338–342.

Napolitano, C.A., Cooke, P., Segalman, K. and Herbette, L. (1983) Organization of calcium pump protein dimers in the isolated sarcoplasmic reticulum membrane. Biophys. J. 42, 119–125.

Pates, R.D. and Marsh, D. (1985) Spin label ESR studies of lipid chain ordering and mobility in rod outer segment disc membranes. to be published.

Post, J.F.M. and Dijkema, C. (1983) An electron spin resonance spin-label study of lipophilin in oriented phospholipid bilayers. Arch. Biochem. Biophys. 225, 795–801.

Rousselet, A., Devaux, P.F. and Wirtz, K.W. (1979) Free fatty acids and esters can be immobilized by the receptor-rich membranes from *Torpedo marmorata*, but not phospholipid acyl chains. Biochem. Biophys. Res. Commun. 90, 871–877.

Sandermann, H. (1978) Regulation of membrane enzymes by lipids. Biochim. Biophys. Acta 515, 209–237.

Sandermann, H. (1982) Lipid-dependent membrane enzymes. A kinetic model for cooperative activation in the absence of cooperativity in lipid binding. Eur. J. Biochem. 127, 123–128.

Sardet, C., Tardieu, A. and Luzzati, V. (1976) Shape and size of bovine rhodopsin: a small-angle X-ray scattering study of a rhodopsin-detergent complex. J. Mol. Biol. 105, 383–407.

Seelig, J., Seelig, A. and Tamm, L. (1982) Nuclear magnetic resonance and lipid-protein interactions. In: *Lipid-Protein Interactions*, Vol. 2, pp. 127–148. Editors: P.C. Jost and O.H. Griffith. Wiley, New York.

Silvius, J.R., McMillen, D.A., Saley, N.D., Jost, P.C. and Griffith, O.H. (1984) Competition between cholesterol and phosphatidylcholine for the hydrophobic surface of sarcoplasmic reticulum Ca^{2+}-ATPase. Biochemistry 23, 538–547.

Smith, R., Cook, J. and Dickens, P.A. (1984) Structure of the proteolipid protein extracted from bovine central nervous system myelin with non-denaturing detergents. J. Neurochem. 42, 306–313.

Thomas, D.D., Bigelow, D.J., Squier, T.C. and Hidalgo, C. (1982) Rotational dynamics of protein and boundary lipid in sarcoplasmic reticulum membrane. Biophys. J. 37, 217–225.

Träuble, H. and Sackmann, E. (1972) Studies of the crystalline-liquid crystalline phase transition of lipid model membranes. III. Structure of a steroid-lecithin system below and above the lipid-phase transition. J. Am. Chem. Soc. 94, 4499–4510.

Watts, A., Volotovski, I.D. and Marsh, D. (1979) Rhodopsin-lipid associations in bovine rod outer segment membranes. Identification of immobilized lipid by spin labels. Biochemistry 18, 5006–5013.

Watts, A., Davoust, J., Marsh, D. and Devaux, P.F. (1981) Distinct states of lipid mobility in bovine rod outer segment membranes. Resolution of spin label results. Biochim. Biophys. Acta 643, 673–676.

Watts/De Pont (Eds.)
Progress in Protein-Lipid Interactions

CHAPTER 5

Translational diffusion of proteins and lipids in artificial lipid bilayer membranes. A comparison of experiment with theory

ROBERT M. CLEGG and WINCHIL L.C. VAZ

Max Planck Institut für Biophysikalische Chemie, D-3400 Göttingen-Nikolausberg, F.R.G.

I. Introduction

The indeterminate lateral transport of a molecule, or molecular aggregate, within a lipid bilayer is ultimately attributable to random statistical fluctuations of the density and energy within the immediate environment of the molecule. Several models have been invoked to describe the lateral diffusion within lipid membranes, and due to the complexity of the lipid molecules all of these models are necessarily simplified theoretical idealizations of the physical situation. The models differ primarily in the physical description of the surrounding environment of the molecule under observation and consequently stress different properties of the lipid bilayer. Another approach to the topic is given in Chapter 1 of this volume.

Before discussing these details it is useful to estimate the order of magnitude of some physical parameters from simple considerations of random walks in two dimensions. The translational diffusion coefficient is determined by measuring the averaged square displacement, $\langle r^2 \rangle$*, of molecules within a time, t, and applying the relationship, $\langle r^2 \rangle = 4D_T t$, where D_T is the translational diffusion constant. In pure lipid bilayers the lipids have D_T of about 10^{-7} cm^2/s (most proteins diffuse about 5-times slower). Thus, the lipid molecule will scan a circular area with a diameter of 100 Å within microseconds. However, seconds will be required to cover a circular area with a diameter of 10 μm. If a lipid molecule

* See list of abbreviations on p. 221.

were to be dragged through the membrane with a constant velocity of 10 μm per second, the required energy to cover 10 μm would be an order of magnitude larger than kT, and only a linear distance would be traced, whereas the thermal motions of the membrane allow the lipid molecules to scan the total circular area within the same time, and require no energy dissipation. In this sense, diffusion is an effective mechanism to distribute molecules in a membrane. The goal of detailed physical studies of lateral diffusion in lipid bilayers is to discover the controlling molecular dynamics responsible for this effective dispersive mobility, especially considering that lipid bilayers are very often thought of as static structural elements of cells.

The theories of lateral transport in lipid bilayers can be divided into two main categories, those which treat the underlying processes from the point of view of continuum mechanics and those which attempt to treat the molecular events within a modelistic framework. We first discuss the intrinsic features of the models which have been developed to describe the diffusion of molecules in lipid bilayers. Then the applicability of the models to experimental diffusion data will be examined. Each model emphasizes disparate physical characteristics of the lipid bilayer and the application of each model presupposes the appropriate conditions. Our goal is to clearly define the framework of each model and discuss their important defining characteristics.

II. Theoretical considerations

1. HYDRODYNAMIC THEORIES

a. General

The hydrodynamic theories treat the membrane as a two dimensional continuum fluid, and the diffusion coefficient of a particle is found by solving the appropriate Navier-Stokes hydrodynamic equations, where the boundary conditions must include the aqueous phase external to the bilayer. Thus, for this approach the molecular structure of the membrane is not considered and the properties of the membrane are defined solely by a viscosity coefficient, the thickness and the two dimensionally anisotropic nature of the bilayer. The size of the diffusing species is considered to be much larger than the size of the solvent molecules. Because the diffusing particles to be considered are so small (on the molecular scale) the viscous forces dominate the inertial forces at all times, and the hydrodynamic equation governing the movement of the fluid *inside and outside* of the membrane sheet is the low Reynolds number limit of the Navier-Stokes equation (Landau and Lifshitz, 1959).

$$-\nabla P + \eta_0 \Delta \mathbf{v} + \mathbf{F} = 0 \tag{1}$$

The terms ϱ_0 $(\partial \mathbf{v}/\partial t)$ and ϱ_0 $(\mathbf{v} \cdot \nabla)\mathbf{v}$ have been neglected on the right side of Eqn. 1, which means that steady-state solutions are sought and that the inertial terms of the Navier-Stokes equations have been ignored. **v** is the fluid velocity, η_0 is the viscosity, ϱ_0 is the density, and P is the pressure. In addition to Eqn. 1, we will always assume the incompressibility equation, div $\mathbf{v} = 0$. **F** is the external force exerted on the fluid per unit volume. Throughout the discussion this is assumed to be zero in the aqueous phase, however, **F** will be defined within the membrane sheet. The translational diffusion coefficient of a molecule within this fluid is calculated from the Einstein-Stokes relation (Landau and Lifshitz, 1959, Sommerfeld, 1964),

$$D = kT/f \tag{2}$$

where f is the friction coefficient of the molecule. This is in complete analogy with the *three*-dimensional case where, according to Stokes' law, $f = 6\,\pi\eta_0 r$, for a sphere of radius, r.

However, at the low Reynolds number limit of the Navier-Stokes equations in two dimensions, which describe the hydrodynamic 'fluid sheet', it is impossible to define a friction coefficient in an *infinitely extending two-dimensional* fluid (Stokes' paradox, Birkhoff (1950), Landau and Lifshitz (1959), Lamb (1932), Saffman and Delbrueck (1975)). This is a peculiarity of the hydrodynamic equations in a two dimensional space and arises from the logarithmic divergence of the general solution of Eqn. 1 in two dimensions if $\mathbf{F} = 0$. There are several ways to circumvent this situation as we will discuss below.

It is interesting, at this point, to note why the friction coefficient in the two dimensional situation defined for the conditions of Eqn. 1 cannot be defined. A useful interpretation of the final equations representing the hydrodynamic models can be gained by applying dimensional arguments to understand Stokes' paradox (see Happel and Brenner, 1973). The important physical parameters defining the fluid flow are **v**, ϱ, η and some linear dimension l (see also Chapter 6). At the low Reynolds number limit, ϱ is not important and the only combination of the parameters with the dimensions of the friction coefficient is $\eta_0 l$. Therefore, dimensional arguments indicate that $f_T \propto \eta_0 l$. However, for an infinite cylinder moving perpendicular to the cylindrical axis, the force per unit length is important, rather than just the force. Thus, by analogous dimensional arguments a constant force per unit length would give a steady-state velocity of the cylinder which is dependent only upon the viscosity, and independent of any dimensions of the cylinder. This is physically unreasonable. Thus, another characteristic length dimension must be introduced into the problem in order to obtain a reasonable solution.

b. Three definitions of f_T

i. Including inertial terms. Oseen (1910), showed that the inertial forces cannot be neglected at large distances from the object moving relative to the fluid, that is, the Reynolds number of the low field is not low throughout the entire two dimensional space (this applies to two and three dimensions). By adding the inertial contributions to the hydrodynamic equations (see the sentence following Eqn. 1), Oseen (1910) showed that the friction coefficient, per length h, of an infinite cylinder moving perpendicular to its cylindrical axis can be described by (Landau and Lifshitz, 1959).

$$f_T = 4\pi\eta_0 h\,(1/2 - \gamma - \log\,(UR_c/4\nu))^{-1} \tag{3}$$

$$\gamma = 0.5772$$

U is the velocity of the fluid relative to the particle at infinity, R_c is the radius of the cylinder and ν is the kinematic viscosity (η_0/ϱ). The Reynolds number, R_{Rey}, is defined by Ul/ν where l is some characteristic length of the problem, so that $-\log(UR_c/4\nu) = \log(l/R_c) - \log(R_{Rey}/4)$. If the characteristic length for the Reynolds number, l, is chosen to be R_c, which is the only dimension available in an infinitely extending two-dimensional space, f_T will be logarithmically inversely proportional to R_{Rey}. However, Eqn. 3 cannot be used in the Einstein relation of Eqn. 2, since this requires a friction coefficient which is independent of the velocity.

ii. Circular boundary. If the spatial extent of the two-dimensional liquid is limited to an area within a circle of radius R (Lamb, 1932), the friction coefficient per length h of the infinite cylinder becomes,

$$f_T = 4\pi\eta_0 h(\log(R/R_c) - 1/2)^{-1} \tag{4}$$

Equation 4 was obtained by explicitly introducing a new dimension, which is the extent of the boundaries limiting the flow, and simply integrating Eqn. 1 with $\nabla P = 0$ and $\mathbf{F} = 0$. This is the appropriate friction coefficient to use for lower values of R, as will be discussed later.

iii. *Effect of exterior fluid.* The following hydrodynamic treatments remove the two dimensional constraint of the total problem, and use the third dimension to limit the extent of the velocity correlations of the fluid in the two-dimensional membrane sheet. Figure 1a schematically defines the models for all the hydrodynamic theories. Equation 1 can be simplified by assuming that $\nabla P = 0$. This is either evident in the solution, see below (Saffman, 1976; Hughes et al. 1981),

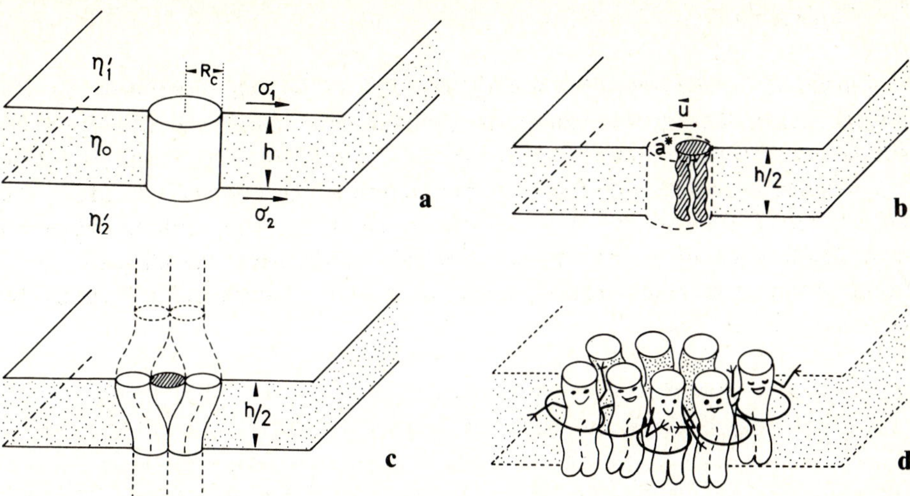

Fig. 1. Schematic representations of the main models considered in the text. h is the thickness of the bilayer in all cases. (**1a**) *Hydrodynamic model.* The hydrodynamic description for a diffusing cylinder (radius R_c) spanning the bilayer. The surface stress (force/area) at the aqueous-membrane interface, $\boldsymbol{\sigma}_1$ and $\boldsymbol{\sigma}_2$, are caused by the fluid velocity gradients in the aqueous phase. The membrane *fluid* is not allowed any velocity gradients perpendicular to the sheet, so the $\boldsymbol{\sigma}$'s produce a *body* force (force/volume) within the membrane sheet which is tangential to the membrane. The energy dissipation within the sheet is modelled by hypothesizing a three-dimensional viscosity which is uniform and constant within the sheet (see text). (**1b**) *Free-volume model.* The free volume of a molecule (in this case a lipid with two chains) is considered to be the volume in which the molecule can move unhindered by the adjacent molecules. The corresponding free area of the two-dimensional membrane is indicated. We have pictured the average *cage* as existing within a continuum fluid and given it a cylindrical shape (circular area). The free volume (free area) around any molecule fluctuates and if sufficient space is available, i.e., $a_f > a^*$, the molecule may translocate to a neighbouring site (the site is not depicted). The figure is obviously an idealization. (**1c**) *Flexible rod model.* The solid lines depict the separating chains for the monolayer part of the bilayer, and the dashed lines are the appropriate extensions for the infinitely long polymer model. The dark circle represents either the penetrant for the polymer model or the polar head group of a neighbouring lipid for the lipid bilayer model. The configuration shown is an activated state where the maximum separation is sufficient to allow the entrance of a neighbouring head group (or penetrant). The distance along the chains from maximum separation to where the chains resume their equilibrium separation is the 'x_0' of the text. (**1d**) Perhaps a more realistic representation of the thermal motions of the lipid molecules in a membrane.

or can be assumed as an *Ansatz* (Wiegel, 1979a, b, c, 1980). Saffman, Hughes and Wiegel neglect the molecular structure of the lipid membrane and model the two-dimensional structure of the membrane by forcing the viscous stress in the water phase, which is applied across the membrane interface to act as an external volume force within the membrane sheet, and to have components only in the direction parallel to the membrane surface (see Fig. 1a). Thus, the **F** of Eqn. 1 within the membrane sheet is related to the shear force per unit area in the aqueous phase, $\boldsymbol{\sigma}$, acting *across the aqueous-membrane interface*.

$$\boldsymbol{\sigma} = \eta'(\partial \mathbf{v}/\partial z)_{z=0} \tag{5}$$

z is the cartesian dimension perpendicular to the plane of the sheet, and the position, $z = 0$, is defined to be at one of the membrane interfaces. An analogous stress is defined for $z = -h$. Note that the origin of $\boldsymbol{\sigma}$ is within the fluid external to the membrane sheet. This stress, at the interface of the aqueous and membrane phases, must be divided by the length within the membrane sheet over which the stress acts like a body force in order to transform it into a force per volume within the membrane sheet. This is usually assumed to be the thickness of the membrane sheet, h. That is, the viscous forces of the external fluid phase are assumed to act like a body force in the membrane sheet. Within a bilayer, the stress, $\boldsymbol{\sigma}$, on the membrane at an interface is distributed at most over a thickness of $h/2$ if each of the lipid monolayers is assumed to undergo lateral displacement individually. This results in a volume force, $\mathbf{F}$ of Eqn. 1, within the membrane of $\boldsymbol{\sigma}/(h/2)$, assuming that the liquid on both sides of the membrane is the same. In this case each monolayer reacts only to the stress on *its* aqueous interface. If the stress at each interface could be distributed over the entire thickness of a membrane, h, the volume force in the membrane contributed by *each* interface would be $\boldsymbol{\sigma}/h$, and the *total* volume force would then be $2x(\boldsymbol{\sigma}/h) = \boldsymbol{\sigma}/(h/2)$, as we just found for separate monolayer components. All the theoretical hydrodynamic treatments assume that the surface stresses on both sides of the membrane act across the entire thickness, h, of the membrane and both contributions sum to give a total volume force within the membrane of $(\boldsymbol{\sigma}_1 + \boldsymbol{\sigma}_2)/h$. The value of the body force is the same for both of the above interpretations provided that the viscosities on both sides of the membrane are equal. If this is not the case, and if it is assumed that the membrane consists of two layers (as it does), the body force on the membrane fluid sheet in the top monolayer could be different than in the bottom monolayer. This would require some type of averaging throughout the membrane, or a redefinition of the problem in terms of two parallel, possibly interacting, two-dimensional fluids to replace the single membrane sheet. Such a situation is not considered in any of the models to be discussed here and would probably be important mainly for particles which do not span the membrane.

At this point we note that, when applying Eqn. 1 to the two-dimensional fluid in the membrane sheet, η_0 refers to a three-dimensional viscosity in the portion of the membrane where the body force acts. Also, the fluid movement within the membrane sheet is restricted to two dimensions by reducing the Laplacian to a two-dimensional form (no gradients in the z direction), even though the three-dimensional viscosity within the sheet is retained. For this reason the viscous stress in the external fluid at the interface is defined as a three-dimensional body force within the sheet. Equivalently one could define the hydrodynamic equations of the membrane sheet in terms of a surface viscosity, $\eta_s \cong \eta_0 h$ (section II.1.d.ii, point 4), and the surface stress. This would avoid the introduction of a body force.

In addition, the velocity of the fluid within the sheet is assumed not to have any gradients in the direction perpendicular to the plane of the membrane, and the boundary conditions at the surface of the membrane are set by equating the tangential movement of the aqueous fluid with that of the 'two-dimensional' lipid fluid. The viscosities of the fluids external to the membrane sheet are the normal three-dimensional values, and the viscosity of the membrane sheet, η_0, is usually found by fitting the data assuming a value of h. The fluids external to the membrane retard the fluid particles of the membrane sheet due to the viscous dissipation of the adjacent external fluids. The drag force exerted directly on the diffusing particle by the fluid external to the membrane, is considered to be very small in Saffman's (1976) analysis. Hughes et al. (1981) have retained this effect, and as ε of Eqn. 7 becomes very large ($\cong 1$) the third term within the brackets of Eqn. 7 contributes significantly.

(1). Saffman's model. Saffman (1976) carried through an analysis which included the hydrodynamic drag of aqueous phases on both sides of the bilayer membrane (but he did not simultaneously consider the inertial effects or the effects of a solid boundary). Saffman's (1976) treatment is a valid approximation whenever the viscosity of the aqueous phase is much smaller than that of the membrane, and his result is,

$$f_T = 4\pi\eta_0 h \, (-\gamma + \log(\eta_0 h/\eta' R_c))^{-1} \tag{6}$$

where η' is the viscosity of the water phase, and h is the thickness of the membrane sheet. Alternative possibilities for h will be discussed later (Section II.1.d.ii). At the moment it is only important to realize that η_0 always appear in the equations as a product, $\eta_0 h$, and that h refers to a dimension of the *membrane* and not of the diffusing particle. Saffman (1976) only considers the case for equivalent viscosities on both sides of the membrane, however, if the viscosities of the fluids on both sides of the membrane sheet are different, η' must only be replaced by $(\eta_1' + \eta_2')/2$, where η_1' and η_2' are the viscosities of the fluids on opposite sides of the sheet.

(2). The model of Hughes et al. Hughes et al. (1981) have extended this treatment by solving the equations for any ratio of the aqueous and membrane viscosities, and their result is,

$$f_T = 4\pi\eta_0 h(\log(2/\varepsilon) - \gamma + (4/\pi)\varepsilon - (\varepsilon^2/2)\log(2/\varepsilon))^{-1} \tag{7}$$

$$\varepsilon = (R_c/h)\,((\eta_1' + \eta_2')/\eta_0)$$

(3). Wiegel's model. Wiegel (1979a, b, c, 1980) has extended the treatment of Saffman (1976) to include the possibility that the diffusing particle is porous, and

is permeable to the fluid molecules of the membrane sheet. This treatment is also applicable to the movement of large protein aggregates which allow some flow of the lipid molecules within the aggregate. His analysis is based upon introducing another body force, **F** of Eqn. 1, (force per unit volume) exerted on the fluid of the membrane sheet. This force is a frictional force equal to $-(\eta_0/k_0)\mathbf{v}$ where k_0 is the permeability. $(k_0)^{1/2}$ represents the depth that a fluid flow penetrates a permeable medium. An expression for the permeability, k_0, can be written (Felderhof and Deutsch, 1975) provided that the space density of the permeable medium, $\nu(r)$, is known, and the translational coefficient of each mass point comprising the material (sphere of radius a) can be approximated by $f = 6\pi\nu_0 a$; then $k_0(r) = \{6\pi a\nu(r)\}^{-1}$. The permeability is generally considered a function of r, where r is the radius dimension in cylindrical coordinates, however, only the case of a uniform disk is considered by Wiegel (1980). That is $k(r) = k_0$ if $r < R_c$, and $k(r) = \infty$ if $r > R_c$, where R_c is the radius of the permeable cylinder. Under the same conditions as Saffman, Eqn. 6, the frictional coefficient can be derived to be:

$$f_T = 4\pi\eta_0 h\{-\gamma + \log(\eta_0 h/(\eta' R_c)) + 2/\sigma^2 + I_0(\sigma)/\sigma I_1(\sigma)\}^{-1} \quad (8)$$

$$\sigma = R_c/(k_0)^{1/2} \quad (k \text{ has dimensions of cm}^2)$$

I_0 and I_1 denote modified Bessel functions.

Equation 8 should be compared to Saffman's result, Eqn. 6. The additive term is such that, as $\sigma \rightarrow 0$, the free draining limit is $f_T \simeq \pi\eta_0 h\sigma^2$, and as $\sigma \rightarrow \infty$ where the disc becomes impermeable, the limit of Eqn. 8 is Eqn. 6. Equation 8 has not been used in the literature to interpret transport data in membranes, but should be considered for the movement of large clusters of proteins in membranes.

c. Interpretation of the expressions for f_T

Comparing the logarithmic terms of Eqns. 4, 6, we see that $\eta_0 h/\eta'$ and R represent numerically similar terms. It was discussed above that the viscous stress, $\boldsymbol{\sigma}$, acts predominantly on the fluid of the membrane sheet and not upon the diffusing particle itself. This is reflected by the fact that the viscous stress in the exterior fluid retards the velocity of the 'fluid' in the membrane sheet. These dissipative forces within the exterior fluid can be regarded as simulating an effective area for fluid motion around the diffusing particle, which is equivalent to a circular area with a radius of $R = \eta_0 h/\eta'$. Thus, the additional linear dimension which is needed to define a friction coefficient for a two-dimensional fluid can be entered, either explicitly as a boundary (Eqn. 4), or by considering an additional mode of momentum dissipation (Eqn. 6). The model for Eqns. 5, 6 and 7 have included the viscous forces in the three-dimensional exterior fluid, thereby limiting the distance over which the velocities are correlated in the two-dimensional membrane sheet. This dissipation has an equivalent effect on the diffusion coefficient as a circular impenetrable boundary around the particle with a radius of $R = \eta_0 h/\eta'$. For a discussion of the term $\eta_0 h$, which is termed the surface viscosity, see point 4 of Section II.1.d.ii.

d. Critique of the hydrodynamic approach

Here we discuss some of the conditions required for the application of the hydrodynamic theory to transport phenomena in lipid membranes. First we consider the positive and negative aspects.

i. Positive aspects. (1) The hydrodynamic theory is based upon fundamental laws of continuum physics and, provided that the requirements for the application are met, a high degree of confidence can be invested in the resulting description.

(2) The general properties of fluids such as viscosity, friction coefficients, shape factors, boundary effects and pressure are intuitively understandable in terms of a hydrodynamic description.

(3) A vast amount of literature in many disciplines is available for consultation, and innumerable problems have been worked out in detail. The mathematical techniques are often difficult, but descriptions usually exist on many different levels of sophistication. The hydrodynamic approach has been applied to many specific cases of direct interest to membrane research, such as liquid crystals, polymers and interacting systems.

(4) The concepts required to describe the basic existing hydrodynamic theories of transport in membranes are simple and straightforward, even if somewhat subtle (Saffman, 1976).

(5) According to the above models the membrane has been modelled as a hydrodynamic fluid. This may even be a valid description for molecules with a size comparable to the lipid molecules. One of the major requirements for linear hydrodynamics to be a valid approximation is that the momentum changes of the particle should be small on a collisional time scale. Due to the rotations about the C-C bonds in the hydrocarbon chain the lipid molecules are quite flexible along the axis perpendicular to the membrane surface. It is likely that in the hydrocarbon region of the bilayer the neighbouring lipid molecules interact by nudging each other (Fig. 1d) due to rapid local intramolecular movements at independent positions along the chain (Edholm and Blomberg, 1980). This dynamic interaction is caused by movements which are much more rapid and considerably smaller in amplitude than if the lipid molecules were to collide as hard cylinders. This means that many small consecutive random interactions effect the movement of even a lipid molecule, and that the environment of any particular molecule may resemble relatively small molecules, rather than cylinders of comparable size, at least in the hydrocarbon phase.

ii. Negative aspects. (1) The conditions for the applicability of hydrodynamic theory require that the fluid volume, for which velocity and density are defined, be small compared with the particle under consideration, but large compared to the size of the molecules comprising the fluid. This condition is surely not valid within the membrane for molecules with a size less than or equal to the lipid molecules, especially if the lipid molecules are modelled as hard cylinders. It may be possible to avoid this difficulty by appropriate averaging over a small space-time region around the particle of interest, in a manner similar to the discussion of point (5) of section II.1.d.i, but it is not obvious how to do this. The concept of a generalized frequency dependent viscosity (Alder and Alley, 1984, Alder et al., 1981) may be useful in this context to reduce the scale for which classical hydrodynamics applies to lipid bilayers, but there is presently no way to formulate *a priori* the form of the frequency dependence of the viscosity. In order to describe the motion of a spherical particle immersed within a simple three-dimensional fluid in hydrodynamic terms it has been estimated that the Brownian particles must be at least three-times larger than the solvent molecules and that the time scale of a dynamic event should be longer than 20 collisions (Alder et al., 1981). The same size restriction has been proposed be Vaz et al., (1982) to interpret the size dependence of the diffusion coefficients of proteins in lipid membranes.

(2) The parameters which are derived from Eqns. 4–8 may be misleading, since the models assume that the physical structure of the lipid sheet is independent of the position along the axis perpendicular to the plane of the membrane. However, the 'fluidity' of the microenvironment is known to depend on the vertical position in the membrane (Seelig and Seelig, 1980).

(3) There is little molecular information which can be derived from the hydrodynamic approach. The viscosity of the membrane sheet represents an averaged dynamic property, and there is no simple way to correlate the chemical structure of the lipids to their effective viscosity in the membrane sheet.

(4) The concept of a viscosity within the membrane sheet is not well defined, and actually the problem can be formulated in terms of a 'surface viscosity', η_s, which for some purposes can be represented by $\eta_0 h$ (Evans and Hochmuth, 1978). For a discussion of surface viscosity see Joly (1972) and Goodrich (1973). The surface viscosity has units of poise-cm and arises from the two-dimensional character of the bilayer. The volume, force/area and viscosity of the three-dimensional case is replaced by the area, force/length and surface viscosity. It can be seen from Eqns. 4–8 that the distance parameter h and the membrane viscosity, η_0, do not enter the expressions independently, but as a product. This is to be expected since the dynamic stress in a two-dimensional environment can only act over a length, not over an area as in three dimensions. It is possible that the actual stresses in membranes cannot be expressed in terms of this strictly two-dimensional picture, but the model chosen has imposed this restriction and the

corresponding parameters must be adhered to when using Eqns. 5–8. The model of Saffman (1976) assumes that the lipid molecules which make up the membrane sheet are rigid entities, extending through the membrane, which do not relax their structure due to imposed forces and that their cylindrical axis remains perpendicular to the membrane sheet. This is not the expected behaviour for the lipid molecules comprising the bilayers (van der Ploeg and Berendsen, 1982). Therefore, to express η_s as $\eta_0 h$ a model is required which defines an averaged viscosity throughout the effective membrane thickness over which the dissipative forces act. It may be misleading to construe the surface viscosity in terms of $\eta_0 h$ since it is not clear over which distance, h, the surface stress, $\boldsymbol{\sigma}$ (Eqn. 3), acts as a body force. Consequently the derived values of η_0 can differ by large factors depending upon the value chosen for h (see also section II.4.e). Since $\eta_0 h$ always appears as one parameter in the equations representing the diffusion constants, the quality of fits of data to the models is not affected by the choice of h, but the resulting estimate of η_o is, of course, dependent upon h. There are indications that possibly only the phosphate headgroups contribute to the surface viscosity (Evans and Hochmuth, 1978). At any rate, the dissipative mechanisms contributing to the viscosity, η_0, of the membrane sheet are almost certainly different for the polar-headgroups than for the hydrocarbon tails. However, the hydrodynamic models developed so far do not consider this.

(5) The transport of lipids or any other particles that do not span both monolayers of the bilayer certainly cannot be represented by the equations of Saffman (1976) since the fluid dynamic description of the interface at the bilayer midplane is surely different from that of a simple fluid extending indefinitely in a halfplane. The solution of Hughes et al. (1982) suffers from the same deficiency in this case, even though the magnitude of either exterior viscosity can be large. One can probably use the equation of Hughes et al. (1982) as a reasonable approximation in this case, but it is not rigorous.

e. Application of the hydrodynamic models, and some general comments

The hydrodynamic theory of Saffman (1976) and Hughes et al. (1981) should be applicable to large integral proteins with a radius larger than about 15 Å and which span the lipid bilayer, or to porous aggregates within the bilayer (Wiegel, 1980). A sufficiently large homogeneous lipid area (greater than about 4000 Å in radius) must be available for the application of Eqns. 6–8, otherwise the parallel effect (i.e., additive friction coefficients) of limiting boundaries must be considered, see Eqn. 4. Restricting ourselves, for simplicity, to Saffman's result (Eqn. 6), we see that in this respect the important parameters to compare are $\eta_0 h/\eta'$ of Eqn. 6 and R of Eqn. 4, and that a more general expression for the friction coefficient would be

$$f_T^{total} = f_T \text{ (of Eqn. 4)} + f_T \text{ (of Eqn. 6)} \qquad (9)$$

as suggested by Saffman and Delbrueck (1975). Also, Eqns. 6–8 are derived only for motion which is unlimited within the membrane *and* in the external fluids. If there are other boundaries in the fluid exterior to the membrane sheet and if they are close to the membrane sheet (as in multiple bilayers) the effective shear force of Eqn. 5 will be increased. To a first approximation, any interaction of the membrane particles with the exterior medium in addition to that already incorporated into the theory will be similar to increasing the η' in Eqn. 5. However, it is not clear how the derivations of Eqns. 6–8 will be affected by close parallel bilayers since the exterior medium then no longer extends infinitely in a direction perpendicular to the membrane. We would expect (Section II.1.c, above) that the effective radius over which the membrane fluid velocities are correlated would decrease due to the interactions with the neighbouring boundary, and that the frictional coefficient would increase as the reciprocal logarithm of this radius of dissipation, as Eqn. 4.

If the concentration of the diffusant or another component (non-interacting particles) is large enough the diffusion constant will be changed due to the increase in the effective viscosity of the membrane sheet. This effect could possibly be approximated by an expression, due to Einstein (1906, 1911, 1956), which expresses the viscosity of the fluid as a function of the fractional volume (ϕ) of dilute solute molecules, i.e., $\eta/\eta_s = 1 + 2.5\,\phi$, where η and η_s are the viscosities of the solution and pure solvent, respectively. See Hiemenz (1977) for a further discussion of this viscosity effect, including the possibility of interactions between the solvent and solute and the effect of higher concentrations. One should be cautioned, however, that this effect has not been calculated for two dimensions, as far as we know. Since Einstein's derivation assumes that the fluid is unperturbed by the movement of the particles at a sufficient distance from them, the derivation for two dimensions may not be directly analogous to that of three dimensions for reasons similar to the Stokes' paradox mentioned in Section II.1.a.

The influence of the lateral compressibility of the membrane sheet upon the diffusion constant has never been considered, although, the lateral compressibility of a lipid bilayer is relatively high, approximately 0.01 cm/dyne for DMPC (Evans and Hochmuth, 1978). General hydrodynamic arguments (Sommerfeld, 1964) indicate that the compressibility does not need to be taken into account in the hydrodynamic equations until the fluid velocities are on the order of the speed of sound in the fluid. Therefore, we do not expect this to contribute to the hydrodynamic description of diffusion in lipid bilayers.

2. FREE-VOLUME THEORY

a. General

The concept that molecular transport occurs primarily as a consequence of the 'free volume' available to the mobile molecules was introduced originally for solids and crystals, and later by analogy to liquids where the displacement of any particular molecule can only take place if a vacancy exists adjacent to the molecules or atoms. Extensions of these ideas have been developed for liquids, glasses and polymers where the concept of a necessary void volume is retained, however, the statistical nature of the fluctuating molecular environment is included. The major difference among the free-volume theories is the mechanism whereby the free volume arises, since the translocation into an existing void volume is usually considered to present no conceptual difficulty. The probability that a diffusional jump will take place is, as a first approximation, considered to be proportional to the probability that an empty space of sufficient size is available. Thus, the problem is reduced primarily to the calculation of the number of void volumes which have a size large enough to accommodate the diffusing molecule. We will consider only one of the free-volume diffusion models, that of Cohen and Turnbull (1959), which was originally suggested to explain diffusion in glass-forming materials, and was adapted by Galla et al. (1979) for diffusion in lipid bilayer membranes.

The free-volume model was first applied by Fox and Flory (1950) to explain the fact that the viscosity of many liquids decreases markedly with increasing pressure. The free volume is defined as

$$v_f = v_{av} - v_0 \tag{10}$$

where v_{av} is the average volume per molecule and v_0 is the van der Waals volume of the molecule. The suggestion that the fluidity of a fluid increases linearly with the specific volume of the liquid was made and extensively tested by Batschinski (1913). It was already proposed by Warburg and Babo, in 1882, that the denisty, rather than the temperature of a liquid is the important factor controlling the viscosity. Fox and Flory (1950, 1951, 1954) suggested that the glass transition temperature is related to a critical value of the free volume below which the freedom of movement of the molecules is drastically limited. Doolittle (1951) found an empirical expression relating the fluidity ($\phi = 1/\eta$) of hydrocarbon liquids to the free volume

$$\phi = \phi_0 \exp(-bv_0/v_f) \tag{11}$$

and Williams et al. (1955) proposed the following expression for the free volume,

$$v_f = v_g\,(\beta + \alpha(T - T_g)) \tag{12}$$

where v_g is the volume per molecule at the glass transition temperature, T_g, and α is the difference between the thermal expansion coefficient of the liquid and glass phase. β is the fractional free volume at the glass transition temperature, and Williams et al. (1955) gave the value 0.025. This has been extended to include the pressure and the concentration of other components (Ferry, 1980).

Cohen and Turnbull (1959) developed a theory of molecular transport which is based upon the movements of molecules into void volumes greater than some critical size. The distinguishing feature of the Cohen-Turnbull theory is that diffusion results from the *redistribution* of the total free volume in a liquid, and an *activation is not required* for the creation of local free volume regions. The average free volume is constant at any temperature or pressure, and the distribution function representing the probability that a molecule will have a free volume, $p(v_f)$, is derived under the constraint of constant total free volume. This probability was found by Cohen and Turnbull (1959) to be,

$$p(v_f) = (\gamma/v_f)\exp(-\gamma v^*/v_f) \tag{13}$$

where γ is a factor to correct for overlapping free volumes ($\simeq$ 0.5–1) and v^* is the critical free volume per molecule which must exist so that transport occurs. For simple liquids this is usually approximately equal to the volume of the molecules. The crucial assumptions on which the validity of Eqn. 13 depends are (Grest and Cohen, 1981): (1) a local volume associated with each molecule, or with each translocatable segment of the molecule, exists; (2) part of this volume is free, i.e., molecular motion within this volume is possible without interaction of the surrounding molecules; (3) a critical free volume, v^*, exists as discussed for Eqn. 13; (4) the distribution of the free volume within the liquid can be readjusted without energy requirements. Provided that each molecule can move only within a 'cage' defined by the adjacent molecules, assumption 1 is valid. The other assumptions are discussed in detail by Grest and Cohen (1981), where they review the success of the free-volume theory to account for the liquid-glass transition of simple liquids. They point out that assumption 4 limits the free volume interpretation of mobility to cases where the 'fluid' is sufficiently liquid-like so that an activation energy is not required for a redistribution of the free volume, and that the material must be sufficiently dense so that the life time of the cage is long compared to the time scale of motions within the cage.

b. Definition of the diffusion constant

Cohen and Turnbull (1959) defined a diffusion coefficient for every value of the free volume, $D(v_f)$, and therefore, the average diffusion constant is a normalized sum of these,

$$D = \int_{v^*}^{\infty} D(v_f) \cdot p(v_f) \cdot dv_f \tag{14}$$

The lower limit of integration is v^* because free volumes smaller than this do not contribute to the diffusion coefficient. At this point the assumption of a critical free volume enters the derivation. To obtain a useful relationship for the diffusion constant an expression for $D(v_f)$ must be available, and Cohen and Turnbull (1959) chose a form identical to that for the gas kinetic theory, that is,

$$D(v_f) = g \cdot l(v_f) \cdot u \tag{15}$$

where g is a geometric factor $\simeq 1/6$, $l(v_f)$ is the free distance of travel for the diffusant in the free volume, v_f, and u is the average velocity of the molecule. $u = [3kT/m]^{1/2}$, where m is the mass of the molecule, k is Boltzmann's constant and T is the temperature. We note that Eqn. 15 is not necessary, but, within the context of the Cohen-Turnbull free-volume theory, Eqn. 14 is essential. Equation 13 is a basic assertion of Cohen and Turnbull (1959), and Galla et al. (1979), however, an alternative can be proposed which also includes energy fluctuations (Section II.2.c).

In order to integrate Eqn. 14, Cohen and Turnbull (1959) reasoned that $p(v_f)$ decreases rapidly as v^* increases, so that $D(v_f)$ will be essentially constant over the region where $p(v_f)$ is significant, that is, at $v_f = v^*$. The resulting expression for D is thus, from Eqn. 14.

$$D = D(v^*) \int_{v^*}^{\infty} p(v_f) \cdot dv_f = D(v^*)\exp(-\gamma v^*/v_f) \tag{16}$$

$$D(v^*) = gl_c u, \quad g \simeq 1/6$$

Galla et al. (1979) redefined the appropriate terms in an obvious way to obtain an expression for two-dimensional motion as follows,

$$D = D(a^*)\exp(-\gamma a^*/a_f) \tag{17}$$

$$D(a^*) = gl_c u, \quad g \simeq 1/4$$

a^* and a_f are the area expressions for the respective volumes, v^* and v_f, see Fig. 1b. In Eqns. 16 and 17, l_c is the average length of free travel of the diffusant within the free volume, which is roughly the diameter of the cage with volume v^* or of area a^*. Equation 18 in Galla et al. (1979) has been misprinted; it should be our Eqn. 17.

MacCarthy and Kozak (1982), have proposed a similar expression which they have derived directly for the two-dimensional case, referring to random walk arguments coupled with the appropriate expression for the probability of the free area being larger than a_0, $P(a_0)$, where they have made the assumption that $a^* = a_0$, a_0 being the van der Waals area of the molecule. Their expression for D has, referring to Eqn. 17. $D(a^*) = l_c u/4$ and $\gamma = 1$.

c. Extension to activated diffusion
Macedo and Litovitz (1965), (see also Ricci et al., 1977) have proposed a theory for the viscosity of liquids which combine the Eyring rate-theory and the Cohen-Turnbull free-volume approach. This was more rigorously derived by Chung (1966) using statistical-mechanical arguments, and he arrived at the identical expression for the liquid viscosity. Since, as mentioned below, we expect the diffusion constant to be directly proportional to the fluidity ($\phi = 1/\eta$) of the environment, a similar expression exists for the diffusion constant which includes an energy barrier between two equilibrium positions before and after the step. As in the above theories for the viscosity, such an expression is required whenever an activation energy is involved in the formation of the free volume (or area) so that the process is not completely entropic. Equation 17 can then be written as,

$$D = A_0 D(a^*) (\exp(-\gamma a^*/a_f - E^*/kT)) \qquad (18)$$

A_0 is a constant dependent only on the temperature and E^* is the activation energy per molecule. Such an expression was not considered by Galla et al. (1979). Equation 18 should be considered whenever energy fluctuations are assumed to be important for the diffusion process in addition to the density fluctuations.

d. Appropriate application of the free-volume theory
The Einstein relation of Eqn. 2 can be written as, $D = kTb$, where b is the mobility of the particle, and for most situations of interest is proportional to the product of the fluidity, ϕ, and the inverse of some length, a_0, i.e. $D \simeq kT\phi/a_0$. It was specifically stated by Cohen and Turnbull (1959) that they only considered the case where D is proportional to ϕ, and where the Einstein relation is valid.

The free-volume theory is strictly valid only for molecules with sizes on the order of the solvent molecules or smaller, and its application to larger molecules is not consistent with the basic assumptions of the theory. The relative size of the diffusant and the solvent molecules for which the hydrodynamic approach is relevant is the opposite extreme, at least as far as the fundamental assumptions of the derivations are concerned. The free-volume theory is probably more appropriately applied to the diffusion of lipids than is the hydrodynamic approach. However, there are deficiencies of the free-volume theory which we consider in the following section.

e. Critical examination of the two-dimensional free-volume formulation
The adaption which Galla et al. (1979) made to the Cohen-Turnbull theory to describe two-dimensional transport in bilayer membranes, Eqn. 17, requires no additional basic concepts, and the strengths and weaknesses of the three-dimensional case also apply. The most basic objection is that the theory is

phenomenological and modelistic. However, this objection is partially compensated for by the attempt to account for at least a minimum of the contributing physical mechanisms on a molecular scale. The free-volume theory is a cell model and as such suffers from the deficiency that an attempt is made to describe a many-body situation with a one-body model (Hoover et al., 1972). This becomes strikingly evident by observing that Eqns. 16 and 17, which represent the free-volume diffusion model, are essentially a product of a gas-kinetic diffusion constant, which is governed by a single particle motion, and a distribution function for the free volume which is derived from a quasi-lattice structure, i.e., a structure with solid properties. Thus, the process of transport has been artificially divided into two independent steps, the formation of the free volume and the translational movement. The translocation step does not involve any cooperative movements of the liquid molecules and this neglect has been criticized and discussed by Ricci et al. (1977) for hard spheres. They conclude, as was already pointed out by Frenkel (1955), that the fluctuations of the local free volume are a result of correlative motions of many particles, and these correlations are important for the disappearance of the void volume, as well as for the formation of it. After the translocation of the diffusant the solvent molecules are assumed to fill the void left behind, but the free-volume theories do not consider this event quantitatively. The objection that the Cohen-Turnbull theory has neglected energy requirements for the formation of the free volume, has been alleviated by the suggestion of Macedo and Litovitz (1972) as discussed above for Eqn. 18. Estimates for the value of the activation energy are available from the lipid diffusion model of Pace and Chan (1982), which we consider in the next section.

It is not obvious which values of the critical volume and the free distance of travel are important for the transport of molecules with geometrical shapes other than spherical or more complicated molecules which have segmental motion. Vogel and Weiss (1981a) have considered extensions of the Cohen-Turnbull theory for the case of disk-like molecules and almost spherical molecules, by introducing geometrical relations for the distance of free travel and the magnitude of free volumes necessary for movement in different directions. They have also investigated the range of sizes of the diffusants relative to the size of the solvent for which the free-volume model is applicable (Vogel and Weiss, 1981b). The theory is unable to account for the larger diffusant molecules for which, as expected, the theory predicts an exaggerated hindrance to the rate of diffusion. This is attributed to the fact that such large molecules do not need a free volume corresponding to their size, as is proposed by the free-volume model. A molecule with the flexibility of a linear hydrocarbon chain, such as a lipid, might be expected to undergo large displacements by executing rapid segmental motions into free volumes, or areas, with sizes on the order of the segment size, rather than the size of the whole molecule. This has been suggested by Jost (1964, 1972).

f. Assumptions of the free-volume theory

In addition to the limitations of the model already mentioned, a number of approximations and assumptions are made when applying the free-volume model to the diffusion of lipids and other small molecules in lipid bilayers. First, it should be realized that the free volume model is not expected to represent the diffusion of a molecule which is very large compared to the lipid solvent. The major applications of the model in other fields, for instance polymers, glasses and liquids, are limited to a description of viscosity, diffusion of molecules small relative to the solvent or to molecules of comparable size with the solvent molecules (self-diffusion). It is obvious from the discussion of the derivation that these limits apply to the model.

All internal motions of the molecules have definitely been neglected and the segmental motion mentioned above has never been considered. The molecules are considered to act as hard cylinders which are not allowed to wobble when undergoing translation in the plane of the membrane, and the motion into the free volume takes place as a concerted motion perpendicular to the cylindrical axis. The ultimate hope is that the most important physical contribution to the diffusion process has been realized, and that the other motions and interactions play only a minor role. The electrostatic interactions of the polar head groups and forces between the hydrocarbon chains or entanglements between them have been completely neglected. The high dielectric constant for water will diminish electrostatic interactions provided that the headgroups are fully hydrated, however, at low values of hydration the electrostatic forces will probably become significant. Also, dissipative interactions of the lipids with the aqueous solvent and of the hydrocarbon regions of opposing bilayers have not been considered. Due to the rather large size of the lipid molecules, the stipulation that the diffusant moves as a rigid object, and that this single particle motion can proceed as a gas over a distance comparable to the diameter of the diffusing particle, seems improbable. It is possible to include a *hydrodynamic* interaction between the diffusing particle in the membrane sheet and the aqueous phase by replacing Eqn. 15 by $D(v_f) = kT/f$, where f is the friction coefficient due to the particle-aqueous and particle-midplane contacts (Clegg and Vaz, 1984; Vaz et al., 1984a). This takes into account the fact that the particle does not move freely within the free volume if it interacts with the external fluid. Another energy requirement which has been neglected is due to the surface tension, γ_s. This energy must be expended to form the free area at the interface between the polar head groups and the aqueous phase. The surface tension is the energy per unit area so that this energy contribution will be proportional to the free area at the polar head groups.

Many of the approximations made when applying the free-volume model to membrane diffusion are *also* asserted for the hydrodynamic model, however, hydrodynamic considerations are not really justified for particles with sizes similar to the solvent.

3. *Molecular models: flexible rod activated jump model for lipid diffusion*

a. General

Pace and Chan (1982) have presented a theory for lateral self-diffusion in lipid membranes which is a modification of an earlier model by Pace and Datyner (1979) for penetrant diffusion in bulk polymers. This is the most detailed theory yet proposed for lateral self-diffusion in lipid bilayers. Estimates of the parameters of the model are either taken from previous work on polymers involving polyethylene chains in the all-*trans* configuration, or are calculated from assumptions which can be considered as part of the model. Pace and Chan (1982) adjust only one parameter to simulate the diffusion constant, and the variance of this parameter is only a factor of two. The agreement with the data which is presented is quite good considering that no attempt is made to 'fit' the data, that is, the values of the parameters are assumed to be identical to, or very close to, those determined for the appropriate polymers. The model will be carefully examined here, since the fundamental concepts are probably unfamiliar to many researchers not dealing directly with polymers, and this makes it difficult to assess the assumptions, limitations and success of the model. We will first describe the general polymer model, stressing the concepts important for the lipid bilayers and then we examine the approximations and limitations. We will give a relatively detailed account of the background material dealing with penetrant diffusion in polymers since the concepts introduced here are not only helpful to understand thc model but are also suggestive for future work.

The model of Pace and Chan (1982) for self-diffusion in lipid bilayers is based entirely upon the model by Pace and Datyner (1979) which is a statistical mechanical model for diffusion of simple penetrants (such as O_2 or N_2) in polymers. No new concepts are developed for lipid diffusion, however, a few assumptions were made which we will discuss below. For this reason it is advantageous to discuss the penetrant diffusion model directly, since the assumptions which apply to the lipid bilayer can then be easily introduced and evaluated.

All the theoretical treatments of this section calculate the diffusion constant by referring to the following equation from random walk statistics,

$$D = (1/2d)\bar{L}^2 \nu \tag{19}$$

d is the dimension of the space, $\bar{L}$ is the root mean square displacement per step, and ν is the frequency of random steps. We will be mainly concerned with calculating the frequency term, since the choice for $\bar{L}$ is straightforward within the Pace and Chan (1982) theory.

The goal of many theories of diffusion is to develop an Arrhenius-type equation for the diffusion constant, i.e., $D = D_0 \exp(-E_d/RT)$, where D_0 is a frequency factor which is only weakly dependent upon the temperature (Glasstone et al.,

1941). It is often found experimentally that the apparent activation energy, E_d, is also temperature dependent. ν of Eqn. 19 is expressed as $\nu = A\exp(-E_d/RT)$, where A may be a function of E_d and T.

b. Diffusion involving concerted movements of chain segments

The Eyring theory of diffusion (Glasstone *et al.*, 1941) does not account for the presence of different segment lengths of polymer structures. For instance, the probability of diffusion 'jumps' which involve only short segments will be limited by the stiffness of the chain due to the strain required to displace a short segment of the chain, and the diffusion jumps involving longer segments will depend upon the probability that several segments will cooperate to perform a concerted movement of long segments. Barrer (1943, 1957) has proposed a zone theory which accounts for the multiple degrees of freedom involved in coordinated movements of molecules which are in the immediate vicinity of a diffusant. Statistical arguments show that the probability of finding a total energy greater than or equal to E distributed among all of the possible subsets of the total number, n, of the degrees of freedom constituting an arbitrary region is (Bueche, 1953).

$$P = \sum_{f=1}^{n} (E/RT)^{f-1}(1/(f-1)!)\exp(-E/RT) \tag{29}$$

For any E/RT there is a maximum in P for a particular $f = f_{max}$ which corresponds to the average energy $E = fRT$. If $f < f_{max}$, the local energy per degree of freedom which is distributed over f degrees of freedom will be greater than the thermal average and the local region will be activated. If the energy distribution is such that the energy per degree of freedom will be less than the thermal average there would be no activation energy at all. Thus, the summation of Eqn. 20 only needs to be at most up to f_{max}. The terms of the summation in Eqn. 20 which are important for the diffusion process, or viscosity, will not be that involving f_{max}, but will lie somewhere between $f = 1$ and $f = f_{max}$.

Another postulate of Barrer's theory is that there is a need for synchronization among the rotations and vibrations to result in a successful unit process. Each term of the summation must then be multiplied by a term, ϱ_f, which is the probability that f degrees of freedom will cooperate in a diffusion step. ϱ_f is often assumed equal to $(\varrho^*)^f$ where ϱ^* is the probability of a group (representing one degree of freedom) moving in a certain direction. This assumes independence of the degrees of freedom. Barrer (1943) also considered replacing the sum in Eqn. 20 by the maximum term, under the assumption that the significant terms are likely to be limited in number, and that often a single selection of degrees of freedom is involved in the important movement rather than a spectrum. Whether one term of the summation is singled out or not, the expression for the diffusion constant according to Barrer is given by combining Eqns. 19 and 20,

$$D = (1/2d)\bar{L}^2 \{\nu_0 [\exp(-E/RT)] \sum \varrho_f (E/RT)^{f-1}/(f-1)!\} \quad (21)$$

ν_0 is the vibration frequency characteristic of a diffusion unit. The form of this equation appears to have been first been proposed by Wheeler (1938) and a more recent derivation is given by Bueche (1953).

c. Selecting one degree of freedom important for diffusion

Pace and Datyner (1979) point out that, according to the treatment of Barrer (1957), the relevant motions in terms of the coordinates contributing to the chain displacements are considered to be loosely coupled. This is actually the opposite of what would be expected of concerted movements of connected chain elements as in polymers. However, if this is the case it is difficult to give a general expression for D without a specific model for the movement of the chains. Provided that the chains are quite stiff, that is, the bending energy per unit length is very large compared to the interaction energies between the chains per unit length, Pace and Datyner (1979) define the only degree of freedom important for diffusion of a penetrant to be that corresponding to a *maximum chain center displacement* which is a separation between two adjacent chains to make room for the penetrant (see Fig. 1). The positions of all the polymer segments are always considered predictable provided that this one degree of freedom is specified. Limiting the possible configurations to this one type is an important assumption and they only justify it by appealing to the uncertainty in the $\bar{L}^2$ term of Eqn. 19 for penetrant diffusion in polymers. This should be kept in mind since the confidence in the parameters derived from the Pace and Datyner expression are dependent upon this basic assumption. In this respect it should also be noted that $\bar{L}$ is well defined for the case of lipid diffusion (Pace and Chan, 1982), in contrast to penetrant diffusion in polymers. Pace and Datyner (1979) consider only interactions between *two* chains and assume that the penetrant diffusion is controlled solely by the synergistic dynamics of neighbouring pairs of chains. Their model is a combination of previous theories by Barrer (1943), Brandt (1959) and DiBenedetto (1963a, 1963b) to which they have added a calculation of the bending energies of the chains.

d. Calculation of the diffusion constant

Pace and Datyner (1979) assume that the energy that corresponds to the chain separation between two chains has a different functional dependence for small and large chain center displacements. If $E(d')$ is the energy required for a maximum chain center displacement of d', they define the following energies for a chain separation of $d' = y_1 + y_2$, where y_1 and y_2 are the distances of the maximum center separation of the two neighbouring chains from their minimum energy positions.

$$E(d') = \Gamma' \ (y_1 + y_2) \text{ for large } d' \tag{22a}$$

$$E(d') = \Lambda' \ (y_1^2 + y_2^2) \text{ for small } d' \tag{22b}$$

d' is approximately the diameter of the penetrant (see below, Eqn. 29). To calculate the probability of a large chain separation they calculate the chain conformation (representing the only degree of freedom for the participating segments) which corresponds to a *minimum* expenditure of the total energy. This consists of intramolecular (bending) and intermolecular components (we discuss below how they define the minimum energy configuration). The constant Γ' is identified by taking the derivative with respect to d' of the non-linear expression for the activation energy of a symmetric separation which they derive (see below, Eqns. 24 and 25). The constant Λ is determined as follows. The average interchain potential per unit length is expanded in a power series involving the lateral displacement of a chain from its equilibrium position. This expansion is then suitably averaged, considering the pairwise interactions of the nearest neighbours, to give a mean cylindrically symmetric potential, as was done by DiBenedetto (1963a). The approximation is made that the adjacent chains are fixed at cell lattice sites, each chain having four nearest neighbours. The interaction potential which is averaged is a Lennard-Jones potential derived by DiBenedetto (1963a). The total effective potential to be used for the low amplitude 'vibrations' (small d'), including the intramolecular bending energy, is then obtained by inserting this averaged interchain potential into the expression which they derived for the energy of a symmetric separation of two chains. The result for small displacements is a quadratic dependence of the energy upon the displacement of the chains from the equilibrium positions (see Eqn. 22b).

To calculate the probability of a chain separation, normal statistical mechanical expressions are used. The integration over all separation distances (which is the normalizing factor) is done with only the quadratic energy dependence, Eqn. 22b, and the integration involving the activation region with distances $\geq d'_c$ is done using only the linear energy expression, Eqn. 22a. Pace and Datyner (1979) assumed that the movement of pairs of chains which is selected as important for diffusion is independent of all the other energy configurations available to the polymer. This statistical mechanical procedure defines an activation energy (Eqn. 23) related to the probability of having a chain separation $\geq d'$. This activation energy $\Gamma' d'$ is identified with the full expression for the activation energy derived by them, which we discuss below. The probability distribution then has to be multiplied by the characteristic frequency. ν_0, for the libration (i.e., the symmetrical chain separation), to give an expression for ν of Eqn. 19 which determines the diffusion constant. ν_0 is calculated by assuming that the chain vibrates like a rigid rod which is fixed at one end (either $+x_0$ or $-x_0$ from the center of maximum displacement, section II.3.e.v) with a hinge joint and has a spring with a linear restoring force (Λz) attached at the other end, $x = 0$. The movable end of the rod (at $x = 0$) corresponds to the position of maximal displacement of the chain ($z = d'$). The final expression in terms of the parameters of Eqn. 22 for the diffusion of a penetrant within a polymer matrix is then,

$$D = (1/6)\bar{L}^2 \nu_0 (\Lambda d'/\pi \Gamma')\exp(-\Gamma' d'/RT) \qquad (23)$$

$$\nu_0 = (1/2\pi)(3\Lambda\lambda/m^* x_0)^{1/2}$$

where λ is the distance between chain segments and m^* is the mass per chain segment.

e. Interchain interaction energy

i. Interaction potential per unit length. The above discussion assumes that an equation exists for the interaction energy between two parallel neighbouring chains as a function of the distance between them. DiBenedetto (1963a) has derived such an expression by assuming that the two chains are parallel and *rigid*. The potentials between the subunits of apposing chains are assumed to be Lennard-Jones (6–12) potentials, and the total interaction energy between a single subunit of one chain and the total neighbouring chain is expressed as a sum, which is arbitrarily limited to ten polymer segments on either side of the center in question. The resulting potential per backbone element divided by the distance between the elements on a chain, λ, is the average interchain potential per unit length, and is given by,

$$f(z) = \varepsilon^* \varrho^* / \lambda^2 \, [0.77(\varrho^*/z)^{11} - 2.32\,(\varrho^*/z)^5] \qquad (24)$$

ϱ^* and ε^* are the Lennard-Jones distance and energy parameters and z is the perpendicular distance between the *parallel* chains. The equilibrium separation between the chains is at a distance $z = \varrho$, and Pace and Datyner (1979) often use the approximation that $\varrho = \varrho^*$; $f(z) = 0$ if $z = 0.83\,\varrho^*$ (DiBenedetto, 1963a). The potential represented by Eqn. 24 is only between two chains, and does not constitute the total potential which would involve at least the other adjacent chains.

ii. Minimum energy for the 'maximum chain center displacement'. Pace and Datyner (1979) have combined features of two previous theories of penetrant diffusion in polymer matrices. Brandt (1959) described a molecular model for diffusion which takes into account intramolecular resistances of the chains to bending motions and intermolecular repulsions of the bent molecular chains by the internal pressure of the system. Brandt (1959) calculated the diffusion constant according to the zone theory of Barrer (1943), see Eqn. 21. DiBenedetto and Paul (1964) considered the diffusion of the penetrant molecules parallel to the axes of bundles or polymer molecules and, as mentioned above, they related the activation energy of the transport process to intermolecular interactions of the polymers (Eqn. 24). We have already described the diffusion model by Pace and Datyner (1979), and now outline their calculation of the activation energy, i.e., the full expression for the linear approximation given in Eqn. 22a.

As an approximation to the energy of symmetric separation of two chains they consider the expression,

$$E = \int_{-\infty}^{+\infty} [f(z) - f(\varrho) + (\beta/2)(d^2z/dx^2)]dx \tag{25}$$

where β is the 'average' effective single chain-bending modulus per unit length, x is the dimension along the average polymer axes, all other parameters have been described above. The expression for $f(z)$ is that proposed by DiBenedetto (1963a) in Eqn. 24. They reason that the contour of the chains will be predicted by the minimum value of E sufficient for the inclusion of the penetrant molecule at $x = 0$ between the two infinite chains. E of Eqn. 25 is then minimized by solving the appropriate Euler-Lagrange equation subject to reasonable boundary conditions. The reader is referred to the paper for the details of the calculation. Their result is Eqn. 30, below.

iii. Simple derivation and interpretation of the activation energy. We will present a much simpler derivation of the total energy required for two chains to separate by a distance ϱ' at the point of maximum separation which results in essentially the same expression for E as Pace and Datyner (1979). This derivation is not as rigorous as that of Pace and Datyner (1979) but it emphasizes the approximations and essentials of their approach, and leads to a simple interpretation of their model for the activation energy. Pace and Datyner (1979) have introduced three main assumptions to make the problem more tractable. First, they assumed that the interchain potential per unit length, $f(z)$ (Eqn. 24), can be approximated by a linear function

$$f(z) = \Gamma z + \text{constant} \tag{26}$$

similar to Eqn. 22a for E. Second, they introduced an approximate length x_0, which is approximately the distance along the chains from the point of maximum separation at which the two chains resume their average separation, ϱ. Thirdly, they assumed that $f(z)$ does not depend upon the contour shape of the two chains at successive positions along their common average axes. DiBenedetto's (1963a) derivation of $f(z)$ is only strictly valid provided that the chains are always parallel within the regions where they are separated, and Pace and Datyner (1979) use DiBenedetto's expression for $f(z)$. This last assumption is expected to be valid for very stiff chains since the local deviation from parallelism will then be very small. Using these assumptions we now approximate a symmetric separation of two chains by a contour shown in Fig. 2. Here we have replaced a continuous curvature of the two chains by straight lines from the point of maximum separation ($z = d + \varrho$ at $x = 0$) to the points of average separation ($z = \varrho$ at $x = x_0$). Now we make the *Ansatz* that the energy required to separate the chains is derived only from the Lennard-Jones potentials. Therefore,

$$E = \int_{-x_0}^{x_0} (f(z) - f(\varrho))dx = 2 \int_{0}^{x_0} \Gamma[d - dx/x_0]dx \tag{27}$$

$$E = \Gamma x_0 d$$

where we have used Eqn. 26 for $f(z)$. The balance between the bending forces and the interchain interactions determine the limits of integration, x_0 (Section II.3.e.v).

iv. Activation energy of Pace and Datyner, and interpretation. Pace and Datyner (1979) derived the following expression for the activation energy.

$$E = (16/15)\Gamma^{3/4}\, d'^{5/4}\, (72\beta)^{1/4} \tag{28}$$

They chose the boundary conditions such that

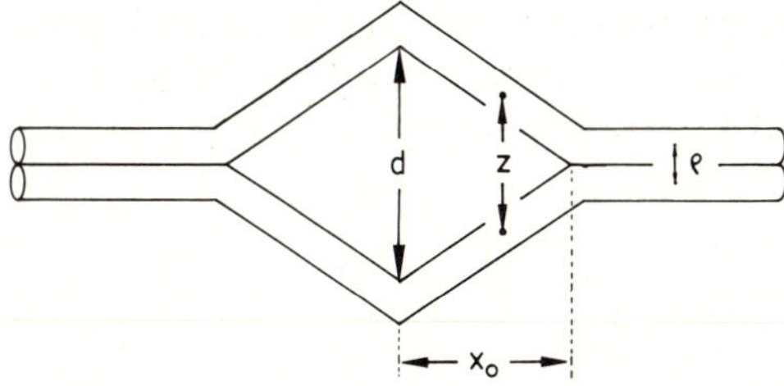

Fig. 2. Simple approximation of the flexible rod model (Section II.3) for calculating the energy of two separating chains (Eqn. 27). x_0, the distance from the maximum chain displacement to the equilibrium chain center separation, can be calculated by either Eqn. 29 (Pace and Datyner, 1979) or by equations given in Section II.3.e.v. x_0 is essentially the distance that a flexible rod must be bent to accommodate a maximum separation of 'd'. The lipid model is derived from this figure by simply considering only the left or the right half of the symmetrical figure.

$$x_0 = [72(\beta/\Gamma)d')]^{1/4} \tag{29}$$

where $d' = d + \varrho^* - \varrho$

so that we can write the Pace and Datyner (1979) expression as

$$E = (16/15)\Gamma x_0 d' \tag{30}$$

This is essentially the same as Eqn. 27, especially considering that $\varrho \cong \varrho^*$; Pace and Datyner (1979) actually set $\varrho = \varrho^*$ throughout their derivation. The constant Γ is chosen by Pace and Datyner (1979) such that the work done to separate two chains by a certain distance at the point of maximum displacement will be the same whether Eqn. 24 or Eqn. 26 is used for $f(z)$. This approximate approach of Eqn. 27 gives a simple physical interpretation to the model proposed by Pace and Datyner (1979). To separate two parallel infinitely extending chains at one point by a distance, d, work must be done to overcome the intermolecular attractions of the chains. The extent of the relative displacements along the lengths of the chains is limited by the parameter, x_0, which is ultimately attributed to a balance between the resistance to bending of the chains and their attraction for each other. By including the stiffness of the chains as a parameter, Eqn. 25, Pace and Datyner (1979) were able to calculate the position of the 'hinges' (x_0) referred to in the derivation of ν_0 in Eqn. 23. The exact shape of the chain is immaterial for the calculation of the activation energy, and only the extent of the chain separation along the chain axis is required.

v. Alternative estimate and meaning of x_0. We can also give a good estimate for x_0 by considering simply the mechanics of bending beams. If a uniform force per unit length, p, (such as gravity) is applied to a rod of length, l, which is fixed at the ends it will bend with a maximum displacement at the middle of the rod equal to $y_{max} = (1/384)(p/\beta)l^4$. The uniform force, p, is equivalent to Γ. Since $l = 2\,x_0$ we have $x_0 = (1/2)(384\beta y_{max}/p)^{1/4}$ (Eshbach, 1966), which is only 35% smaller than the estimate of Eqn. 29 if we remember that $y_{max} = d'/2$, and this is well within the limits of the theory. This simple analogy provides a clear demonstration of the important roles of the bending and intermolecular interactions. If one lifts a flexible rod, such as a water hose, the energy required will depend upon the length of the rod to be lifted, and this will in turn depend upon the stiffness. The major portion of the work is expended in overcoming the attraction of the two chains, not in bending them. The stiffness determines the length of the two chains along which they must be displaced from their equilibrium separations. Our estimate for x_0, substituted into either Eqn. 30 or 27 will give essentially the same value of the activation energy.

f. Application of the polymer diffusion model for self-diffusion in lipid bilayers

i. Assumptions and definitions asserted specifically for lipid self-diffusion. Pace and Chan (1982) have adapted the model of Pace and Datyner (1979) of penetrant diffusion in polymers to describe self-diffusion within lipid bilayers. The polymer model is taken over with only minor variations, only one additional assumption and the necessary definitions to adapt the model to self-diffusion within a bilayer. The polar head group of the phospholipids plays the role of the penetrant diffusant and the diffusion step is the insertion of a head group between two adjacent head groups. Motion at the head group of the lipid molecule is assumed to be the controlling factor for diffusion since the two chains of the hydrocarbon end will be less likely to have a concerted effect. The free chain ends are assumed to immediately follow the motion of the head groups, offering no resistance either statistically or mechanically. The jump length, $\bar{L}$, is assumed to be $(\sqrt{3}/2)\varrho$ for the two chain lipids and the jump frequency is $8/3\nu_{L0}$ where ν_{L0} is the frequency of chain separation. The hydrocarbon chains are modelled as flexible rods as in Pace and Datyner (1979) with a bending coefficient, β. The activation energy is assumed to be represented by Eqn. 28 (or Eqn. 30) but these expressions are to be divided by 2, since, relative to the positions of the head groups (the penetrant), only half of the integral of Eqn. 25 is to be performed. That is, the chain extends only on one side of the penetrant (polar head group) position. This assumption will overestimate the activation energy since the expression by DiBenedetto (1963a) for $f(z)$ at the end of a chain is about 1/2 of the value in the middle of a pair of chains. The dispersion forces of the glycerol and polar head groups are neglected and only the intermolecular forces of the CH_2 groups are considered. Even though for a two chain lipid molecule there are two acyl chains per molecule which must move concertedly, the dispersion forces of only a single pair of acyl chains is included because Pace and Chan (1982) have defined the conformation of the separating pair of molecules such that only one of the two acyl chains per lipid molecule are adjacent within the intermolecular pair. The bending potential per molecule, β, is doubled for the two chain species of lipids but, as noted from Eqn. 29, the activation energy is relatively insensitive to the value of β. The expression for D of Eqn. 23 is multiplied by the probability that a diffusion step will be completed following a chain separation, and this factor is varied from 0.5 to 1 to simulate the diffusion data. This description, together with the previous discussion of the polymer-penetrant diffusion, completes the description of the Pace and Chan (1982) model of lateral lipid self-diffusion.

ii. Temperature dependence and parameterization of D_t. The diffusion constant of Eqn. 23 can be expressed in terms of the basic parameters, ϱ, ϱ^*, ε^*, β, λ, d' and m^* introduced earlier by substituting the equations for Γ', Λ and ν_0 (see Eqn. 4a of Pace and Chan (1982)). If the assumption is made that $\bar{L} = \varrho^*$, all the parameters have been previously estimated for polyethylene chains in the all-*trans* configuration. Pace and Chan (1982) give an estimate for the lipid chains of 10.9 kJ · nm/mol for β of one chain (rather than 12 kJ · nm/mol for the all-*trans* configuration) and $\varepsilon^*\varrho^*/\lambda = 1.12$ kJ/mol ($\varepsilon^* = 229$ kJ/mol). The temperature dependence of the activation energy requires knowledge of the temperature dependence of ϱ^*/ϱ ($\cong 1$) and they assume that $d/dT(\varrho^*/\varrho) = -1/2\alpha_v$, where α_v is the thermal volume expansivity of the bilayer ($\cong 8 \times 10^{-4}$ K^{-1}). With this value of α_v the diffusion constant shows an Arrhenius dependence.

The prediction of the temperature dependence of the self-diffusion coefficient definitely shows the general trend of the selected data, which means that the estimate of the activation energy is of the correct order of magnitude. The temperature dependence of the activation energy does not seem to be pronounced, as the Arrhenius plots are linear over a 50 K temperature range. This was to be expected from the earlier work of Pace and Datyner (1979). This would mean that the data of Vaz et al. (1984) would not be represented as well by the model since they find a nonlinear temperature dependence (Section III). The magnitude of the apparent activation energy is compatible with an activation process involving a separation of two acyl chains by 4 Å over a chain distance of about 1 nm (see Figs. 1 and 2).

iii. General critique of the lipid self-diffusion model. Pace and Chan (1982) adopt, almost entirely, the premises and approximations of the penetrant-polymer diffusion model of Pace and Datyner (1979), which is one of the most detailed molecular descriptions for penetrant diffusion in polymers. We do not have the space to discuss the success of the polymer model but refer the reader to a recent review (Frisch and Stern, 1983). We have tried to present the polymer model in sufficient detail so that the reader can judge for himself the basic assumptions underlying the model. It is of primary interest here to inquire whether the Pace and Chan (1982) model describes correctly the situation for self-diffusion in lipid bilayers. The model has not, to our knowledge, been referred to or treated to interpret lipid diffusion data, except by Pace and Chan (1982), and we also know of no previous critical discussion of the theory. We have given a relatively detailed account of the theoretical background because we believe that it is necessary that future accounts of movement in bilayer membranes, both rotational and translational, be based on more molecular details, and despite any shortcomings of the model, it is the first attempt since that of Traeuble (1971) and Galla et al. (1979) to account for diffusion in lipid bilayers from a molecular point of view. As should be obvious from the discussion above the polymer model is not an isolated proposition but integrates many concepts which have been previously developed (see Frisch and Stern, 1983; Crank and Park, 1968). This reflects the enormous amount of research directed toward penetrant diffusion in polymer substances over many years, however, there is a need for much more data and different approaches to better define the important dynamic events contributing to lateral movement in lipid bilayers.

We will now examine several of the critical assumptions of Pace and Chan (1982).

(1) They have used the expression for the energy corresponding to a chain separation for infinite chains developed by Pace and Datyner (1979) and have simply divided this expression by two to account for the fact that, relative to the head group (penetrant), only half of the chain is present, see Fig. 1c. This will definitely overestimate the energy. The intermolecular interactions per unit length will be reduced in the vicinity of the polar headgroups since there is no chain extending on one side of the maximum displacement. According to the procedure of DiBenedetto (1963a), $f(z)$ of Eqn. 24 will thereby be halved at the polar head groups. This reduction will apply for several polymer segments away from the polar head groups. Although this neglect may be partially cancelled by their neglect of the dispersion forces of the glyceryl group, it is amplified exponentially when estimating the diffusion constant.

(2) They have assumed without justification that the bending conformation will be the same whether the chain is free at the point of maximal separation, or not (see Fig. 1c). Referring to the simple approximation to their model described above in Section II.3.e.iii (Fig. 2), we can estimate the effect of this approximation by comparing the length over which a rod must deform for the two cases of a rod fixed at both ends (which corresponds to the infinite polymer) or a rod fixed only at one end with the other end free (which corresponds to the lipid bilayer). To produce a maximal separation of y_{max}, with uniformly distributed loads per unit length, w, the length of the rod, l, must be $l_1 = (8y_{max}\beta/w)^{1/4}$ for a single fixed end and $l_2 = (384y_{max}\beta/w)^{1/4}$ for two fixed ends (Eshbach, 1966; Sommerfeld, 1964). These l's should allow estimates for the x_0 of the Pace and Chan (1982) model as shown above in Section II.3.e.v. The relations of the l's to x_0 are, for the infinite chains, $l_2 \cong 2x_0$, and for the lipids, $l_1 \cong x_0$. We find that $l_2/l_1 = (384/8)^{1/4} = 2.2$, which means that the two estimates of x_0 are very close. According to the simple model of Section II.3.e.iii, this corresponds to decreasing the energy by a factor of two compared to the infinite chain polymers. This is what was assumed by Pace and Chan (1982), and justifies their assumption.

(3) Pace and Datyner (1979), and consequently Pace and Chan (1982), have neglected the effect of the surrounding chains upon the energy of large separations, Eqn. 22a. It is difficult to estimate the magnitude of this error, and it would certainly depend upon the model chosen to describe the local environment of the

chain pairs. In favor of this assumption is the large lateral compressibility of the lipid bilayer. The polymer model of Brandt (1959) calculated the diffusion coefficient of penetrants assuming that the energy to expand a vacancy due to the internal pressure is a large contribution to the activation energy. It would be valuable to estimate the magnitude of such an effect and compare it to the energy of separation for isolated chains. For the smaller displacements, Eqn. 22b, Pace and Datyner (1979) did estimate a mean field contribution of the adjacent molecules assuming a lattice model.

(4) Their model assumes that a vacancy is produced first and then the neighbouring polar head group 'jumps' into the hole. No contribution to the energy from the acyl chain of the incoming lipid molecule is considered. This would lower the required energy, since such a large separation of two chains would no longer be required if the third chain moves simultaneously into the 'hole'. Although this is a concerted motion, the energy may be considerably lower.

(5) The activated configuration assumes that the acyl ends of the lipid chains are anchored and do not move until the polar head groups have translocated. No justification is given for this assumption and it does not seem to be completely consistent with their assumption that the acyl ends offer no resistance to subsequent lateral movement. One would not except the acyl ends to act like rigid rods since the order parameters for the CH_2 segments below the C_{10} chain position fall off very rapidly (Seelig and Seelig, 1980), and it is not expected that the chain ends can resist lateral stresses. If it is really necessary to consider bending, some attempt should be made to justify this assumption. x_0, which is the effective length over which the chains bend, is about 1 nm, and the order parameter of the chains is high only for the first 1.2 nm from the polar head groups (Seelig and Seelig, 1980). Thus, the assumption of infinite chains becomes questionable.

(6) The entropic constraints, discussed in Section II.2 on free-volume theory, are not considered. The model is purely an energetic model where the free energy of activation is assumed to be local. The effect of a free-volume redistribution could be combined with the model of Pace and Chan (1982) as discussed above for the Macedo and Litovitz (1965) model (Section II.2.c).

4. RELATED TOPICS, COMPARISON OF THEORIES AND SUGGESTIONS FOR FUTURE WORK

a. Scope

We have tried to give a generally understandable but still relatively detailed account of the present theories which are available for interpreting diffusion coefficients of particles embedded in lipid bilayers. Several other treatments of transport properties are pertinent to our discussion of lipid bilayers, since the principles used to describe the dynamic features of these other systems may be useful and certain data can be compared to those of the lipid bilayers. These approaches are not usually referred to in studies dealing primarily with membranes.

b. Polymers

The theory of Pace and Chan (1982) (see Section II.1.a above) is based upon ideas developed for diffusion within polymer systems and this calls attention to the vast literature existing on polymers. We cannot enter a detailed discussion of related topics in this field, but note that it may be fruitful to compare the dynamic properties of lipid bilayers to concepts which have been profitable in correlating the viscoelastic properties of polymers, especially if the diffusion is controlled by the hydrocarbon portions of the lipids. For instance, the effective viscosity for a foreign molecule within a polymer matrix is very dependent upon the relative size of the molecule and the polymer segments which define the free-volume elements in the matrix. Thus, by analogy, one should be very careful in defining the viscosity of a lipid bilayer, since situations may arise where the dominant dissipative forces are exerted within the polar head groups (suggested by Evans and Hochmuth (1978) for protein diffusion), the hydrocarbon chains of the lipids (rotational diffusion of DPH, Shinitzky and Barenholz, 1978), or both. The local segments of the lipids which characterize the mobility in each of these cases are different and the effective viscosity will probably depend upon the distances over which the configurational rearrangements are coordinated.

The viscosity is operationally defined as the force required to move two opposing parallel unit areas, one unit dimension apart, one velocity unit relative to each other. If this concept is correct at the level of molecular dimensions, the viscosity will probably be dependent upon the size of the system under consideration, and the appropriate value will approach a maximum limit as the size of the diffusant becomes large compared to the segmental elements. Also, the viscosity and diffusion within a polymer matrix demonstrate a strong correlation with the free volume. The presence of a foreign penetrant perturbs the local free volume, and at higher concentrations the total free volume of the polymer system can be affected considerably. Analogous effects within lipid bilayers may be expected. The temperature and pressure dependence of the free volume within polymer matrices has been investigated extensively, and there exists a considerable amount of theoretical discussion concerning diffusion and viscosity measurements. The interested reader can obtain further information related to this subject from two excellent monographs, Crank and Park (1968) and Ferry (1980).

c. Monolayers, surface viscosity and Eyring's two-dimensional theory

The first measurements of surface viscosities were made by Myers and Harkins (1937), where they reported values for monolayers on water of a series of saturated fatty acids, from tetradecanoic to eicosanoic acids (14–20 carbon atoms), and of 1-hexadecanol and pentadecanoic acid, both on 0.01 M HCl. The surface viscosity was found to be linearly proportional to the two-dimensional film pressure well below the 'plastic' region. Similar experiments were carried out in more detail on the corresponding series of alcohols (Fourt and Harkins, 1938) and a detailed discussion of these and additional results was given by Boyd and Harkins (1939). The values of the surface viscosities for the acids were found to be an order of magnitude less than the ones for the corresponding alcohols, and the values were strongly dependent upon the film pressure and temperature. The values lie between 5×10^{-4} and 5×10^{-2} dyne · s/cm for the acids in the 'liquid' region where the viscosity is Newtonian. If the surface viscosity is approximated by $\eta_s = \eta d$, where d, the thickness of the surface layer (see above discussion of the hydrodynamic model) is 10 Å, the corresponding three-dimensional viscosities are 500–50000 poise, which is considerably larger than those for the corresponding hydrocarbons or water!

The temperature dependence of the surface viscosities were also investigated. Boyd and Harkins (1939) analyzed their data in terms of the description of Moore and Eyring (1938). Moore and Eyring (1938) adapted a previous description of the fluidity of a liquid (Eyring, 1936; Ewell and Eyring, 1937; Glasstone et al., 1941) to two-dimensional liquid films, and compared the theory to the results of Meyers and Harkins (1937, 1938) and Langmuir and Schafer (1937). In the first approximation this approach results in the following equation,

$$\phi = (a/h)\exp(-\Delta F^{\ddagger}/kT) \tag{31}$$

where ϕ is the fluidity ($1/\eta$), h is Plancks constant, $\Delta F^{\ddagger}$ is the free energy of activation for viscous flow, and a is the cross-sectional area of the units which flow. This model assumes, in complete contrast to the model of Cohen and Turnbull (1959), that the controlling process for transport is the energy required to produce a hole in the liquid rather than the redistribution (entropy) of the free volume in the liquid. Also, the theory of viscosity developed by Eyring assumes that all the free space in the liquid is within holes of identical sizes, whereas Cohen and Turnbull (1959) have taken a much more realistic view that the free volume (or area) fluctuates, but that the transport of material only happens if the space is large enough. However, the viscosities of many liquids are found to follow an exponential dependence upon the temperature, especially at temperatures far enough above the glass temperature. In this model the pressure dependence is calculated from the free energy required to make a hole, which for a mole of holes becomes, $(\alpha/\beta)T\Sigma$, where Σ is the total area of one mole of holes ($\Sigma = N\delta s$, δs is the area of one hole), and α and β are the coefficients of thermal expansion and compressibility of the liquid films (Moore and Eyring, 1938). If one assumes that this term is the only contribution of the free energy which depends upon temperature and pressure (per unit length) and, if the variation of α and β is known, then the variation of the viscosity should be predictable if the theory is correct. Moore and Eyring (1938), Glasstone et al. (1941) and Davies and Rideal (1961) state that the bulk liquid adjacent to monolayers (usually water or low concentrations of inorganic acids) probably contributes significantly to the surface viscosity. This is assumed to be due to the solvation of the polar groups of the acids or alcohols, and it is suggested by Davies and Rideal (1961) that the film of water at the surface forms a hydration layer of 'soft ice'. The latter conclusion is based on the fact that the hydrocarbon chains do not cohere in the oil phase and should not contribute much to the surface viscosity which, as noted above, is quite large. The reader is referred to the following more recent references, Joly (1972), Goodrich (1973) and Harkins (1952).

d. Significant structure theory

Another theory of the liquid state which has been able to explain a large body of equilibrium and transport data for a wide variety of liquids, is the 'significant structure' theory of Eyring et al. (1958) (see Jhon and Eyring (1971) for references). This theory has been criticized for being too modelistic and for requiring too many parameters. However, since a two-dimensional account has been presented (Wang et al., 1965) and a treatment accounting for transport properties is available (Ree et al., 1964) which combines the concepts of free volume and of an energy of activation, we will discuss it briefly. The name of the model signifies the attempt to identify the significant structures of a medium which are fundamental for describing it as a liquid. Three significant structures are chosen: (1) molecules which are solid-like, i.e., restrained to an equilibrium position; (2) a positional degeneracy of the solid-like molecules due to the additional vacancies in the liquid; and (3) the molecules which have escaped the potential well of their neighbours are gas-like. It was explicitly stated by Eyring and Ree (1961) that they do not regard the liquid state as a mixture of solid and gas. The state of any particle is ephemeral and there are no large solid or gaseous molecular arrangements such as nucleation sites. This division into solid- and gas-like properties is reminiscent of the Cohen-Turnbull free-volume theory discussed above, however, the basic principles of both theories are quite different. The equilibrium properties of the liquid are derived from the partition function, which is a weighted hybrid of the solid-like and gas-like partition functions. The degeneracies are expressed in terms of the volume fractions of vacancies (holes) and molecules in the liquid, and this is where the free-volume enters. The transport properties are calculated using the earlier theory of Eyring (Glasstone et al., 1941) where the activation energy is inversely proportional to the free volume for the solid-like portion and the viscosity is the weighted sum of the gas- and solid-like viscosities. The diffusion coefficient can be calculated by methods described by Eyring and Ree (1961).

This model could provide an additional framework for describing the self-diffusion of lipids, where the activation energy of Pace and Chan (1982) could be included and the two-dimensional character of the viscosity taken into account. This would provide another description of the viscosity within lipid bilayers including the free volume (or area), which appears to be necessary to describe the data for the diffusion of small molecules.

e. Interfacial diffusion

The properties of interfaces, which are laminae or films of characteristic thickness separating two homogeneous phases, have been extensively studied, and the characteristics of the surface phase show properties differing from the bulk phases on either side of the surface (see, for instance, Davies and Rideal, 1961). Especially interesting for the present discussion are the investigations involving additives which adsorb at the interface between two dissimilar fluids forming monolayers, and the measurements of the surface viscosities (Davies and Rideal, 1961; Boyd and Harkins, 1939) also mentioned above.

Brenner and Leal (1977, 1978, 1982) have proposed a microscopic theory of the vicinity of the interface from which the surface constitutive laws of a macroscopic theory can be derived rather than guessed at. Thus, the surface diffusivity, which is a macroscopic term, is given a rigorous derivation in terms of the bulk (three-dimensional) transport parameters. Terms, such as the surface diffusion, which are normally introduced as new concepts when dealing with interfaces between two bulk phases, are expressed as integrals involving the micromechanical *bulk* properties over distances close to the interface. This avoids the *ad hoc* introduction of jump, or discontinuous conditions, and gives a more fundamental molecular view of the physical processes involved. They present a novel discussion of surface diffusion, whereby the diffusive flux of particles parallel to the fluid-fluid interface is greater in the presence of an attractive force potential than with an inert surface. If the potential force is strong enough this excess diffusive flux will become considerable, and if the potential has a deep narrow minimum (in a direction perpendicular to the interface) in the neighbourhood of the surface the integrals defining the average excess surface diffusivity will result in a surface diffusivity equal to the bulk diffusivity of the particle which is constrained at the surface. This limit corresponds to the situation treated by the hydrodynamic theories above, however, without the anisotropic hydrodynamics within the thickness of the bilayer. That is, only the interactions of the external solutions with the particle determine the transport properties and the theory would apply to bilayers only if the membrane sheet offers no resistance to flow.

A related model by Jones et al. (1975) treats the effect of the particle geometry upon the surface diffusion under such conditions. This is obviously not the correct picture for proteins which are permanently integrated in the bilayer, however, this approach is useful for describing the excess surface transport of proteins which associate with the membrane externally. The macroscopic equations governing the interface effects can be derived from a detailed microscopic view of the association process. In this case the membrane offers no direct frictional resistance to the transport, and the movement along the surface can be much more rapid than expected if the protein were to be integrated into the membrane.

f. Computer molecular dynamics

It has recently become feasible (Hoover et al., 1983; Hoover, 1984) to carry out realistic polyatomic deterministic simulations of molecular dynamics. The work dealing with monolayers (Kox et al., 1980, and references therein) and in bilayers (van der Ploeg and Berendsen, 1982, Edholm et al., 1983) have indicated the potential information which is available from these calculations. This has not yet been applied to investigate the transport properties within mono- or bilayers; however, this is definitely a worthwhile future goal.

5. THEORETICAL CONCLUSIONS

The only models which have been extensively applied to lipid bilayer membranes are the hydrodynamic model as solved by Saffman (1976) and Hughes (1981), and the free-volume model of Cohen and Turnbull (1959) which was adapted to the two-dimensional membrane bilayer system by Galla et al. (1979) and Clegg and Vaz (1984). We have suggested some specific extensions of these theories and indicated several approaches which have been successfully applied in related fields in the hope that some of these ideas will be helpful in future interpretations, or as suggestive guides for experiments. The application of the hydrodynamic model to describe diffusion of particles with a size larger than three-times that of the lipid molecules seems reasonable from a theoretical point of view and this restriction is in agreement with simple liquid theory (Alder et al., 1981). The viscosity of the membrane sheet which is referred to in the hydrodynamic expressions for the diffusion coefficient is not well defined. Expressions for the surface viscosity derived from the free-volume model, or from Eyrings model, are more desirable than simply assuming an Arrhenius dependence of a three-dimensional viscosity coefficient. The free-volume or free-area model is more appropriate for the diffusion of particles with diameters comparable to those of the lipids, or smaller. However, we hoped to make it clear in the discussion above that all these representations are very modelistic and that each viewpoint expresses an idealized opinion of the important factors which control the dynamic behavior.

The situation is similar to that described by Hildebrand (1971) in his evaluation of the "motions of molecules in liquids: viscosity and diffusivity", except that the present problem is more complicated and the necessary experimental techniques are unfortunately more sophisticated and less direct. Hildebrand emphasizes the importance of experiments on well defined systems which can be compared to the available theories, and notes the inability of these theories to convincingly account for the data. He stresses the simple correlation of the viscosity of many liquids with the relative free volume and shows that the fluidity of many liquids is directly proportional to the relative free volume and has the same linear constant. However, even though such simple relationships exist, also for long chain hydrocarbons, the theories are all unsatisfactory since they incorporate concepts of different material phases (solids and gases) to the liquid. There is no justification for doing this other than the fact that the more fundamental approaches (Gray, 1968) which seem to be more satisfactory *a priori* are not amenable to solution without drastic simplifications. One is forced to apply models which greatly reduce the difficulty of the problem but involve concepts which do not necessarily correspond to reality. This state of affairs, however, makes the subject attractive from both a theoretical and experimental point of view since there is obviously much to be done.

III. Experimental results

In this section we shall restrict our attention to studies of translational diffusion in model phospholipid membrane systems. Since these systems are, in principle, completely under the control of the investigator with regard to their compositional complexity, they are most ideally suited to test the applicability of the theoretical models to diffusion in membranes. Most model membrane studies have an ultimate interest in elucidating the diffusive behaviour in biological membranes and acquiring information about their effective 'viscosity'. From this point of view, translational diffusion in the phospholipid bilayer in its liquid crystalline, L_α, phase is of greatest interest and will be emphasized here. Several recent reviews (Webb, 1981; Schlessinger and Elson, 1981; Peters, 1981; Vaz et al., 1982a) have covered the literature on studies of translational diffusion in biological and model membranes up to the end of 1981. Here we shall briefly review the literature on studies of translational diffusion in model systems that have appeared in the literature since then and proceed to compare the results with the predictions of the theoretical models.

1. TRANSLATIONAL DIFFUSION OF INTEGRAL PROTEINS

Several laboratories have reported studies on the translational diffusion of integral membrane proteins which have been reconstituted into phospholipid bilayers. Although most of the proteins whose diffusion has been studied have a measurable function in their native membranes, practically none of the studies on diffusion have attempted to first verify the retention of this function in the recombinants. We shall assume here that the proteins have essentially retained their native conformations in the reconstituted membranes notwithstanding the treatments they have undergone in the process of reconstitution.

Following the earliest reports on the diffusion of reconstituted integral proteins in artificial phospholipid bilayers (Smith et al., 1979, 1980; Vaz et al., 1981), several laboratories have reported studies on the translational diffusion of larger and more complex proteins incorporated into lipid bilayers. Chang et al. (1981) have reported on the translational diffusion of human erythrocyte Band 3 protein reconstituted into DMPC bilayers at a lipid/protein molar ratio of 200. The translational diffusion of this protein was examined between about 10 and 35°C. A rapid diffusion ($D_t \sim 1\text{–}2 \times 10^{-8}$ cm²/s) was seen in the lipid L_α phase and a slow diffusion ($D_t < 10^{-9}$ cm²/s) was observed in the gel phase. The change from fast to slow diffusing states occurred at ~ 22°C (midpoint of the transition), which is somewhat below the phase transition temperature of DMPC (23.9°C). This is also below the transition from fast to slow diffusing states seen in the same work for surface-attached proteins. Peters and Cherry (1982) examined the translational and rotational diffusion of bacteriorhodopsin reconstituted into bilayers of DMPC at lipid/protein molar ratios of 210, 140, 90 and 30. Rapid translational diffusion of the reconstituted protein ($D_t \sim 1\text{–}4 \times 10^{-8}$ cm²/s) was seen in the first three recombinants above 25°C. In the recombinant with a lipid/protein molar ratio of 30, D_t was ~ 1×10^{-9} cm²/s above 25°C. In all cases D_t dropped gradually to a valuc bclow 10^{-10} cm²/s below 25°C. Vaz and co-workers (Vaz et al., 1982b; Criado et al., 1982) have examined the translational diffusion of bovine rhodopsin, the acetylcholine receptor protein of *Torpedo marmorata*, and the sarcoplasmic reticulum Ca^{2+}-activated adenosine triphosphatase in L_α phase phospholipid bilayers at lipid/protein molar ratios of 3000 or higher. D_ts for all these proteins were found to be in the range of $1\text{–}4 \times 10^{-8}$ cm²/s in the L_α phase. In the case of bovine rhodopsin which was also examined in the gel phase of DMPC, D_t was found to drop from a value of ~ 2×10^{-8} cm²/s at 25°C to a value below 10^{-9} cm²/s at 23°C. The recovery of fluorescence after photobleaching in the latter case was, however, incomplete, indicating a non-homogeneous diffusing protein population.

Tank et al. (1982) have reported on the translational diffusion of a gramicidin C dimer (referred to as gramicidin hereafter) in DMPC and egg yolk PC bilayers at lipid/protein molar ratios between 5000 and 7. No effect of gramicidin concentration upon D_t for gramicidin was observed in DMPC bilayers at 30°C between lipid/protein molar ratios of 5000 and 100, but D_t, dropped from about 4×10^{-8} cm²/s at a molar ratio of 100, to about 1×10^{-8} cm²/s at a molar ratio of 7. D_t as a function of temperature in DMPC was shown to be biphasic with rapid diffusion ($D_t \geqslant 2 \times 10^{-8}$ cm²/s) above 23°C and a sharp transition at about 23°C to a slow diffusion ($D_t < 10^{-10}$ cm²/s) below this temperature. In egg yolk PC gramicidin translational diffusion was rapid with a monotonic decrease in D_t from about 5×10^{-8} cm²/s at about 35°C to about 2×10^{-8} cm²/s at about 5°C. The presence of 20 mol% cholesterol in the DMPC bilayers abolished the sharp transition of the diffusion behaviour of gramicidin at 23°C. In egg yolk PC bilayers a monotonic decrease in D_t for gramicidin was observed with increasing cholesterol concentration.

All of the above intrinsic peptides and proteins have radii in the membrane plane of between 0.6 nm for gramicidin (Tank et al., 1982) and 3.0 nm for the sarcoplasmic reticulum adenosine triphosphatase (Vaz et al., 1982b). In an attempt to increase the radius of the diffusing protein particle in bilayers, Vaz and Criado (1984) examined the translational diffusion of a covalently cross-linked tetramer of the acetylcholine receptor protein in bilayers of soybean phospholipids. D_t for the tetrameric protein was found to be about equal to D_t for the monomer between 10 and 35°C. A similar attempt was made with glycophorin which can be cross-linked by the lectin, wheat germ agglutinin (Vaz, W.L.C., unpublished results). In this case there is no good estimate of the size of the agglutinated glycophorin but extensive cross-linking was inferred from triplet anisotropy decay measurements (Matayoshi, E., and Vaz, W.L.C., unpublished results). D_t for the agglutinated glycophorin in DMPC bilayers was found to be only 20% lower than that for non-agglutinated glycophorin under similar conditions. Table 1 lists some results of the translational diffusion of several integral membrane proteins in reconstituted phospholipid bilayers.

TABLE 1
Translational diffusion coefficients for some proteins in reconstituted phospholipid bilayers

Protein/peptide diffusant	Host bilayer lipid	Lipid/protein molar ratio	Temp. (°C)	D_t (cm²/s)	Ref.
Membrane-spanning (Integral) Proteins					
Gramicidin C	DMPC	>100	21	6.6×10^{-11}	1
			25	3.0×10^{-8}	
			34	4.3×10^{-8}	
	egg PC	not specified	15	2.9×10^{-8}	1
			34	4.9×10^{-8}	
M-13 Viral coat protein	DMPC	2000	25	3.0×10^{-8}	2
			35	4.8×10^{-8}	
Glycophorin	DMPC	4500	15	$<5.0 \times 10^{-11}$	3
			20	1.1×10^{-8}	
			25	1.3×10^{-8}	
			35	1.7×10^{-8}	

Bacteriorhodopsin	DMPC	210	25	1.4×10^{-8}	4
			32	3.3×10^{-8}	
		140	25	1.1×10^{-8}	
			32	2.2×10^{-8}	
		90	25	8.0×10^{-9}	
			32	1.3×10^{-8}	
		30	25	1.1×10^{-9}	
			32	1.4×10^{-9}	
Bovine rhodopsin	DMPC	3000	24	1.6×10^{-8}	5
			36	3.3×10^{-8}	
SR Ca^{2+}-adenosine triphosphatase	SR Lipids	5000	13	9.9×10^{-9}	5
			36	1.8×10^{-8}	
Human erythrocyte Band 3 protein	DMPC	not specified	12	1.5×10^{-10}	6
			20	5.4×10^{-10}	
			25	1.2×10^{-8}	
			35	1.8×10^{-8}	
Acetylcholine receptor (monomer)	DMPC	>5000	25	1.0×10^{-8}	7
			36	2.3×10^{-8}	
	Soybean Lipids	>5000	15	1.5×10^{-8}	7
			37	3.3×10^{-8}	
Acetylcholine receptor (dimer)	Soybean Lipids	>5000	15	1.6×10^{-8}	7
			37	3.1×10^{-8}	
Acetylcholine recepter (tetramer)	Soybean Lipids	5000	15	1.3×10^{-8}	8
			35	3.1×10^{-8}	
Surface-attached (Extrinsic) Proteins					
Apolipoprotein ApoC-III	DPPC	1000	16	1.0×10^{-9}	9
			30	1.0×10^{-8}	
			36	6.7×10^{-8}	
			47	9.7×10^{-8}	
	egg PC	1000	13	2.0×10^{-8}	9
			34	5.4×10^{-8}	
Human erythrocyte spectrin	DMPC	not specified	10	1.2×10^{-10}	6
			22	1.6×10^{-10}	
			25	4.1×10^{-8}	
			35	6.9×10^{-8}	
Human erythrocyte Band 4.1 protein	DMPC	not specified	9	1.2×10^{-10}	6
			17	1.8×10^{-10}	
			25	4.2×10^{-8}	
			35	6.2×10^{-8}	

Refs.: 1, Tank et al. (1982); 2, Smith et al. (1979); 3, Vaz et al. (1981); 4, Peters and Cherry (1982); 5, Vaz et al. (1982); 6, Chang et al. (1981); 7, Criado et al. (1982); 8, Vaz and Criado (1984); 9, Vaz et al. (1979).

2. TRANSLATIONAL DIFFUSION OF EXTRINSIC PROTEINS

Following the initial studies of Vaz et al. (1979) on the translational diffusion of an apolipoprotein, ApoC-III, bound to phosphatidylcholine bilayers with and without cholesterol, the only reported studies on the translational diffusion of surface-attached or extrinsic proteins bound to artificial lipid bilayers have been those of Chang et al. (1981), who have examined the diffusion of human erythrocyte spectrin and Band 4.1 protein bound to DMPC bilayers. These workers showed that D_t for both spectrin and Band 4.1 protein on DMPC was about the same as D_t for lipid probe diffusion in the same bilayers. Both proteins show $D_t \geqslant 5 \times 10^{-8}$ cm²/s above the lipid phase transition temperature, with a sharp drop at 23°C to values of D_t of about 10^{-10} cm²/s below this temperature. These results are similar to those reported earlier for ApoC-III diffusion at the surface of DPPC and egg yolk PC bilayers (Vaz et al., 1979). A summary of the results on the diffusion of surface-attached proteins in artificial phospholipid bilayers is presented in Table 1.

3. TRANSLATIONAL DIFFUSION OF LIPIDS AND LIPID DERIVATIVES

The study of the translational diffusion of lipids in bilayer membranes can be done using magnetic resonance techniques (Traeuble and Sackmann, 1972; Devaux and McConnell, 1972; Lindblom and Wennerstroem, 1977; Kuo and Wade, 1979) or optical techniques (Vaz et al., 1982a). The latter techniques particularly the 'fluorescence recovery after photobleaching' (FRAP) method (Axelrod et al., 1976) has been by far the more productive with regard to information on the translational diffusion of lipid-like molecules in lipid bilayers. Consequently, all the results we shall discuss below have been obtained by the FRAP method. This requires the use of a fluorescent diffusant which is usually a fluorescent derivative of a phospholipid such as NBD-PE or some other amphipathic fluorophore such as diI or diO. From structural considerations (see, for example, Vaz et al. 1982a) it is clear that NBD-PE is likely to mimic the behaviour of the phospholipids of the bilayer more closely than diI and diO.

In Fig. 3 we show the temperature dependence of the translational diffusion of some lipid probes in bilayers of DMPC in excess water in the gel and L_α phases. The data are from different laboratories. As seen in Fig. 1, the L_α phase is characterized by a rapid translational diffusion of the lipids (we assume here that the diffusion of the host bilayer lipid molecules is equivalent to that of the probes) with D_t in the range of $\sim 5 \times 10^{-8}$ cm²/s. D_t drops by a factor of ~ 100-fold at the phase transition temperature of the host bilayer and the gel phase is characterized by a lipid $D_t \leqslant 10^{-10}$ cm²/s. The same sort of behaviour is seen in other phosphatidylcholine bilayers and in bilayers of bovine brain phosphatidylserine, other lipid systems not having been studied. There seems to be no disagreement between the results from different laboratories as to the value

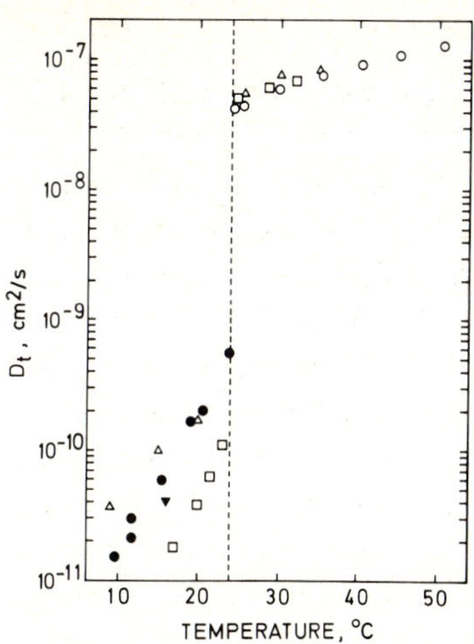

Fig. 3. Temperature dependence of the translational diffusion of lipid derivatives and lipid-like amphipathic probes in bilayers of DMPC in excess water. The broken lines indicate the main lipid bilayer phase transition temperature according to Mabrey and Sturtevant (1976). Data are for: (○) NBD-DMPE in DMPC multibilayers at a DMPC/NBD-DMPE molar ratio of 2000 (Vaz et al., 1984a); (□) dioctadecyl-diO in DMPC multibilayers at a DMPC/diO molar ratio of 100 (Peters and Cherry, 1982); (△) NBD-egg PE in DMPC multibilayers (Chang et al., 1981); (•) NBD-egg PE in DMPC multibilayers at a DMPC/NBD-PE molar ratio of 200 (Smith and McConnell, 1978); (▼) dioctadecyl-diI in DMPC multibilayers at a DMPC/diI molar ratio of $\leqslant 10^4$ (Schneider et al., 1983). The last results are attributed by the authors to diffusion along linear disordered defect structures in the $P_{\beta'}$ phase DMPC bilayer, the value of D_t in ordered structures of this phase being estimated at $\leqslant 10^{-16}$ cm²/s.

of D_t in the L_α phase but considerable disagreement as to its value in the gel phase. It has been shown (see, for example, Derzko and Jacobson, 1980) that lipid translational diffusion in the L_α phase is isotropic in the plane of the membrane but that diffusion in the gel phase may not be. Specifically, in the $P_{\beta'}$ phase (the gel phase existing between the main and pretransition temperatures) Derzko and Jacobson (1980) were able to describe their fluorescence recovery curves by the assumption of two diffusing components: a fast component with D_t between 10^{-8} and 10^{-10} cm²/s and a slow component with $D_t \leqslant 5 \times 10^{-11}$ cm²/s depending upon the host bilayer, the diffusing probe, and the temperature. The fast diffusion was attributed to one-dimensional diffusion along linear defects in the gel phase bilayers. The slow diffusion was suggested to be representative of the lipid diffusive behaviour in bulk gel phase domains. More recently Schneider et al. (1982) have examined the translational diffusion of dioctadecyl-diI in $P_{\beta'}$ phase bilayers of DMPC. Their results were interpreted in terms of a submicroscopic disordered band model for $P_{\beta'}$ phase bilayers. In this model the 'rippled' bilayer has bands of disordered lipid separated by bands of ordered lipid and occurring with a repeat distance of 7.0 nm. The disordered bands are coincident with the regions of high curvature in the rippled $P_{\beta'}$ structure. Rapid diffusion (in this case $D_t \sim 4 \times 10^{-11}$ cm^2/s in DMPC at 16°C) was attributed to diffusion along the disordered bands, whereas the diffusion in the ordered bands was not measurable and estimated to have an upper limit of $\leqslant 10^{-16}$ cm^2/s. Schneider et al., (1982) did not encounter the diffusive process wih $D_t \sim 10^{-8}$ cm^2/s seen by Derzko and Jacobson (1980).

The earlier work on lipid diffusion in bilayer membranes has been reviewed elsewhere (Vaz et al., 1982a). Most of this work was concerned with obtaining reproducible values for D_t in lipid bilayers. More recent results have been obtained with the explicit intention of evaluating the applicability of theoretical models for diffusion in membranes (Vaz and Hallmann, 1983; Vaz et al., 1982b, 1984a; Peters and Cherry, 1982; Peters and Beck, 1983) or evaluating the effect of protein concentration on the diffusive behaviour of proteins and lipids in artificial membranes (Jacobson et al., 1981; Peters and Cherry, 1982; Tank et al., 1982). These results are discussed below.

In an attempt to evaluate various predictions of the different models described earlier, we have undertaken a systematic study on the translational diffusion of NBD-PE analogous with varying acyl chain lengths in several phosphatidylcholine bilayers in the liquid crystalline phase (Vaz and Hallmann, 1983; Vaz et al., 1984a). It was shown (Vaz and Hallmann, 1983) that D_t had, at the most, a weak dependence upon acyl chain length of the diffusant. It was also shown (Vaz et al., 1984a) that the values of D_t for NBD-PEs whose acyl chain lengths and structures were matched to those of the host bilayer lipid were indistinguishable within experimental error on an absolute temperature scale in L_α phase bilayers of DLPC, DMPC, DPPC and DSPC. D_t for NBD-POPE in POPC bilayers was, however, about 30% lower than D_t at the same absolute temperature for matched NBD-PEs in saturated acyl chain PC bilayers. The data for diffusion in saturated acyl chain PC bilayers plotted on a 'reduced temperature' scale (the reduced temperature is defined as $T_r = (T - T_m)/T_m$ where all temperatures are given in degrees Kelvin) showed that at the same reduced temperature the magnitude of D_t followed the order DSPC > DPPC > DMPC > DLPC = POPC, *i.e.*, at equivalent reduced temperatures diffusion was fastest in the thickest bilayers. In other work (Vaz et al.,1984b) we examined the translational diffusion of an NBD derivative of a membrane-spanning diglycerophosphoethanolamine tetraether lipid synthesized from a membrane-spanning tetraether lipid extracted from archaebacteria (De Rosa et al., 1983) in L_α phase bilayers of POPC. D_t for the membrane-spanning lipid was found to be 2/3 of D_t for NBD-POPE in the same bilayers at every temperature examined between 10 and 50°C.

In a recent study (Beck and Peters, 1982; Peters and Beck, 1983) the translational diffusion of NBD-egg PE was examined in monolayers of DLPC at the air-water interface. In this study the lateral surface pressure was varied at a constant temperature of $\sim$ 21°C. It was shown (Peters and Beck, 1983) that log

(D_t) varied linearly with $1/a_f$, where a_f is the free area per lipid molecule (for a discussion, see Section *II.2.b*) in the monolayer. The translational diffusion in these monolayers was found to be considerably faster than in bilayers of DLPC (D_t in the monolayers at 21°C and lateral surface pressures of 10 and 30 dyne/cm was found to be $\sim 50 \times 10^{-8}$ and $\sim 25 \times 10^{-8}$ cm^2/s, respectively, compared with D_t of $\sim 5 \times 10^{-8}$ cm^2/s in DLPC bilayers at 20°C (Vaz et al., 1984a). Such rapid diffusion in monolayers at an air-water interface had been also been reported earlier for anthroyl fatty acids in monolayers of DPPC and DPPG (Teissie et al., 1978). Von Tscharner and McConnell (1981) have examined the translational diffusion of NBD-DMPE in monolayers of DPPC adsorbed onto an alkylated glass surface. This experimental system is similar to a lipid bilayer except that the area per lipid molecule in the adsorbed lipid monolayer can be controlled by the investigator simply by adsorbing the lipid monolayer at different lateral surface pressures. The results of von Tscharner and McConnell (1981) differ from those obtained on monolayers at an air-water interface in two important respects: (1) D_t in the adsorbed monolayers is at least 10-fold lower than D_t in the monolayers at an air-water interface under comparable conditions of molecular area. This difference may be due to the additional frictional drag forces acting upon the adsorbed monolayer due to its contact with the alkylated glass. It could also be due to increased ordering of the acyl chains in the adsorbed monolayers relative to the monolayers at the air-water interface. Perhaps the value of D_t in this case is also dependent upon the 'fluidity' of the alkylated surface. It is also interesting to note in this regard that D_t in the adsorbed monolayers is about 5-fold lower than D_t in lipid bilayers. (2) The adsorbed monolayers show no dependence of D_t upon molecular area (von Tscharner and McConnell, 1981) unlike D_t for monolayers at the air-water interface (Peters and Beck, 1983; see, however, Beck and Peters, 1982).

4. COMPARISON OF EXPERIMENT WITH THEORY

First we compare the experimental results for protein diffusion with the hydrodynamic model. Hydrodynamic theory (Section *II.1.*) provides the only appropriate model for the diffusion of particles, such as proteins in membranes, which are considerably larger in size than the molecules which make up the fluid. For the case of lipids diffusing in lipid membranes other models may be more appropriate.

Perhaps the most striking aspect of the model of Saffman and Delbrueck (1975) is the weak dependence of D_t upon the diffusant radius: $D_t \propto \ln(1/R_c)$, Eqn. 6. This prediction has, in fact, been experimentally borne out for the translational diffusion of integral membrane proteins in artificial lipid bilayers (Vaz et al., 1982b). A problem in this work was that the range of protein radii examined (~1.5 nm to ~3.0 nm), while supporting a conclusion that the Saffman-Delbrueck model adequately describes integral protein diffusion in lipid bilayers, was not sufficiently large to allow a critical evaluation of the theory. However, to our knowledge, there are few if any integral membrane proteins with radii larger than 3.0 nm which do not spontaneously aggregate to form ordered structures in bilayers. One way around this problem is to examine the translational diffusion in membranes of covalently linked integral protein oligomers. This was done in the case of the acetylcholine receptor protein (Vaz and Criado, 1984) and glycophorin (Vaz, W.L.C., unpublished results). The fact that the acetylcholine receptor tetramer diffuses at about the same rate as the monomer (see Table 1) and that an aggregated form of glycophorin has D_t only 20% lower than D_t for the non-aggregated glycophorin, clearly can be considered as a critical evaluation of the Saffman-Delbrueck model. In Fig. 4 we show the agreement of the experimental results for integral protein diffusion (see Table 1) with the theoretical predictions of Saffman and Delbrueck (1975). As seen, the agreement is reasonably good if the membrane viscosity is assumed to be between 1 and 2 poise.

So far we have considered only the large integral membrane proteins whose radii in the plane of the membrane $\geqslant$ 1.5 nm. The translational diffusion of several smaller proteins has also been examined (Smith et al., 1979; Vaz et al., 1981; Tank et al., 1982). If these proteins are considered as single peptide helices that span the membrane, their radii in the membrane plane is ~ 0.6 nm. This is not very much larger than the radius of a lipid molecule. It is, therefore, not clear whether a discussion of their diffusion in membranes should involve the same hydrodynamic considerations as used above for the larger integral proteins. We have included the data for gramicidin and the M-13 viral coat peptide in Fig. 4. As seen, the experimentally obtained values of D_t for gramicidin and the M-13 viral coat peptide are about the same as the hydrodynamic theory would predict for proteins of this size. D_t for glycophorin (not included in Fig. 4) is considerably lower but this may be due, among other things, to oligomerization of the protein in the membrane or due to interactions that the large hydrophilic portions of this protein may have with the polar head groups of lipids in the bilayer.

Another way to test the hydrodynamic model is by comparing the translational and rotational mobilities of a diffusant in a membrane (Saffman and Delbrueck, 1975):

$$D_t/D_r = b_t/b_r = R_c^2\{\ln(\eta_0 h/\eta' R_c) - \gamma\} \tag{32}$$

where D_r, b_r and b_t are the rotational diffusion coefficient, the rotational mobility, and the translational mobility, respectively. For the case of a sphere of radius R_s diffusing in an unbounded three-dimensional fluid, the corresponding Stokes-Einstein equations give

$$D_t/D_r = b_t/b_r = 4R_s^2/3 \tag{33}$$

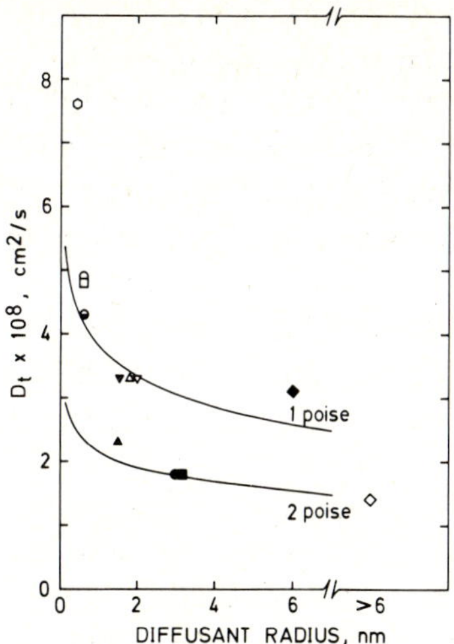

Fig. 4. A comparison of experimental results for the translational diffusion of membrane-spanning proteins in reconstituted liquid crystalline phase lipid bilayers with the size dependence for such diffusion predicted by the model of Saffman and Delbrueck (1975). The lines are theoretical size dependences predicted for η_0 values of 1.0 and 2.0 poise at 35°C. The values of h and η' were assumed to be 5×10^{-7} cm and 0.01 poise, respectively. Values of the protein radii were taken from the literature. The results are for: gramicidin C in DMPC bilayers at 34°C (◒), and in egg PC bilayers at 34°C (○) (Tank et al., 1982); M-13 viral coat peptide in DMPC bilayers at 35°C (□) (Smith et al., 1979); bacteriorhodopsin in DMPC bilayers at a lipid/protein molar ratio of 210 and at 32°C (△) (Peters and Cherry, 1982); bovine rhodopsin in DMPC bilayers at 36°C (▽) (Vaz et al., 1982b); sarcoplasmic reticulum Ca^{2+}-adenosine triphosphatase in a total lipid extract of sarcoplasmic reticulum at 36°C ((●) (Vaz et al., 1982b); erythrocyte Band 3 protein in DMPC bilayers at 35°C (□) (Chang et al., 1981); acetylcholine receptor protein monomer in DMPC bilayers at 36°C (▲) and in soybean phospholipids at 37°C (▼) (Criado et al., 1982); acetylcholine receptor protein tetramer in soybean phospholipids at 35°C (◆) (Vaz and Criado, 1984); and wheat germ agglutinin-aggregated glycophorin in DMPC bilayers at 36°C (⬡) (Vaz, W.L.C., unpublished results). For comparison, the D_t for NBD-DMPE in DMPC bilayers at 35°C (◇) (Vaz et al., 1984a) has been included with the assumption that the lipid molecule has a radius of 0.4 nm.

The difference between the Saffman-Delbrueck equations and the Stokes-Einstein equations is, therefore, that the mobility ratio is proportional to the logarithm of the membrane and bounding fluid viscosity ratio. Since the viscosity ratio is about 100, the theory predicts $D_t/D_r \sim 4R_c^2$. Whether this is a good test of the hydrodynamic model or not will depend upon whether the values of η_0 are identical for translational and rotational motions. A deviation from the expected viscosity ratios may be an indication that the dissipative mechanisms are different for translation and rotation. In carrying out this test it should also be remembered that large errors are inherent in mobility measurements in membranes. This is seen in the following examples. Saffman and Delbrueck (1975) used the measured rotational and translational diffusion coefficients for rhodopsin in rod outer segment membranes (Cone, 1972; Poo and Cone, 1974) for a comparison of their model with experiment. Assuming R for rhodopsin to be about 2.0 nm (Yeager, 1976; Osborne et al., 1978), h, η_0, and η' to be 5.0 nm, 1.0 poise, and 0.01 poise, respectively, the theoretical value of D_t/D_r is $\sim 20 \times 10^{-14}$ cm^2. The experimental mobility ratio, given that $D_t = 3.5 \pm 1.5 \times 10^{-9}$ cm^2/s (Poo and Cone, 1974) and $D_r = 5 \times 10^4$ s^{-1} within a factor of 2 (Cone, 1972), is between 2 and 20 $\times 10^{-14}$ cm^2. A similar test can be attempted for results on reconstituted systems. D_t for bacteriorhodopsin at a lipid/protein molar ratio of about 210 in DMPC bilayers at 32°C has been given by Peters and Cherry (1982) to be $3.4 \pm 0.2 \times 10^{-8}$ cm^2/s and D_r for this protein under identical conditions has been given by Cherry and Godfrey (1981) to be $7.5 \pm 2.5 \times 10^4$ s^{-1}. Bacteriorhodopsin has dimensions of 4.5 (height) $\times$ 25 $\times$ 35 nm (Henderson and Unwin, 1975). If we assume the maximum radius (R = 1.75 nm), the membrane thickness equal to the height of the protein (4.5 nm), and η_0 and η' to be 1 poise and 0.01 poise, respectively, the theoretical mobility ratio is $\sim 15 \times 10^{-14}$ cm^2 and the experimental mobility ratio lies between 32 and 72 $\times 10^{-14}$ cm^2. Still another system that can be tested is that of erythrocyte Band 3 protein reconstituted into DMPC bilayers. At 35°C, Chang et al. (1981) give D_t for this protein as 1.8×10^{-8} cm^2/s, and the same laboratory (Sakaki et al., 1982) estimates D_r for Band 3 in DMPC at 37°C to be 3.9×10^4 s^{-1}. Assuming this protein to have a radius of 3.0 nm (Weinstein, 1974) and using the same values of h, η_0, and η' as before, D_t/D_r (theoretical) = 41 $\times$ 10^{-14} cm^2 and D_t/D_r (experimental) = 46 $\times 10^{-14}$ cm^2. Peters and Cherry (1982) have attempted to use this test further by adding sucrose to the aqueous phase in their reconstituted multibilayers. However, X-ray diffraction and differential scanning calorimetric studies in our laboratory (Vaz, W.L.C., and Stuempel, J., unpublished results) indicate that the addition of sucrose to the aqueous phase of multibilayer lipid-water systems decreases the interlamellar spacing, slightly increases the main lipid bilayer phase transition temperature, and reduces the difference between the main and pretransition temperatures. This is characteristic of dehydration of the multibilayers, a phenomenon that is known to increase the order and decrease D_t (at least for lipids) in these systems (McCown et al., 1981). We, therefore, suggest that the sucrose does not actually penetrate the interlamellar aqueous space but simply dehydrates the multilamellar system. In this case, it may be only coincidental that the results obtained by Peters and Cherry (1982), by addition of sucrose to their reconstituted system, are about what would be expected if the sucrose had increased the aqueous phase viscosity.

A case worthy of analysis is that of the acetylcholine receptor protein in soybean lipid bilayers. This protein was studied in its monomeric, dimeric (Criado et al., 1982), and oligomeric (Vaz and Criado, 1984) forms. The oligomers (predominantly tetramers) were prepared by *in situ* covalent cross-linking of the protein in native membranes (Criado and Barrantes, 1984), the reaction occurring in the hydrophilic head groups of the protein. From neutron-scattering studies it is known that the membrane-spanning portion of this protein can be approximated to a cylinder with a radius of ~ 1.5 nm (Wise et al., 1979). Similar studies from the same laboratory on a naturally occurring covalent dimer of this protein give a center-to-center distance between the two monomer units of ~ 8.5 nm (Wise et al., 1979). If we assume the most compact form of a tetramer, it must have roughly the same center-to-center separation distance between the monomer units. We therefore estimate that the tetramer has a radius of ~ 6.0 nm. In this structure the space between the membrane-spanning portions of the monomer units of the tetramer must be filled with lipid molecules. Using the approach of Wiegel (1980) for the diffusion of porous discs in thin fluid sheets, we consider the acetylcholine receptor tetramer to be a porous disc and estimate that the solvent penetration length in the disc is ~ 2.6 nm or roughly half the radius of the tetramer.

We now turn our attention to the diffusion of lipids in phosphatidylcholine bilayer membranes. *A priori* there is no good reason to describe the diffusive behaviour of lipids in a fluid sheet composed of other lipids by a hydrodynamic treatment (see, for example, Landau and Lifshitz (1959) and discussion in Section *II.1.* above). There are several indications that lipid diffusion in bilayers may not be adequately described by the hydrodynamic model. (1) The experimental lipid D_t is considerably higher than the hydrodynamic equations predict, based upon the diffusion results for membrane-spanning proteins (see, for example, Fig. 4). Assuming $\eta_0 = 1$ poise, the equations of Saffman and Delbrueck (1975) are consistent with the experimental lipid D_t only when a frictionless sliding of the two monolayers of a bilayer lipid membrane over each other is assumed (Vaz et al., 1982b). It is expected that the order and 'viscosity' of the bilayer midplane are considerably lower than elsewhere in the bilayer (Seelig and Seelig, 1980; Tilley et al., 1979; Vincent et al., 1982), however, they are probably not zero. (2) The lipid D_t values in the L_α phase for different thicknesses of lipid bilayer (DLPC to DSPC) are the same (± 10%), indicating an insensitivity of D_t to h, contrary to the prediction of the hydrodynamic model (Vaz et al., 1984a). (3) On a reduced temperature scale, the magnitude of D_t in saturated acyl chain PC bilayers follows the order DSPC > DPPC > DMPC > DLPC. This means that at the respective reduced temperature the fastest diffusion is occurring in the thickest bilayer which is exactly the opposite of what might be expected if the hydrodynamic model were applicable to lipid diffusion in lipid bilayers. Some years ago, Sackmann and co-workers (Traeuble and Sackmann, 1972; Galla et al., 1979) proposed that diffusion in membrane could be understood in terms of a free area model for two-dimensional fluids based upon the three-dimensional free-volume model of Cohen and Turnbull (1959). We (Vaz et al., 1982a, b) also proposed that this model may be able to describe lipid diffusion behaviour in bilayers better than the model of Saffman and Delbrueck (1975). More recently, we (Vaz et al., 1984a) have provided experimental results which agree fairly well with a two-dimensional free-area model for diffusion of lipid-like particles in membranes, taking into account the frictional drag forces acting at the bilayer-water interface and the bilayer midplane (Clegg and Vaz, 1984). These conclusions are strengthened by the finding that the logarithm of D_t is linearly proportional to the reciprocal free area in dehydrated bilayers (McCown et al., 1981) and in PC monolayers at the air-water interface (Peters and Beck, 1983).

IV. Concluding comments

1. "If the reader has by now concluded that little is known about the prediction [of diffusivities in lipid bilayer membranes], he is correct. There is an urgent need for experimental measurements, both for their own value and for the development of future theories" (Bird et al., 1960). This citation, which was true for diffusivities in simple liquids in 1971 (Hildebrand, 1971), is certainly valid for lipid bilayers today.

2. We have attempted here to present the premises and assumptions upon which the theoretical models for diffusion in lipid bilayers are based. An interpretation of the recent data on translational diffusion in artificial bilayer systems has been undertaken with reference to the different models.

3. Diffusion of membrane-spanning proteins with large radii (> 1.0 nm in the plane of the membrane) seems to be described well by the hydrodynamic models. These models predict that, for lateral transport of large particles, the motion of the fluid elements of the membrane sheet are correlated over large distances within the membrane. The radius of dissipation is dictated by the dissipative interactions of the membrane sheet with the bounding fluid. The fact that this distance is large has consequences relevant to studies on biological membranes where extra- and intramembranal structures may interfere with the validity of application of the theory.

4. For protein diffusion in membranes it is not clear that the dissipative mechanisms for translational and rotational diffusion are the same. This means that membrane viscosity values derived from diffusion measurements should be considered as estimates, at best.

5. Lipid translational diffusion may be more appropriately described by either the free volume (or area) models or by the 'flexible rod activated jump' model. It should be experimentally testable which of these models provides a better description of lipid transport in a lipid bilayer and future experiments should provide an answer. Since lipid diffusion may emphasize properties of a bilayer that are not as important for protein diffusion, one should be careful in the use of viscosity values for membranes.

List of abbreviations

a_f	= free area.
a^*	= critical free area.
A	= preexponential factor.
A_0	= see Eqn. 18.
b	= see Eqn. 11.
b_r	= rotational mobility.
b_t	= translational mobility.
D_r	= rotational diffusion coefficient.
D or D_t	= translational diffusion constant (cm^2/s).
$D(v_f)$	= particle diffusion constant within the free volume v_f.
d'	= $y_1 + y_2$.
diI	= indocarbocyanine iodide.
diO	= oxacarbocyanine iodide.
DHPE	= dihexanoylphosphatidylethanolamine.
DLPC	= dilauroylphosphatidylcholine.
DLPE	= dilauroylphosphatidylethanolamine.
DMPC	= dimyristoylphosphatidylcholine.
DMPE	= dimyristoylphosphatidylethanolamine.
DPPC	= dipalmitoylphosphatidylcholine.
DPPE	= dipalmitoylphosphatidylethanolamine.
DSPC	= distearoylphosphatidylcholine.
DSPE	= distearoylphosphatidylethanolamine.
E^*	= activation energy/molecule.
E_d	= activation energy for diffusion.
f or f_T	= friction coefficient.
F	= force (dynes).
FRAP	= fluorescence recovery after photobleaching.
g	= geometric factor ($\simeq 1/6$, Eqn. 15, $\simeq 1/4$, Eqn. 17).
h	= thickness of the membrane sheet.
I_0, I_1	= modified Bessel functions, Eqn. 8.
k	= Boltzmann's constant.
k_0	= permeability (cm^2), Eqn. 8.
l	= characteristic length.
$l(v_f)$	= free distance of travel within free volume, v_f, Eqn. 15.
l_c	= approximate diameter of 'cage' with the volume v^*.
$\bar{L}$	= root mean square displacement/step.
m or m^*	= mass.
NBD	= 4-nitrobenz-2-oxa-1,3-diazole.
$p(v_f)$	= probability to have a free volume v_f, Eqn. 13.
P	= pressure ($dynes/cm^2$).

PC = phosphatidylcholine.
PE = phosphatidylethanolamine.
POPC = 1-palmitoyl-2-oleoyl phosphatidylcholine.
POPE = 1-palmitoyl-2-oleoyl phosphatidylethanolamine.
r = radius.
R = radius of circular area for diffusion, or gas constant.
R_c = cylinder radius.
R_{Rey} = Reynolds' number (Ul/v), dimensionless.
R_s = sphere radius.
$\langle r^2 \rangle$ = average square displacement.
t = time.
T = temperature.
T_g = glass transition temperature.
u = average velocity of molecule.
u = particle velocity.
U = fluid velocity at infinite distance from particle.
v = velocity (cm/s).
v_{av} = average volume/molecule, Eqn. 10.
v_g = volume/molecule at the glass transition temperature.
v_f = free volume = $v_{av} - v_0$, Eqn. 10.
v_0 = van der Waals molecular volume.
x_0 = half-distance along polymer where chains are separated see Eqn. 22.
y_i = maximum separation of the ith chain, Eqn. 20.

Greek

α = difference of thermal expansion coeff. of liquid and glass Eqn. 12.
β = fractional free volume at the glass transition temperature Eqn. 12; or, bending modulus. Eqs. 21 and 22.
Γ = linear constant for interaction potential per unit length, Eqn. 21.
Γ' = linear constant for chain separation energy, Eqn. 20.
γ = 0.5772, Euler's constant, or correction factor for overlapping free volume (or area), Eqn. 13.
Δ = Laplacian operator.
∇ = gradient operator.
ε = quantity defined in Eqn. 7.
η = viscosity.
η_0 = membrane viscosity.
η'_i = viscosity of the external liquid phase.
η_s = surface viscosity $\cong \eta_0 h$.
Λ = effective 'spring constant', Eqn. 20.

λ = distance between chain segments, Eqn. 20.
ν = kinematic viscosity (η/ϱ), or frequency of random steps.
ν_0 = fundamental frequency, Eqn. 20.
$\nu(r)$ = space density (cm^{-2}), Section II.1.b.iii (3).
ϱ = density (g/cm^3).
σ = $R_c/(k_0)^{1/2}$, dimensionless, Eqn. 8.
$\boldsymbol{\sigma}$ = shear force/unit area ($dyne/cm^2$).
ϕ = fractional volume (solute/(solvent + solute)), Sect. II.1.e or fluidity ($1/\eta$).
ϕ_0 = see Eqn. 11.

Acknowledgements

We would like to thank Dr. Thomas Jovin for his support of this work. We are also grateful to Drs. Robert Macgregor, Hagen Fueldner, Juergen Stuempel, Erich Sackmann and Evan Evans for stimulating discussions. Dr. Macgregor's careful reading of the manuscript and thoughtful suggestions are especially appreciated. Peter Regenfuss is thanked for his suggestion of Fig. 1D.

References

Alder, B.J., Alley, W.E. and Pollock, E.I. (1981) Validity of macroscopic concepts for fluids on a microscopic scale. Ber. Bunsenges. Phys. Chem. 85, 944–952.

Alder, B.J., and Alley, W.E. (1984) Generalized hydrodynamics. Phys. Today, January, 56–63.

Axelrod, D., Koppel, D.E., Schlessinger, J., Elson, E. and Webb, W.W. (1976) Mobility measurement by analysis of fluorescence photobleaching recovery kinetics. Biophys. J. 16, 1055–1069.

Barrer, R.M. (1943) Viscosity of pure liquids. I. Non polymerized fluids. Trans. Faraday Soc. 39, 48–67.

Barrer, R.M. (1957) Some properties of diffusion coefficients in polymers. J. Phys. Chem. 61, 178–189.

Batschinski, A.J. (1913) Untersuchungen ueber die innere Reibung der Fluessigkeiten. Z. Phys. Chem. 84, 643–706.

Beck, K. and Peters, R. (1982) Translational diffusion in lipid monolayers at the air/water interface. Hoppe-Seyler's Z. Physiol. Chem. 363, 894.

Bird, R.B., Stewart, W.E. and Lightfoot, E.N. (1960) *Transport Phenomena*, pp. 28–29, Wiley, New York.

Birkhoff, G. (1950) *Hydrodynamics, A Study in Logic, Fact, and Similitude*, Princeton University Press, Princeton.

Boyd, E. and Harkins, W.D. (1939) Molecular interaction in monolayers: Viscosity of two-dimensional liquids and plastic solids. V. Long chain fatty acids. J. Am. Chem. Soc. 61, 1188–1195.

Brandt, W.W. (1959) Model calculation of the temperature dependence of small molecule diffusion in high polymers. J. Phys. Chem. 63, 1080–1084.

Brenner, H. and Leal, L.G. (1977) A model of surface diffusion on solids. J. Coll. Int. Sci. 62, 238–258.

Brenner, H. and Leal, L.G. (1978) A micromechanical derivation of Fick's law for interfacial diffusion of surfactant molecules. J. Colloid Interface Sci. 65, 191–209.

Brenner, H. and Leal, L.G. (1982) Conservation and constitutive equations for adsorbed species undergoing surface diffusion and convection at a fluid-fluid interface. J. Colloid Interface Sci. 88, 136–184.

Bueche, F. (1953) Segmental mobility of polymers near their glass temperature. J. Chem. Phys. 21, 1850–1855.

Chang, C.H., Takeuchi, H., Ito, T., Machida, K. and Ohnishi, S. (1981) Lateral mobility of erythrocyte membrane proteins studied by the fluorescence photobleaching recovery technique. J. Biochem. (Tokyo) 90, 997–1004.

Cherry, R.J. and Godfrey, R.E. (1981) Anisotropic rotation of bacteriorhodopsin in lipid membranes. Comparison of theory with experiment. Biophys. J. 36, 257–276.

Chung, H.S. (1966) On the Macedo-Litovitz hybrid equation for liquid viscosity. J. Chem. Phys. 44, 1362–1364.

Clegg, R.M. and Vaz, W.L.C. (1984) Inclusion of interfacial interactions in the free-area model for diffusion in lipid bilayer membranes. Submitted for publication.

Cohen, M.H. and Turnbull, D. (1959) Molecular transport in liquids and glasses. J. Chem. Phys. 31, 1164–1169.

Cone, R.A. (1972) Rotational diffusion of rhodopsin in the visual receptor membrane. Nature New Biol. 236, 39–43.

Crank, J. and Park, G.S., Editors (1968) *Diffusion in Polymers*, Academic Press, New York.

Criado, M., Vaz, W.L.C., Barrantes, F.J. and Jovin, T.M. (1982) Translational diffusion of acetylcholine receptor (monomeric and dimeric forms) of *Torpedo marmorata* reconstituted into phospholipid bilayers studied by fluorescence recovery after photobleaching. Biochemistry 21, 5750–5755.

Criado, M. and Barrantes, F.J. (1984) Conversion of acetylcholine receptor dimers to monomers upon depletion of non-receptor peripheral proteins. Biochim. Biophys. Acta 798, 74–381.

Davies, J.T. and Rideal, E.K. (1961) *Interfacial Phenomena*. Academic Press, New York.

De Rosa, M., Gambacorta, A., Nicolaus,B., Chappe, B. and Albrecht, P. (1983) Isoprenoid ethers: Backbone of complex lipids of the archaebacterium *Sulfolobus solfataricus*. Biochim. Biophys. Acta 753, 249–256.

Derzko, Z. and Jacobson, K. (1980) Comparative lateral diffusion of fluorescent lipid analogues in phospholipid multibilayers. Biochemistry 19, 6050–6057.

Devaux, P. and McConnell, H.M. (1972) Lateral diffusion in spin-labelled phosphatidylcholine multilayers. J. Am. Chem. Soc. 94, 4475–4481.

DiBenedetto, A.T. (1963a) Molecular properties of amorphous high polymers. I. A cell theory for amorphous high polymers. J. Polym. Sci. Part A, 1, 3459–3476.

DiBenedetto, A.T. (1963b) Molecular properties of amorphous high polymers. II. An interpretation of gaseous diffusion through polymers. J. Polym. Sci. Part A, 1, 3477–3487.

DiBenedetto, A.T. and Paul, D.R. (1964) An interpretation of gaseous diffusion through polymers using fluctuation theory. J. Polym. Sci. Part A, 2, 1001–1015.

Doolittle, A.K. (1951) Studies in Newtonian flow. II. The dependence of the viscosity of liquids on free-space. J. Appl. Phys. 22, 1471–1475.

Edholm, O. and Blomberg, C. (1981) Dynamics of kink motion in hydrocarbon chains: Calculation of NMR relaxation times. Chem. Phys. 53, 185–195.

Edholm, O., Berendsen, H.J.C. and van der Ploeg, P. (1983) Conformational entropy of a bilayer membrane derived from a molecular dynamics simulation. Mol. Phys. 48, 379–388.

Einstein, A. (1906) Eine neue Bestimmung der Molekueldimensionen. Ann. Phys. 19, 289–306.

Einstein, A. (1911) Berichtigung zu meiner Arbeit: 'Eine neue Bestimmung der Molekueldimensionen'. Ann. Phys. 34, 591–592.

Einstein, A. (1956) *Investigations on the Theory of the Brownian Movement.* Editor: R. Fuerth. Dover Publications, New York.

Eshbach, O.W. (1966) *Handbook of Engineering Fundamentals*, 2nd Edn. Wiley, New York.

Evans, E.A. and Hochmuth, R.M. (1978) In: *Current Topics in Membrane and Transport*, Vol. 10, pp. 1–64. Editors: F. Brenner and A. Kleinzeller. Academic Press, New York.

Ewell, R.H. and Eyring, H. (1937) Theory of the viscosity of liquids as a function of temperature and pressure. J. Chem. Phys. 5, 726–736.

Eyring, H. (1936) Viscosity, Plasticity, and diffusion as examples of absolute reaction rates. J. Chem. Phys. 4, 283–291.

Eyring, H., Ree, T. and Hirai, N. (1958) Significant structures in the liquid state. Proc. Natl. Acad. Sci. U.S.A. 44, 683–688.

Eyring, H. and Ree, T. (1961) Significant liquid structures. VI. The vacancy theory of liquids. Proc. Natl. Acad. Sci. U.S.A. 47, 527–537.

Felderhof, B.U. and Deutch, J.M. (1975) Frictional properties of dilute polymer solutions. I. Rotational friction coefficient. J. Chem. Phys. 62, 2391–2397.

Ferry, J.D. (1980) *Viscoelastic Properties of Polymers*, 3rd Edn. Wiley, New York.

Fourt, L.G. and Harkins, W.D. (1938) Surface viscosity of long-chain alcohol monolayers. J. Phys. Chem. 42, 897–910.

Fox, T.G. and Flory, P.J. (1950) Second order transition temperatures and related properties of polystyrene. Influence of molecular weight. J. Appl. Phys. 21, 581–591.

Fox, T.G. and Flory, P.J. (1951) Further studies on the melt viscosity of polyisobutylene. J. Phys. Chem. 55, 221–234.

Fox, T.G. and Flory, P.J. (1954) The glass temperature and related properties of polystyrene. Influence of molecular weight. J. Polym. Sci. 14, 315–319.

Frenkel, J. (1955) *Kinetic Theory of Liquids*. Dover Publications, New York.

Frish, H.L. and Stern, S.A. (1983) Diffusion of small molecules in polymers. Crit. Rev. Solid State Mat. Sci. 11, 123–187.

Galla, H.J., Hartmann, W., Theilen, U. and Sackmann, E. (1979) On two dimensional passive random walk in lipid bilayers and fluid pathways in biomembranes. J. Membrane Biol. 48, 215–236.

Glasstone, S., Laidler, K.J. and Eyring, H. (1941) *The Theory of Rate Processes*. McGraw-Hill, New York.

Goodrich, F.C. (1973) Progr. Surf. Membr. Sci. 7, 151–181.

Gray, P. (1968) In: *Physics of Simple Liquids*, pp. 507–578. Editors: H.N.V. Temperley, J.S. Rowlinson, and G.S. Rushbrooke. Elsevier North Holland, Amsterdam.

Grest, G.S. and Cohen, M.H. (1981) Adv. Chem. Phys. 48, 455–525.

Happel, J. and Brenner, H. (1973) *Low Reynolds Number Hydrodynamics*. Noordhoff, Leyden.

Harkins, W.D. and Kirkwood, J.G. (1938) The viscosity of monolayers: Theory of the surface slit viscosimeter. J. Chem. Phys. 6, 53.

Harkins, W.D. (1952) *The Physical Chemistry of Surface Films*. Reinhold, New York.

Henderson, R. and Unwin, P.N.T. (1975) Three dimensional model of purple membrane obtained by electron microscopy. Nature 257, 28–32.

Hiemenz, P.C. (1977) *Principles of Colloid and Surface Chemistry*. Marcel Dekker, New York.

Hildebrand, J.H. (1971) Motions of molecules in liquids: viscosity and diffusivity. Science 174, 490–493.

Hoover, W.G., Ashurst, W.T. and Grover, R. (1972) Exact dynamical basis for a fluctuating cell model. J. Chem. Phys. 57, 1259–1262.

Hoover, W.G., Ladd, A.J.C. and Hoover, V.N. (1983) In: *Molecular-Based Study of Fluids, Advances in Chemistry Series 204*, pp. 29–46. Editors: J.M. Haile, and G.A. Mansoori. American Chemical Society, Washington.

Hoover, W.G. (1984) Computer simulation of many-body dynamics. Phys. Today, January, 44–50.

Hughes, B.D., Pailthorpe, B.A. and White, L.R. (1981) The translational and rotational drag on a cylinder moving in a membrane. J. Fluid Mech. 110, 349–372.

Hughes, B.D., Pailthorpe, B.A., White, L.R. and Sawyer, W.H. (1982) Extraction of membrane microviscosity from translational and rotational diffusion coefficients. Biophys. J. 37, 673–676.

Jacobson, K., Hou, Y., Derzko, Z., Wojcieszyn, J. and Organisciak, D. (1981) Lipid lateral diffusion in the surface membranes of cells and in multibilayers formed from plasma membrane lipids. Biochemistry 20, 5268–5275.

Jhon, M.S. and Eyring, H. (1971) *Physical Chemistry, an Advanced Treatise*, Vol. VIIIA, *Liquid State*, pp. 335–375. Editor: D. Henderson. Academic Press, New York.

Joly, M. (1972) Surf. Colloid Sci. 5, 1–193.

Jones, R.B., Felderhof, B.U. and Deutch, J.M. (1975) Diffusion of polymers along a fluid-fluid interface. Macromolecules 8, 680–684.

Jost, W. (1964) Grundlagen der Diffusionsprozesse. Angew. Chem. 76, 473–483.

Jost, W. (1972) Biomembranes 3, 5–36.

Kox, A.J., Michels, J.P.J. and Wiegel, F.W. (1980) Simulation of a lipid monolayer using molecular dynamics. Science 287, 317–319.

Krueger, G.J. (1982) Diffuysion in thermotropic liquid crystals. Phys. Reports 82, 229–269.

Kuo, A.L. and Wade. C.G. (1979) Lipid lateral diffusion by pulsed nuclear magnetic resonance. Biochemistry 18, 2300–2308.

Lamb, H. (1932) *Hydrodynamics*, 6th Edn. Cambridge.

Landau, L.D. and Lifshitz, E.M. (1959) *Course of Theoretical Physics*, Vol. 6, *Fluid Mechanics*, Ch. 1, p. 1, Pergamon, Oxford.

Langmuir, I. and Schaefer, V.J. (1937) The effect of dissolved salts on insoluble monolayers. J. Am. Chem. Soc. 59, 2400–2414.

Lindlom, G. and Wennerstroem, H. (1977) Amphiphile diffusion in model membrane systems studied by pulsed NMR. Biophys. Chem. 6, 167–171.

MacCarthy, J.E. and Kozak, J.J. (1982) Lateral diffusion in fluid systems. J. Chem. Phys. 77, 2214–2216.

Macedo, P.B. and Litovitz, T.A. (1964) On the relative roles of free volume and activation energy in the viscosity of liquids. J. Chem. Phys. 42, 245–256.

McCown, J.T., Evans, E., Diehl, S. and Wiles, H.C. (1981) Degree of hydration and lateral diffusion in phospholipid multibilayers. Biochemistry 20, 3134–3138.

Moore, W.J. and Eyring, H. (1938) Theory of the viscosity of unimolecular films. J. Chem. Phys. 6, 391–394.

Myers, R.J. and Harkins, W.D. (1937) The viscosity (or fluidity) of liquid or plastic monomolecular films. J. Chem. Phys. 5, 601–603.

Osborne, H.B., Sardet, C., Michel-Villaz, M. and Chabre, M. (1978) Structural study of rhodopsin in detergent micelles by small-angle neutron scattering. J. Mol. Biol. 123, 177–206.

Oseen, C.W. (1910) Ark. Mat. Astron. Fys. 6, no. 29.

Pace, R.J. and Chan, S.I. (1982) Molecular motions in lipid bilayers. III. Lateral and transverse diffusion in bilayers. J. Chem. Phys. 76, 4241–4247.

Pace, R.J. and Datyner, A. (1979) Statistical mechanical model for diffusion of simple penetrants in polymers. I. Theory. J. Polym. Sci., Polym. Phys. Ed. 17, 437–451.

Peters, R. (1981) Translational diffusion in the plasma membrane of single cells as studied by fluorescence microphotolysis. Cell Biol. Int. Rep. 5, 73–760.

Peters, R. and Cherry, R.J. (1982) Lateral and rotational diffusion of bacteriorhodopsin in lipid bilayers: Experimental test of the Saffman-Delbrueck equations. Proc. Natl. Acad. Sci. U.S.A. 79, 4317–4321.

Peters, R. and Beck, K. (1983) Translational diffusion in phospholipid monolayers measured by fluorescence microphotolysis. Proc. Natl. Acad. Sci. U.S.A, 80, 7183–7187.

Poo, M. and Cone, R.A. (1974) Lateral diffusion of rhodopsin in the photoreceptor membrane. Nature 247, 438–441.

Ree, T.S., Ree, T. and Eyring, H. (1964) Significant structure theory of transport phenomena. J. Phys. Chem. 68, 3262–3267.

Ricci, F.P., Ricci, M.A. and Rocca, D. (1977) On the free volume theory and on the Macedo-Litovitz hybrid equation for diffusion in liquids. J. Phys. Chem. 81, 171–177.

Saffman, P.G. (1976) Brownian motion in thin sheets of viscous fluid. J. Fluid Mech. 73, 593–602.

Saffman, P.G. and Delbrueck, M. (1975) Brownian motion in biological membranes. Proc. Natl. Acad. Sci. U.S.A. 72, 3111–3113.

Sakaki, T., Tsuji, A., Chang, C.H. and Ohnishi, S. (1982) Rotational mobility of an erythrocyte membrane integral protein, Band 3, in dimyristoylphosphatidylcholine reconstituted vesicles and effect of binding of cytoskeletal peripheral proteins. Biochemistry 21, 2366–2372.

Schneider, M.B., Chan, W.K. and Webb, W.W. (1983) Fast diffusion along defects and corrugations in phospholipid $P_{\beta'}$ liquid crystals. Biophys. J. 43, 157–165.

Seelig, J. and Seelig, A. (1980) Lipid conformation in model membranes and biological membranes. Q. Rev. Biophys. 13, 19–61.

Schlessinger, J. and Elson, E.L. (1981) In: *Receptors and Recognition*, Ser. B. Vol. 11, pp. 159–170. Editors: S. Jacobs and P. Cuatrecasas. Chapman and Hall, London.

Shinitzky, M. and Barenholz, Y. (1978) Fluidity parameters of lipid regions determined by fluorescence polarization. Biochim. Biophys. Acta 515, 367–394.

Smith, B.A. and McConnell, H.M. (1978) Determination of molecular motion on membranes using periodic pattern photobleaching. Proc. Natl. Acad. Sci. U.S.A. 75, 2759–2763.

Smith, L.M., Smith, B.A. and McConnell, H.M. (1979) Lateral diffusion of M-13 coat protein in model membranes. Biochemistry 18, 2256–2259.

Smith, L.M., Rubenstein, J.L.R., Parce, J.W. and McConnell, H.M. (1980) Lateral diffusion of M-13 coat protein in mixtures of phosphatidylcholine and cholesterol. Biochemistry 19, 5907–5911.

Sommerfeld, A. (1964) *Lectures on Theoretical Physics*, Vol II, *Mechanics of Deformable Bodies*. Academic Press, New York.

Tank, D.W., Wu, E.S., Meers, P.R. and Webb, W.W. (1982) Lateral diffusion of Gramicidin C in phospholipid multibilayers. Effects of cholesterol and high Gramicidin concentration. Biophys. J. 40, 129–135.

Teissie, J., Tocanne, J.F. and Baudras, A. (1978) A fluorescence approach to the determination of translational diffusion coefficients of lipids in phospholipid monolayers at the air-water interface. Eur. J. Biochem. 83, 77–85.

Tilley, L., Thulborn, K.R. and Sawyer, W.H. (1979) An assessment of the fluidity gradient of the lipid bilayer as determined by a set of *n*-(9-anthroyloxy) fatty acids ($n = 2, 6, 9, 12, 16$). J. Biol. Chem. 254, 2592–2594.

Traeuble, H. (1971) The movement of molecules across lipid membranes: A molecular theory. J. Membr. Biol. 4, 193–208.

Traeuble, H. and Sackmann, E. (1972) Studies of the crystalline-liquid crystalline phase transition of lipid model membranes. III. Structure of a steroid-lecithin system below and above the lipid phase transition. J. Am. Chem. Soc. 94, 4499–4510.

Turnbull, D. (1965) In: *Liquids: Structure, Properties, Solid Interactions*. pp. 6–24. Editor: T.J. Hughel. Elsevier, New York.

Turnbull, D. and Cohen, M.H. (1970) On the free volume model of the liquid-glass transition. J. Chem. Phys. 52, 3038–3041.

Van der Ploeg, P. and Berendsen, H.J.C. (1982) Molecular dynamics simulation of a bilayer membrane. J. Chem. Phys. 76, 3271–3276.

Vaz, W.L.C., Jacobson, K., Wu, E.S. and Derzko, Z. (1979) Lateral mobility of an amphiphatic apolipoprotein. ApoC-III, bound to phosphatidylcholine bilayers with and without cholesterol. Proc. Natl. Acad. Sci. U.S.A. 76, 5645–5649.

Vaz, W.L.C., Kapitza, H.G., Stuempel, J., Sackmann, E. and Jovin, T.M. (1981) Translational mobility of glycerophorin in bilayer membranes of dimyristoylphosphatidylcholine. Biochemistry 20, 1392–1396.

Vaz, W.L.C., Derzko, Z.I. and Jacobson, K.A. (1982a) In: *Cell Surface Reviews*, Vol. 8, pp. 83–135. Editors: G. Poste and G. Nicolson. Elsevier, Amsterdam.

Vaz, W.L.C., Criado, M., Madeira, V.M.C., Schoellmann, G. and Jovin, T.M. (1982b) Size dependence of the translational diffusion of large integral membrane proteins in liquid crystalline phase lipid bilayers. A Study using fluorescence recovery after photobleaching. Biochemistry 21, 5608–5612.

Vaz, W.L.C. and Hallmann, D. (1983) Experimental evidence against the applicability of the Saffman-Delbrueck model to the translational diffusion of lipids in phosphatidylcholine bilayer membranes. FEBS Lett. 152, 287–290.

Vaz, W.L.C., Clegg, R.M. and Hallmann, D. (1984a) Translational diffusion of lipids in liquid crystalline phase phosphatidylcholine multibilayers. A comparison of experiment with theory. Biochemistry, in press.

Vaz, W.L.C., Hallmann, D., Clegg, R.M., Gambacorta, A. and De Rosa, M. (1984b) Translational diffusion of a membrane-spanning lipid compared to that of a normal lipid in POPC multibilayers. Eur. Biophys. J., in press.

Vaz, W.L.C. and Criado, M. (1984) A comparison of the translational diffusion of acetylcholine receptor monomers and oligomers in lipid bilayers. Submitted for publication.

Vincent, M., de Foresta, B., Gallay, J. and Alfsen, A. (1982) Nanosecond fluorescence anisotropy decays of *n*-(9-anthroyloxy) fatty acids in dipalmitoylphosphatidylcholine vesicles with regard to isotropic solvents. Biochemistry 21, 708–716.

Vogel, H. and Weiss, A. (1981a) Transport properties of liquids. I. Self diffusion, viscosity and density of nearly spherical and disk-like molecules in the pure liquid phase. Ber. Bunsenges. Phys. Chem. 85, 539–548.

Vogel, H. and Weiss, A. (1981b) Transport properties of liquids. II. Self diffusion of almost spherical molecules in athermal liquid mixtures. Ber. Bunsenges. Phys. Chem. 85, 1022–1026.

Von Tscharner, V. and McConnell, H.M. (1981) Physical properties of lipid monolayers at alkylated planar glass surfaces. Biophys. J. 36, 421–427.

Walton, A.J. (1969) Bulk and surface transport in liquids – an introductory treatment. Contemp. Phys. 10, 489–504.

Wang, Y.L., Ree, T., Ree, T.S. and Eyring, H. (1965) Significant-structure theory and cell theory for two-dimensional liquids of hard disks. J. Chem. Phys. 42, 1926–1930.

Warburg, E. and von Babo, L. (1882) Ueber den Zusammenhang zwischen Viscositaet und Dichtigkeit bei fluessigen, insbesondere gasfoermig fluessigen, Koerpern. Ann. Phys. (3) 17, 390–427.

Webb, W.W. (1981) Luminescence measurements of macromolecular mobility. Ann. N.Y. Acad. Sci. 366, 300–314.

Weinstein, R.S. (1974) In: *The Red Blood Cell*, Vol. I, pp. 213–268. Editor: D.M. Surgenor. Academic Press, New York.

Wheeler, T.S. (1938) The theory of liquids. Trans. Natl. Sci. Inst. India 1, 333–365.

Wiegel, F.W. (1979a) Hydrodynamics of a permeable patch in the fluid membrane. J. Theor. Biol. 77, 189–193.

Wiegel, F.W. (1979b) Rotational friction coefficient of a permeable cylinder in a viscous fluid. Phys. Lett. 70A, 112–113.

Wiegel, F.W. (1979c) Translational friction coefficient of a permeable cylinder in a sheet of viscous fluid. J. Phys. A12, 2385–2392.

Wiegel, F.W. (1980) *Lecture Notes in Physics*, Vol. 121, *Fluid Flow Through Porous Macromolecular Systems*. Editors: J. Ehlers et al. Springer-Verlag, Berlin.

Williams, M.L., Landel, R.F. and Ferry, J.D. (1955) The temperature dependence of relaxation mechanisms in amorphous polymers and other glass-forming liquids. J. Am. Chem. Soc. 77, 3701–3707.

Wise, D.S., Karlin, A. and Schoenborn, B.P. (1979) An analysis by low-angle neutron scattering of the structure of the acetylcholine receptor from *Torpedo californica* in detergent solution. Biophys. J. 28, 473–496.

Yeager, M. (1976) Neutron-diffraction analysis of structure of retinal photoreceptor membranes and rhodopsin. Brookhaven Symp. Biol. 27, 3–37.

Watts/De Pont (Eds.)
Progress in Protein-Lipid Interactions

CHAPTER 6

Lipid dynamics and lipid-protein interactions in intestinal plasma membranes

DAVID SCHACHTER

Department of Physiology, Columbia University College of Physicians & Surgeons, New York, NY 10032, U.S.A.

I. Plasma membranes of the intestinal enterocyte

The enterocyte is the major cell type lining the mucosa of the small intestine and responsible for the primary functions of the organ: absorption of all nutrients; digestion of certain substances such as disaccharides and peptides; synthesis of compounds for transport from the cell to the extracellular spaces (triglycerides, lipoproteins); and secretion of water and electrolytes into the lumen. To perform these varied functions, the cell is highly specialized and the plasma membrane is differentiated into a luminal (microvillus) and contraluminal (basolateral) region separated by tight junctions. In effect these antipodal membranes are portals which regulate the exchange of substances between the organism and its environment. Our understanding of the functions of these membranes owes much to the development of methods for the bulk isolation of reasonably pure suspensions of each type (Hopfer et al., 1973; Schmitz et al., 1973; Murer et al., 1974). The microvillus and basolateral membranes differ from each other in ultrastructure (Bloom and Fawcett, 1968; Oda, 1976), enzyme activities (Douglas et al., 1972;

Abbreviations: $(Na^+ + K^+)$ ATPase, sodium plus potassium-dependent adenosine triphosphatase; CaATPase, calcium-dependent adenosine triphosphatase; MgATPase, magnesium-dependent adenosine triphosphatase; DPH, 1,6-diphenyl-1,3,5-hexatriene; HMG-CoA, 3-hydroxy-3-methylglutaryl coenzyme A; r, fluorescence anisotropy; r_o, maximal limiting fluorescence anisotropy; r_∞, limiting hindered fluorescence anisotropy; $[(r_o/r)-1]^{-1}$, anisotropy parameter; τ_F, fluorescence excited-state lifetime; τ_c, correlation time; PGE_1, prostaglandin E_1.

Lewis et al., 1975; Murer et al., 1974, 1976), transport mechanisms (Murer et al., 1974) electrophysiological properties (Rose and Schultz, 1971; Okada et al., 1977) and protein components (Fujita et al., 1973). Differences in the lipid composition (Forstner et al., 1968; Douglas et al., 1972; Kawai et al., 1974; Lewis et al., 1975; Brasitus and Schachter, 1980a) and fluidity (Brasitus et al., 1980; Brasitus and Schachter, 1980a; Gray et al., 1981) will be described in further detail below.

Several features of the differentiation of small intestinal enterocytes are noteworthy. The cells differ in function in successive segments along the length of the small intestine. For example, proximal as compared to distal segments transport calcium and iron more rapidly across the mucosa (Dowdle et al., 1960; Schachter et al., 1960; Kimberg et al., 1961); absorption of conjugated bile salts and of the vitamin B12-intrinsic factor complex are maximal in the distal small intestine (Weiner and Lack, 1968; Glass, 1963); and transport of D-galactose is most efficient in the proximal 2/3 of the rat intestine (Finkelstein and Schachter, 1962). Along with these functional differences, there are changes in lipid composition and fluidity of the enterocyte plasma membranes along the length of the gut, as described further below. Enterocytes of the intestinal epithelium are constantly regenerated by mitotic division in the crypts of Lieberkühn, followed by progression of the cells to the villus tips. Although it is well established that various enzymatic activities characteristic of the mature enterocyte appear in the course of migration toward the villus tip, concomitant developmental changes in membrane lipid fluidity and lipid-protein interactions have not been characterized.

The flow of dietary lipids across the microvillus membranes is of particular interest from the standpoint of the lipid composition, fluidity and lipid-protein interactions in the membrane. Long-chain triglycerides comprise the bulk of the fat calories ingested, and the major products of luminal triglyceride digestion are the 2-monoglycerides, which are absorbed from bile salt mixed micelles. It is instructive to calculate the quantity of acyl chains per day absorbed across the microvillus membrane in relation to the number of acyl chains in the endogenous phospholipids of the membrane itself. I have estimated that in a 200 g rat the microvillus membrane involved in lipid absorption (i.e., the proximal 2/3 of the intestine) contains approximately 3×10^{18} acyl chains, whereas the triglyceride ingested and absorbed per day (taken as approximately 0.5 g) contains roughly 10^{21} acyl chains. In other words, the daily flow of dietary acyl chains very greatly exceeds the number of acyl chains in the membrane itself, by as much as 300-fold or more. This estimate points to the need for homeostatic mechanisms to control microvillus membrane lipid fluidity, since the intercalation of dietary acyl chains of varying saturation and length could otherwise alter and possibly destabilize the membranes. That such homeostatic mechanisms exist is also indicated by the results of dietary experiments (Brasitus et al., 1984a). We have recently shown that the fluidity of rat enterocyte microvillus membranes was not altered significantly by maintenance for 3 weeks on a diet enriched in linoleic acid (18:2) residues (corn oil), although each day on this diet the intestinal mucosa was exposed to an excess, as compared to control values, of approximately 4×10^{21} linoleic acyl chains, i.e. about 1000-times the number of endogenous acyl chains in the microvillus membrane. As discussed further below, maintenance for 6 weeks on the corn oil diet did increase the fluidity of the enterocyte microvillus and basolateral membranes.

In summary, enterocyte plasma membranes offer a number of advantages for the experimental investigation of membrane lipid-protein interactions. Suspensions of purified microvillus and basolateral membranes can be prepared in bulk for lipid composition, fluidity, enzyme and transport studies. Moreover, the membranes are highly differentiated for a variety of essential functions and, therefore, well-suited for exploring such questions as the influence of lipid-protein interactions on intestinal function and the nature of the mechanisms which control membrane lipid composition and fluidity.

II. Lipid fluidity of intestinal plasma membranes

1. PLASMA MEMBRANES AS COMPARED TO CYTOSOLIC ORGANELLES

It is well established that mammalian plasma membranes, as compared to organellar membranes of the cytosol, contain more cholesterol and higher cholesterol/phospholipid ratios. Hence the lipid fluidity of the plasma membrane is generally decreased relative to that of the other organelles. This pattern has been observed in comparisons of rat enterocyte total homogenates versus purified microvillus membranes and in corresponding fractions prepared from rat liver (Table 1). We also confirmed the pattern by fluorescence polarization microscopy (Schachter and Shinitzky, 1977). Suspensions of rat jejunal mucosal cells or isolated brush borders (microvillus membranes plus their core material) were loaded with DPH and the fluorescence anisotropy, r, was quantified individually in the microscope. Mean values of r for whole cells and brush borders, respectively, were 0.111 and 0.177, indicative of lower fluidity in the brush border. These results emphasize the necessity of assuring the purity and comparability of plasma membrane suspensions used in fluidity experiments.

2. MICROVILLUS AS COMPARED TO BASOLATERAL MEMBRANES

There is considerable evidence that the enterocyte microvillus membrane is much less fluid than the antipodal basolateral membrane. Assessment of the rotational diffusion of lipid fluorophores via estimation of the fluorescence polarization clearly shows this pattern (Brasitus and Schachter, 1980a), and illustrative values are listed in Table 2. In the face of similar excited-state lifetimes in the two membranes, the fluorescence anisotropy, *r*, of each fluorophore is greater, i.e., fluidity is less, in the microvillus membrane. This is evident for both the static (DPH values) and dynamic (anthroyloxystearate values) components of fluidity, and it is observed with probes located both nearer the aqueous interfaces (2-anthroyloxystearate) or more deeply in the lipid core of the membrane (12-anthroyloxystearate). Gray et al. (1981) have also reported electron spin resonance

TABLE 1
Lipid fluidity of preparations of rat jejunal mucosa and rat liver as assessed by the fluorescene anisotropy (*r*) of various fluorophores at 25°C

Tissue	Fluorescence probe	Preparation	Fluorescence anisotropy (*r*)	Ref.[a]
Jejunal mucosa	Diphenylhexatriene	Total homogenate	0.241	1
		Microvillus membrane	0.302	
	12-Anthroyloxy-stearate	Total homogenate	0.108	1
		Microvillus membrane	0.155	
	Retinol	Total homogenate	0.275	1
		Microvillus membrane	0.298	
Liver	Diphenylhexatriene	Total homogenate	0.154	2
		Plasma membranes	0.267	

a. Reference: 1, Schachter and Shinitzky, 1977; 2, Livingstone and Schachter, 1980b.

TABLE 2
Fluorescence anisotropy at 23°C of lipid fluorophores in isolated microvillus and basolateral membranes prepared from the proximal half of the small intestine of the rat (Data from Brasitus and Schachter, 1980a)

	Fluorescence anisotropy (*r*)		Fluorescence lifetime (τ_F) (ns)	
Fluorophore	Microvillus membrane	Basolateral membrane	Microvillus membrane	Basolateral membrane
Diphenylhexatriene	0.285	0.219	11	11
2-Anthroyloxystearate	0.134	0.108	12	12
12-Anthroyloxystearate	0.109	0.081	14	16
Retinol	0.275	0.221	8	8
Dansylphosphatidylethanolamine	0.162	0.142	14	14

studies which indicate that the microvillus membrane is less fluid than the basolateral membrane. Moreover, a similar pattern of lower fluidity in the luminal as compared to the contraluminal pole of the plasma membrane has been observed in human and canine renal tubular cells (Le Grimellec et al., 1982, 1983) and rat hepatocytes (Lowe and Coleman, 1982; Storch et al., 1983). In the latter cell the luminal region is the canalicular membrane, which faces the lumen contiguous with the alimentary canal; the contraluminal side (sinusoidal and contiguous membranes) exhibits a number of additional functional homologies with the enterocyte basolateral membrane. Thus, in each cell type the contraluminal membranes contain the (Na^+ + K^+)-dependent adenosine triphosphatase activity, and calcium ion decreases the fluidity of these intact membranes by inducing metabolic alterations in the lipid composition (Storch et al., 1983).

The fluidity of enterocyte microvillus and basolateral membranes has also been compared in our laboratory by assessment of short-range lateral diffusion via quantification of the intramolecular excimer fluorescence of dipyrenylpropane. At all temperatures in the range of 0–40°C (see Fig. 1) the excimer/monomer intensity ratio was considerably higher in the basolateral membranes, an indication of greater fluidity. At 37°C, for example, the microvillus and basolateral membranes yielded ratios of 0.20 and 0.39, respectively, corresponding to ap-

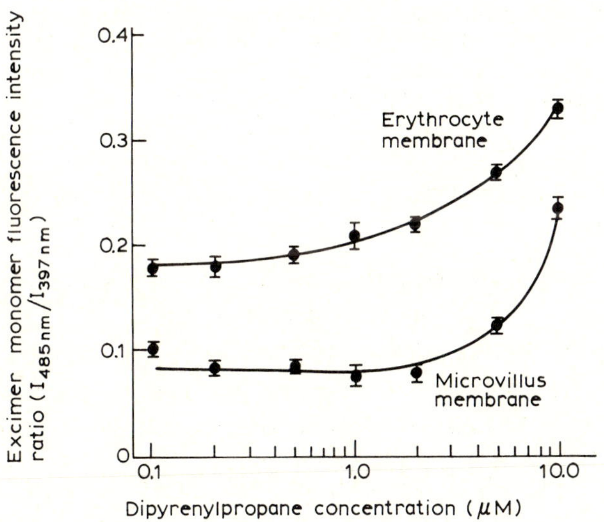

Fig. 1. Ratio of the excimer/monomer fluorescence intensity ratio (I_{485nm}/I_{397nm}) of dipyrenylpropane in rat intestinal microvillus membranes and human erythrocyte ghost membranes as a function of the loading concentration. To achieve plateau values the loading concentration must be less than approximately 0.5–1.0 μM; in the plateau region the probe/lipid molar ratio ranged from 5×10^{-5} to 7×10^{-4}.

parent fluidity (poise^{-1}) values of 0.15 and 0.81, as determined by reference to sorbitol/ethylene glycol solutions of various viscosities.

3. PROXIMAL AS COMPARED TO DISTAL INTESTINAL SEGMENTS

Enterocyte plasma membranes prepared from distal segments of the small intestine are less fluid than those of more proximal regions, as indicated by the values shown in Table 3. Lipid composition data to be discussed in more detail below indicate that one factor in this pattern is a higher cholesterol content and cholesterol/phospholipid molar ratio in the distal preparations. In turn, this compositional difference may result from the higher rate of cholesterol biosynthesis in the distal as compared to the proximal intestinal mucosa (Dietschy and Siperstein, 1965).

4. INTACT MEMBRANES AS COMPARED TO LIPOSOMES

To explore protein-lipid interactions, total lipid extracts of enterocyte plasma membranes were obtained and sonicated to prepare small unilamelllar vesicles (liposomes). The fluorescence anisotropy values of diphenylhexatriene in the intact membranes as compared to the corresponding liposomes are shown in

TABLE 3
Fluidity of rat enterocyte plasma membranes prepared from various segments of the small intestine and assessed by fluorescence polarization of lipid fluorophores at 23–25°C.

Membrane	Fluorophore	Segment	Fluorescence anisotropy (*r*)	Ref.[a]
Microvillus	Diphenylhexatriene	Duodenal[b]	0.282	1
		Jejunal	0.282	
		Ileal	0.293	
	Retinol	Duodenal[b]	0.265	2
		Jejunal	0.290	
		Ileal	0.312	
	12-Anthroyloxy-stearate	Duodenal[b]	0.123	1
		Jejunal	0.133	
		Ileal	0.137	
Basolateral	Diphenylhexatriene	Proximal[c]	0.239	3
		Middle	0.235	
		Distal	0.271	

a. References: 1, Schachter and Shinitzky, 1977; 2, Schachter et al., 1976; 3, Brasitus and Schachter, 1984.

b. Duodenal, proximal 12 cm of intestine; jejunal, middle 25 cm; ileal, distal 15 cm.

c. Entire intestine divided into proximal, middle and distal thirds.

Table 4, along with similar data for rat hepatocyte plasma membranes and human erythrocyte ghost membranes. With the exception of the erythrocyte preparations, the fluorescence anisotropy is characteristically and significantly lower, i.e., fluidity is greater, in the liposomes. A similar pattern has also been observed in studies of rat colonic mucosa basolateral membranes (Brasitus and Keresztes, 1983) and rabbit enterocyte microvillus membranes (Hauser et al., 1982). The evidence points to the existence in the intact membranes of protein-lipid interactions which decrease the mobility of the lipids. Further evidence to confirm this conclusion comes from differential scanning calorimetry experiments, which are independent of exogenous probes, and these results will be described below. Why the erythrocyte membrane does not exhibit protein-lipid interactions to a similar degree is not understood, but it is reasonable to suggest that the marked differences in plasma membrane function between erythrocytes, enterocytes and hepatocytes are significant in this respect. Presumably the organization of the plasma membranes in the last two cell types is essential for certain specific functions, e.g., lipid transport or other translocation mechanisms. It is also noteworthy in Table 4 that liposomes of enterocyte basolateral membrane lipid are considerably more fluid than those of the microvillus membrane, and hence the fluidity difference between the membranes is due mainly to differences in the lipid rather than the protein-lipid interactions.

Evidence of protein-lipid interactions in enterocyte microvillus membranes has also come from other comparisons of the intact membranes and liposomes. The effects of pH, for example, on the DPH fluorescence anisotropy in both kinds of preparations has been examined (Schachter and Shinitzky, 1977) and Table 5 lists the values observed. An increase in hydrogen ion concentration rapidly and reversibly decreases the fluorescence anisotropy of DPH in intact membranes with

TABLE 4

Fluorescence anisotropy of diphenylhexatriene (25°C) in intact membranes as compared to liposomes prepared from the lipid extracts of each membrane

Membrane type	Diphenylhexatriene fluorescence anisotropy (r)		Ref.[a]
	Intact membrane	Liposomes	
Rat enterocyte, microvillus membrane	0.282	0.215	1
Rat enterocyte, basolateral membrane	0.212	0.153	1
Rat hepatocyte, plasma membrane	0.267	0.236	2
Human erythrocyte, ghost membrane	0.270	0.272	3

a. References: 1, Brasitus et al., 1980; 2, Livingstone and Schachter, 1980b; 3, Schachter and Shnitzky, 1977.

TABLE 5
Effects of pH on the fluorescence anisotropy (25°C) of diphenylhexatriene in rat enterocyte microvillus membranes and in liposomes of the membrane lipid (Data from Schachter and Shinitzky, 1977)

	Diphenylhexatriene fluorescence anisotropy (r)	
pH of suspension	Intact membranes	Liposomes
7.3	0.294	0.250
6.3	0.285	0.253
4.7	0.272	0.249
3.9	0.255	0.238

little or no effect in the liposomes. We have suggested, therefore, that pH influences the fluidity via alteration in the protein-lipid interactions. Careful pH titration of this effect revealed inflection points at pH values suggestive of the involvement of glutamic acid (pK, 4.1) and histidine (pK, 6.1) residues of the proteins. Aside from the significance for the molecular organization of the membrane, the change in fluidity with pH is of interest because the microvillus membrane is the interface normally in contact with intralumenal H^+ in the proximal duodenum and charged with protecting the cell from destabilizing effects of acidification. Bloj and Zilversmit (1982) have compared cholesterol dynamics in rabbit intestine brush border vesicle membranes with that in liposomes of the membrane lipid. The exchange of intact membrane cholesterol mediated by a nonspecific lipid transfer protein was biphasic, whereas liposomes exhibited a single phase exchange. From these results and from studies of the availability of the bilayer cholesterol to cholesterol oxidase, the authors concluded that two pools of the sterol exist in the intact membrane, a rapidly exchangeable and enzyme-accessible moiety which amounts to less than 1/3 of the total, and a larger fraction which is inaccessible and exchanges slowly owing to association with membrane protein. It is conceivable that the pH-dependent changes in lipid fluidity mentioned above could be due to altered interactions of cholesterol with one or more membrane proteins.

III. Lipid composition in relation to the fluidity of intestinal plasma membranes

1. MICROVILLUS AS COMPARED TO BASOLATERAL MEMBRANES

The differences in membrane composition which underlie the fluidity differences between enterocyte basolateral and microvillus membranes have been explored, and Table 6 lists at least three compositional parameters which play a role. The protein/lipid ratio of the microvillus membrane considerably exceeds that of the contraluminal membrane, so that protein-lipid interactions are expected to predominate in the former. The cholesterol/phospholipid molar ratio is higher in the microvillus membrane. This results mainly from a significant increase in the total phospholipid content of the basolateral membranes (Brasitus and Schachter, 1980a; Brasitus et al., 1984a). The cholesterol content (percent by weight of total lipid) is usually higher in the microvillus membrane, but the difference may not be statistically significant if the number of determinations is limited. The degree of saturation of the acyl chains is significantly greater in the microvillus membrane, but only in the distal 2/3 of the intestine (Table 6). The differences in fatty acid composition which account for the difference in saturation index include a higher content of stearic acid (18:0) in the microvillus membranes and more arachidonic acid (20:4) in the basolateral membranes (Brasitus and Schachter, 1984). The foregoing changes in composition are all expected to increase the fluidity of the basolateral as compared to the microvillus membrane. Unfortunately, the comparisons remain qualitative, since there is at present no accepted method for quantifying the relative effect of each difference in composition on the fluidity. Moreover, such factors as the asymmetry of hemi-leaflet composition (Rothman and Lenard, 1977; Op den Kamp, 1979), the heterogeneity of microdomains within each leaflet, and the interactions of proteins and lipids, including the cholesterol-protein association mentioned above (Bloj and Zilversmit, 1982), would have to be characterized fully and taken into account.

TABLE 6
Differences in membrane composition of rat enterocyte microvillus as compared to basolateral membranes

Parameter	Intestinal segment	Value of parameter		Ref.[a]
		Microvillus membrane	Basolateral membrane	
Protein/lipid ratio (wt/wt)	Proximal 1/2	1.8	0.6	1
Cholesterol/phospholipid ratio (mol/mol)	Proximal 1/2	0.87	0.62	1
Saturation index	Proximal 1/3	0.49	0.46	2
	Middle 1/3	0.56	0.44	
	Distal 1/3	0.82	0.60	

a. References: 1, Brasitus and Schachter, 1980a; 2, Brasitus and Schachter, 1984.

2. DIFFERENCES IN COMPOSITION ALONG THE LENGTH OF THE INTESTINE

Evidence described above indicates that distal intestinal microvillus and basolateral membranes are less fluid than the proximal preparations in animals maintained on standard laboratory diets (Purina Rat Chow or Camm Maintenance Rodent Diet). As indicated by the results in Table 7, the lipid of the distal microvillus and basolateral membranes in such animals has a higher content of cholesterol and a higher cholesterol/phospholipid molar ratio. In addition, the saturation index of both the microvillus and basolateral membrane lipids is higher in the distal region (Table 6), owing to a marked decrease in linoleic acid (18:2) and a less marked increase in palmitic acid (16:0). The stearic acid (18:0) content of the microvillus membrane lipid is also greater in the distal and middle thirds as compared to the proximal third of the intestine (Brasitus and Schachter, 1984).

3. DIFFERENCES IN COMPOSITION OF RAT MICROVILLUS MEMBRANES WITH AGE

A number of investigators have reported that the lipid fluidity of enterocyte microvillus membranes isolated from suckling rats or rabbits is greater than that of the adult preparations (Schwarz et al., 1982; Pang et al., 1983). We have recently observed that microvillus membranes prepared from 6-week-old rats (2 weeks post-weaning) are also more fluid than the corresponding preparations from animals aged 17 or 117 weeks (Brasitus et al., 1984b). At least three differences in membrane composition contribute to the higher fluidity in the youngest age group. Owing mainly to a higher content of phosphatidylethanolamine and of total phospholipid, the cholesterol/phospholipid molar ratio is lower in the youngest preparations. Further, these preparations contain more arachadonic acid (20:4) residues and have a higher double bond index than the membranes of the older animals. Lastly, the lipid/protein ratio is higher in the 6-week-old animals.

TABLE 7
Differences in cholesterol content and cholesterol/phospholipid molar ratio along the length of the rat small intestine

Membrane	Intestinal segment	Cholesterol content (% by wt of total lipid)	Cholesterol/phospholipid molar ratio	Ref.[a]
Microvillus	Proximal 1/4	14.9 ± 1.7	0.83 ± 0.10	1
	Distal 1/4	18.8 ± 1.7	1.01 ± 0.10	
Basolateral	Proximal 1/3	14.4 ± 1.7	0.40 ± 0.06	2
	Middle 1/3	17.4 ± 1.7	0.51 ± 0.04	
	Distal 1/3	22.3 ± 2.1	0.68 ± 0.06	

a. References: 1, Brasitus and Schachter, 1982; 2, Brasitus and Schachter, 1984a.

IV. Thermotropic transitions

1. DIFFERENTIAL SCANNING CALORIMETRY AND FLUORESCENCE POLARIZATION STUDIES

The characterization of lipid thermotropic transitions has provided a useful tool for the study of lipid-protein interactions in a number of membrane types (Tourtellotte, 1972; Linden et al., 1973; Fox, 1975; Lee, 1975; 1977a, b; Razin, 1975; Melchior and Steim, 1976; Bach et al., 1977). Thermotropic transitions in enterocyte microvillus and basolateral membranes have been detected and characterized by both differential scanning calorimetry and Arrhenius studies of fluorescence polarization parameters (Schachter and Shinitzky, 1977; Brasitus et al., 1980; Brasitus and Schachter, 1980a, b; Brasitus et al., 1984a). Differential scanning calorimetry reveals broad lipid transitions of low enthalpy in the enterocyte plasma membranes (Brasitus et al., 1980) and in hepatocyte plasma membranes (Livingstone and Schachter, 1980b). Steady-state fluorescence polarization experiments yield Arrhenius plots (fluorescence anisotropy parameter against 1/absolute temperature) which deviate from linearity at the lower critical temperature of the broad transition seen by calorimetry. Table 8 summarizes the transition temperatures observed by fluorescence polarization in enterocyte microvillus and

TABLE 8
Thermotropic transition temperatures of rat enterocyte plasma membranes and liposomes of the membrane lipid as detected by fluorescence polarization studies

Preparation	Intestinal segment	Fluorophore	Transition temperature (°C)	Ref.[a]
Microvillus membranes	Middle 25 cm	Diphenylhexatriene	26 ± 2	1
		12-Anthroyloxystearate	27	
	Proximal 1/2	Diphenylhexatriene	23 ± 1.1	2
	Proximal 1/2	Diphenylhexatriene	24 ± 0.6	3
Microvillus membrane liposomes	Middle 25 cm	Diphenylhexatriene	28	1
	Proximal 1/2	Diphenylhexatriene	23 ± 2.2	2
Basolateral membranes	Proximal 1/2	Diphenylhexatriene	26 ± 1.5	2
Basolateral membrane liposomes	Proximal 1/2	Diphenylhexatriene	25 ± 1.7	2

a. References: 1, Schachter and Shinitzky, 1977; 2, Brasitus et al., 1980; 3, Brasitus et al., 1984b.

basolateral membranes and in the liposomes prepared from the membrane lipids. The transition occurs at approximately 26 ± 2°C in both the membranes and liposomes, confirming that it is a lipid transition. The foregoing transition was observed in rats aged 10 weeks or older; in younger animals aged 6 weeks we noted the transition at 17.5 ± 1.3°C in the microvillus membrane (Brasitus et al., 1984b), consistent with increased fluidity of the membrane as compared to the older animals.

The results of differential scanning calorimetry of microvillus and basolateral membranes are summarized in Table 9. Both of the intact membrane preparations exhibited broad transitions with lower critical temperatures similar to those observed in the fluorescence polarization experiments. Of particular interest is the observation that the upper critical temperature, 39–40°C, slightly exceeds physiological body temperature. Hence these membranes function within the thermotropic transition in vivo. The physiological implications of this arrangement are not understood, although Linden et al. (1973) and Lee (1975) have pointed out that lateral compressibility of the membrane lipids is markedly enhanced when the temperature is lowered through the upper extremity of the lipid transition and several phases coexist. Presumably the increased compressibility could facilitate the insertion of substances into the membrane and thereby enhance processes of transport or membrane assembly (Linden et al., 1973). Whether this or other factors are involved, it is worthy of emphasis that the function of rat enterocyte plasma membranes can be modulated by changes in state of the lipids in the range of physiological body temperatures.

TABLE 9
Differential scanning calorimetry of rat enterocyte microvillus and basolateral membranes and of the lipids extracted from the membranes (Data from Brasitus et al., 1980; Enthalpy values of the intact membranes are approximations)

Membrane	Preparation	Transition temperature range (°C)		
		Lower critical temperature	Upper critical temperature	Transition enthalpy (cal/g)
Microvillus	Intact membrane	23	39	~0.1
	Extracted lipids,			
	rehydrated	14	38	~0.4
	dried	15	40	2.0 – 3.0
Basolateral	Intact membrane	27	40	~0.15
	Extracted lipids,			
	rehydrated	13	33	0.55
	dried	5	60	4.5

The enterocyte membrane lipid transitions seen by differential scanning calorimetry are broad and of low enthalpy. The values listed in Table 9 should be compared to values reported for a number of pure phospholipids, 8–15 cal/g (Ladbrooke and Chapman, 1969; Hinz and Sturtevant, 1972b). The intact enterocyte membranes exhibit the lowest transition enthalpies, in the range of ~ 0.1–0.15 cal/g. Such broad transitions of low enthalpy, often attributed to low cooperativity, are frequently observed for complex mixtures of lipids with different transition temperatures and have been reported for membranes of *Mycoplasma* (Steim et al., 1969; Tourtellotte, 1972; Razin, 1975), *Escherichia coli* (Haest et al., 1974; Melchior and Steim, 1976), and other bacterial species (Haest et al., 1974). The factors which account for the relatively low enthalpy of the enterocyte transitions include the presence of cholesterol and of protein-lipid interactions. In aqueous dispersions of defined phospholipids, cholesterol decreases the enthalpy of the major transition (Ladbrooke and Chapman, 1969; Oldfield and Chapman, 1972; Hinz and Sturtevant, 1972a; Estep et al., 1978). In biological membranes, the transition enthalpy observed for *Azotobacter laidlawii* was decreased on increasing the membrane cholesterol to 12% by weight of the membrane lipid (De Kruijff et al., 1972). Comparison of two strains of *Mycoplasma mycoides* revealed a lipid transition in an adapted strain, whose membrane contained 3% cholesterol by weight of the lipids, but no transition was observed in the native strain with 22–25% cholesterol (Rottem et al., 1973). Human erythrocyte membranes contain 24% cholesterol (Farquhar, 1962), and Ladbrooke et al. (1968) observed no transition on differential scanning calorimetry unless cholesterol was removed. The cholesterol content (% by weight of total lipid) of microvillus and basolateral membranes prepared from the proximal half of the rat intestine is approximately 14–16% (Brasitus and Schachter, 1980a), a level at which the transition enthalpy is considerably decreased but not obliterated. It is also noteworthy that a significant moiety of the microvillus membrane cholesterol may be relatively unavailable for sterol-phospholipid interaction owing to cholesterol-protein interaction (Bloj and Zilversmit, 1982).

The higher enthalpies observed for dried as compared to rehydrated lipids (Table 9) are also related to effects of cholesterol. Ladbrooke and Chapman (1969) reported that the interactions of cholesterol and lecithin molecules which underlie the decreases in transition enthalpy require the presence of water, and on removal of water these components 'crystallize separately'. Separation of cholesterol-poor domains in the unhydrated lipid extracts probably account for the higher enthalpy values observed, e.g., 4.5 cal/g for the basolateral membrane lipids. This value is comparable to that of *Mycoplasma* membrane lipids, 3–4 cal/g (Reinert and Steim, 1970; Tourtellotte, 1972). Hydration of pure lecithins is also reported to narrow the temperature range and to lower the midpoint temperature of the transition (Ladbrooke and Chapman, 1969), and similar effects were observed in the enterocyte lipid scans. Hydration of the basolateral membrane lipids decreased the extent of the transition temperature range from 55 to 20°C (Table 9) and lowered the peak transition temperature from 34 to 23°C (Brasitus and Schachter, 1980a).

Protein-lipid interactions also lower the lipid transition enthalpies of the enterocyte plasma membranes, as shown by the values of the transition enthalpy ratios (intact membranes)/(extracted, hydrated lipids), which are in the range 0.2–0.5. Thus, the membrane proteins prevent a significant fraction of the lipids from participating in the cooperative transition. Comparison of these values with the comparable ratios reported for rat hepatocyte plasma membranes, 0.4–0.6 (Livingstone and Schachter, 1980), and for *Mycoplasma* membranes, 0.9 ± 0.1 (Reinert and Steim, 1970), indicates that the protein-lipid interactions are relatively more effective in the mammalian plasma membranes.

2. TEMPERATURE STUDIES OF INTRAMOLECULAR EXCIMER FLUORESCENCE

The thermotropic transition of rat enterocyte microvillus and basolateral membranes can also be detected by following the temperature dependence of the excimer/monomer fluorescence intensity ratio of dipyrenylpropane (Fig. 1, p. 235). As noted previously, this probe provides an assessment of short-range lateral diffusion, as compared to the rotational diffusion studied by fluorescence polarization. Figure 2 illustrates Arrhenius plots of the excimer/monomer intensity ratios in the enterocyte plasma membranes. Major inflection points are ob-

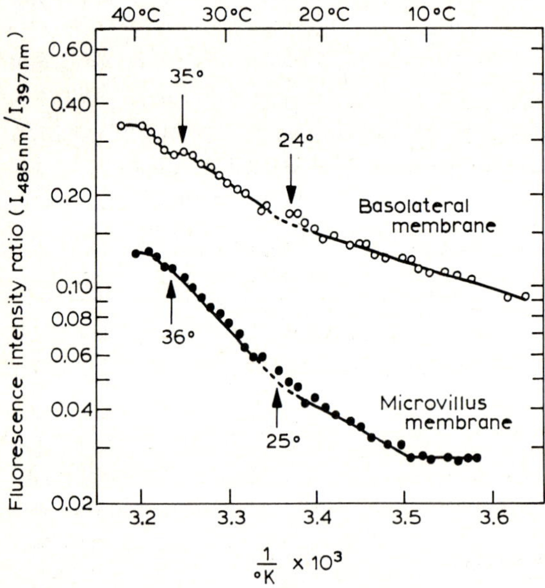

Fig. 2. Arrhenius plot of the dipyrenylpropane excimer/monomer fluorescence intensity ratio (I_{485nm}/I_{397nm}) in rat intestinal microvillus and basolateral membranes.

served at approximately 24–25°C and 35–36°C, in reasonable agreement with the lower and upper critical temperatures detected by calorimetry and fluorescence polarization. The higher excimer/monomer ratio of the basolateral as compared to the microvillus membrane throughout the temperature range examined demonstrates again the greater lipid fluidity of the contraluminal membrane.

3. TEMPERATURE STUDIES OF MEMBRANE ENZYME AND TRANSPORT ACTIVITIES

Protein-lipid interactions in the enterocyte plasma membranes have been explored further by examination of Arrhenius plots of membrane enzyme and transport activities in relation to the lipid thermotropic transitions characterized above. Table 10 lists the Arrhenius plot break points observed for a variety of membrane protein activities. The microvillus membrane activities fall into two groups. In the first group are the digestive enzymes lactase, maltase, sucrase, leucine aminopeptidase and γ-glutamyl transpeptidase, which yield single-slope

TABLE 10

Arrhenius plot break points (means ± S.E.) of enzyme and transport activities in rat enterocyte plasma membranes

Membrane	Activity	Break point temperature (°C)	Ref.[a]
Microvillus	Lactase	none	1
	Maltase	none	
	Sucrase	none	
	γ-Glutamyl transpeptidase	none	
	Leucine aminopeptidase	none	
	p-nitrophenylphosphatase	29 ± 0.7	
	Calcium ATPase	26 ± 0.1	
	Magnesium ATPase	28 ± 1.5	
	Na^+-dependent glucose transport	27 ± 1.3	
	Guanylate cyclase	30 ± 1.0	2
Basolateral	5′-Nucleotidase	28 ± 2.1	3
	Adenylate cyclase		
	Basal	30 ± 1.6	
	NaF-stimulated	30 ± 1.5	
	PGE_1-stimulated	29 ± 0.6	
	(Na^+ + K^+)-dependent ATPase	22 ± 1.5	
	K^+-dependent *p*-nitrophenyl-phosphatase	20 ± 1.0	
	Magnesium-dependent ATPase	20 ± 1.3	

a. References: 1, Brasitus et al., 1979; 2, Brasitus and Schachter, 1980b; 3, Brasitus and Schachter, 1980a.

Arrhenius plots and thus do not appear to experience functionally the effects of the bulk lipid transition. These are appropriately designated 'extrinsic' activities of the membrane, and they are known to mediate digestive reactions at the aqueous, luminal interface of the membrane. Each activity of the second group, including calcium- and magnesium-dependent ATPases, *p*-nitrophenylphosphatase, guanylate cyclase and sodium-dependent transport of D-glucose, shows a change in slope of the Arrhenius plot in the temperature range 26–30°C, i.e., near the lower critical temperature of the lipid transition. These are appropriately designated 'intrinsic' membrane activities.

Biochemical and ultrastructural evidence reinforces the conclusion that the extrinsic activities of the microvillus membrane are less intimately associated with the lipids than are the intrinsic activities. Louvard et al. (1975) reported that treatment with papain elicited rapid release of maltase, sucrase and aminopeptidase, but much slower release of *p*-nitrophenylphosphatase from rat, rabbit and pig microvillus membrane vesicles; and 1% Triton X-100 released 80–100% of the porcine maltase and aminopeptidase, but only 30–50% of the *p*-nitrophenylphosphatase. Boedeker et al. (1976) observed that guinea pig microvillus membranes treated with 0.04% Triton X-100 released 62% of the maltase and sucrase but only 3.9% of the 'alkaline phosphatase'. Papain treatment of human microvillus membranes released considerably more lactase, maltase, sucrase, leucine aminopeptidase and γ-glutamyl transpeptidase than *p*-nitrophenylphosphatase (Maestracci, 1976). Further, there is electron microscopic evidence that digestive enzymes of this extrinsic group are organized in 'knob-like' structures visible on the luminal surface of the microvillus membrane (Johnson, 1966; Nishi et al., 1968; Takesue and Sato, 1968; Maestracci, 1976).

The lipid thermotropic transition of the basolateral membranes influences the activities of 5′-nucleotidase, adenylate cyclase, (Na^+ + K^+)-dependent ATPase and magnesium-dependent ATPase (Table 10), and these are categorized as 'intrinsic' enzymes. The Arrhenius plot break points shown in Table 10 indicate that they can be subdivided further into two groups. A break point at 28–30°C characterizes 5′-nucleotidase and adenylate cyclase (basal, sodium fluoride- and prostaglandin-stimulated). The (Na^+ + K^+)-dependent ATPase, K^+-dependent *p*-nitrophenylphosphatase and Mg^{2+}-dependent ATPase, on the other hand, show a discontinuity at 20–22°C, some 5–7°C below the lower critical temperature of the membrane lipid transition. A similar result was noted by Kimelberg and Papahadjopoulos (1974), who delipidated and subsequently relipidated rabbit rabbit kidney (Na^+ + K^+)ATPase with various phospholipids. Following reconstitution with dipalmitoylphosphatidylglycerol or distearoylphosphatidylgycerol, a break point in the Arrhenius plot of the enzyme activity was observed some 6–8°C below the lower critical temperature of the lipid transition observed by differential scanning calorimetry. The results suggest that in the rat enterocyte basolateral membrane the (Na^+ + K^+)ATPase and K^+-dependent *p*-nitrophenylphosphatase, which represent one enzyme complex (Glynn and Karlish, 1975), and the Mg^{2+}-dependent ATPase, presumably a separate enzyme, function in a lipid microenvironment which is more fluid than the bulk lipid and requires a lower temperature for the transition. In accord with this hypothesis (Kimelberg and Papahadjopoulos, 1974), Grisham and Barnett (1972) noted that the lipid associated with purified lamb kidney (Na^+ + K^+)ATPase was more fluid than that of the crude microsomal starting material. The ability of an intrinsic membrane enzyme to determine its lipid microenvironment is reasonably ascribed to strong and relatively specific protein-lipid interactions. Such interactions could also affect the conformation or aggregation of the protein, and the resulting changes in protein structure with temperature might also determine Arrhenius plot break points, as some authors have emphasized (Thorneley et al., 1975; Anzai et al., 1978; Gómez-Fernández et al., 1979).

The influence of membrane lipid fluidity on the activity of the basolateral membrane adenylate cyclase was also explored by treatment with benzyl alcohol, a compound which is known to increase the fluidity and lower the thermotropic transition temperature of artificial lipid bilayers and biological membranes (Colley and Metcalfe, 1972; Seeman, 1972; Dipple and Houslay, 1978). Treatment of basolateral membranes with 50 mM benzyl alcohol decreased the DPH fluorescence anisotropy and lowered the break point temperature observed by fluorescence polarization of DPH from control values of 26.1 ± 1.2°C to 20.1 ± 1.4°C, a decrease of approximately 6°C (Brasitus and Schachter, 1980a). Concomitantly, the benzyl alcohol decreased the break point temperatures observed by Arrhenius plots of the basal adenylate cyclase by 4°C, of the NaF-stimulated adenylate cyclase by 7°C, and of the PGE_1-stimulated adenylate cyclase by 6°C. The specific activities of the basal and stimulated enzyme were also increased considerably, particularly at temperatures above the break point temperature. The results thus demonstrate that the enzyme is very responsive to the fluidity and physical state of the membrane lipids.

4. DELIPIDATION AND RELIPIDATION STUDIES OF MICROVILLUS MEMBRANE *p*-NITROPHENYLPHOSPHATASE

To explore protein-lipid interactions further, Arrhenius studies of the microvillus membrane *p*-nitrophenylphosphatase were undertaken. Microvillus membranes were treated with butanol-1 to remove membrane lipids, a procedure which also solubilizes the enzyme (Saini and Done, 1972). Progressive delipidation with removal of >95% of the phospholipids and >93% of the cholesterol completely eliminated the Arrhenius plot break point observed at 29 ± 0.7°C in the native membranes (Brasitus and Schachter, 1980a). Of interest is an increase in specific activity of the enzyme which we observed on partial delipidation (removal of 77 and 86% of the phospholipids and cholesterol, respectively) with butanol. The lipid of the native membrane thus appears to restrain the maximal activity of the enzyme. The delipidated preparations were then relipidated by the procedure of Racker (1972) with seven different lipids: microvillus membrane endogenous lipid; pure phospholipids (dipalmitoyllecithin, dimyristoyllecithin, phosphatidylethanolamine and phosphatidylinositol); glycerol trioleate; and sodium oleate. In each instance the relipidation reversed the effects described above, i.e., the specific activity of the enzyme decreased and the Arrhenius plot break point returned. Moreover, the restored break point ranged from 25 to 29°C and did not necessarily correspond to the thermotropic transition temperature of the lipid used. For example, in preparations relipidated with dipalmitoyllecithin, studies of DPH fluorescence anisotropy revealed a bulk lipid transition at 40–43°C, as expected, yet the *p*-nitrophenylphosphatase break point occurred at 28°C (Brasitus et al., 1979). Similar results have been observed on relipidation of a number of other enzymes, including *Azotobacter* nitrogenase (Ceuterick et al., 1978), rabbit kidney (Na^+ + K^+)ATPase (Kimelberg and Papahadjopoulos, 1974) and calcium-dependent ATPase of sarcoplasmic reticulum (Anzai et al., 1978). The exact explanation for this behavior is unknown and reflects a more general lack of an accepted picture of how thermally induced changes in the physical state of the lipids influence membrane enzymes. Some authors (Thorneley et al., 1975; Anzai et al., 1978; Gómez-Fernández et al., 1979) emphasize the importance of temperature-dependent conformational changes or alterations in aggregation of the proteins, processes which presumably require a certain lipid environment. Others (Ceuterick et al., 1978) stress the role of vicinal lipids and their interaction with the protein as a determinant of the break-point temperature. We have incorporated these factors into a working hypothesis to explain the observations with microvillus membranes. In the intact membrane the bulk and vicinal lipids appear to be similar in their properties, inasmuch as quite different activities. e.g., cation ATPases, glucose transport, guanylate cyclase, all show a similar break point temperature, which corresponds to the bulk lipid transition. Treatment of the membrane with l-butanol yields a solubilized *p*-nitrophenylphosphatase which retains some vicinal lipid. The Arrhenius plot break point is lost, however, because it requires bulk lipid. On reconstitution with one of a variety of lipids the break point is restored, with the temperature of the break point determined in each instance by the endogenous vicinal lipid.

V. Regulation of enterocyte plasma membrane fluidity

1. ROLE OF CHOLESTEROL BIOSYNTHESIS

The foregoing sections provide considerable evidence that lipid fluidity and lipid-protein interactions can modulate the functions of intrinsic enzyme and transport activities. In itself this would suggest the existence of homeostatic mechanisms to regulate the membrane fluidity, and we have pointed out above that such mechanisms are also likely because marked changes in dietary lipids require many weeks to alter the fluidity. To explore the nature of such regulatory processes, about which little is now known, we initiated studies to test the hypothesis that plasma membrane cholesterol content, and hence fluidity, is modulated by the rate of cholesterol biosynthesis in the enterocyte. This hypothesis is attractive in view of the findings described above that ileal enterocytes synthesize cholesterol more rapidly than do more proximal cells, while ileal plasma membranes have a higher content of cholesterol, higher cholesterol/pospholipid molar ratios and lower fluidity values as compared to more proximal cells. Moreover, Sinensky (1977, 1978) has described a Chinese hamster ovary cell mutant which is defective in the regulation of cholesterol biosynthesis by exogenous cholesterol. When the cholesterol of the culture medium was increased the mutant failed to reduce cholesterol biosynthesis and the cholesterol content of the plasma membrane rose. Under similar growth conditions the parental cell line decreased the biosynthetic rate and maintained normal, control levels of membrane cholesterol. The author suggested that "the role of the cholesterol-synthesizing enzymes of the mammalian fibroblast is to regulate the membrane fluidity" (Sinensky, 1978).

To examine our working hypothesis, we varied the rate of cholesterol biosynthesis of rat mucosal enterocytes in vivo (Brasitus and Schachter, 1982), using procedures described by Andersen and Dietschy (1977). As summarized in Table 11, either feeding rats 2% sodium taurocholate or fasting the animals decreased the synthesis of cholesterol in the ileal mucosa, as monitored by the specific activity of the 3-hydroxy-3methylglutaryl coenzyme A (HMG-CoA) reductase, the rate-limiting enzyme of cholesterol synthesis. Concomitantly, the lipid fluidity of the ileal microvillus membranes, assessed by the fluorescence polarization of DPH, increased significantly, and, in the fasted animals, the cholesterol/phospholipid molar ratio was reduced. The endogenous synthesis of cholesterol was increased, on the other hand, by biliary ligation or by feeding 2% cholestyramine, a resin which binds and sequesters the luminal bile salts. In these cases the cholesterol/phospholipid molar ratio of the ileal microvillus membranes was significantly increased and the lipid fluidity was decreased.

To determine whether the foregoing responses can also be detected in the contraluminal membranes, we fed rats cholestyramine or taurocholate, as described above, and examined the isolated ileal basolateral membranes. No significant changes in lipid fluidity from the control levels were observed. The modulation of ileal microvillus membrane cholesterol content and fluidity by endogenous cholesterol synthesis is therefore relatively specific for the luminal membrane.

Inasmuch as the relative changes in microvillus membrane lipid fluidity owing to variations in cholesterol biosynthesis are in the range −6 to +14% Table 11, the question arises whether such changes are sufficient to influence membrane

TABLE 11
Effects of variations in cholesterol biosynthesis on cholesterol/phospholipid molar ratio and lipid fluidity (diphenylhexatriene anisotropy parameter at 25°C) of ileal microvillus membranes (Values calculated from results of Brasitus and Schachter (1982))

Procedure	% change in HMG-CoA reductase[a] activity	% change in cholesterol/phospholipid molar ratio	% change in DPH anisotropy parameter[b]
Feeding 2% sodium taurocholate, 2 weeks	−20.6 ($P < 0.05$)[c]	n.s.[d]	−12.0 ($P < 0.001$)
Fasting, 4 days	−48.4 ($P < 0.01$)	−10.5 ($P < 0.05$)	−6.3 ($P < 0.025$)
Biliary ligation, 4 days	+66.3 ($P < 0.01$)	+15.2 ($P < 0.05$)	+14.2 ($P < 0.05$)
Feeding 2% cholestyramine, 2 weeks	+30.4 ($P < 0.01$)	+35.8 ($P < 0.05$)	+11.1 ($P < 0.001$)

a. HMG-CoA reductase, hydroxymethylglutaryl coenzyme A reductase.
b. Anisotropy parameter = $[(r_o/r) - 1]^{-1}$, where r_o is the maximal limiting anisotropy, taken as 0.362 for DPH. Lipid fluidity varies inversely with the anisotropy parameter.
c. *P* values are for significance of the differences as compared to controls (Brasitus and Schachter, 1982).
d. n.s., not significantly different.

function. Accordingly, we prepared and tested ileal microvillus membrane vesicles for sodium-dependent influx of D-glucose (Brasitus and Schachter, 1982). Membrane vesicles prepared from the rats fed taurocholate, which decreases cholesterol synthesis and enchances membrane fluidity, transferred D-glucose considerably more rapidly than did controls. Based on a 15-s flux period, the Na^+-dependent D-glucose uptake was increased by approximately 60% ($P < 0.02$); in the absence of a sodium gradient no significant difference from the controls was observed. These results do not necessarily prove that the enhanced flux of glucose results solely from the increase in lipid fluidity. Nonetheless, it is reasonable to suggest that modulation of ileal microvillus membrane cholesterol and fluidity can regulate functions mediated by intrinsic membrane proteins.

2. DIETARY MODULATION OF ENTEROCYTE MEMBRANE COMPOSITION AND FLUIDITY

To explore the effects of diet, we recently maintained groups of rats on a nutritionally complete diet supplemented with either corn oil (which supplies an excess of polyunsaturated acyl residues, largely 18:2) or butter fat (saturated acyl chains). The final lipid content of each diet, 37% by weight, and all other components including cholesterol were identical. As noted in a prior section, after 3 weeks on the diets no consistent difference in lipid fluidity of the enterocyte plasma membranes was detected. After 6 weeks, however, lipid fluidity of enterocyte microvillus and basolateral membranes, as assessed by the fluorescence anisotropy of DPH, was increased significantly, by 12–23%; and similar increments were observed in colonocyte basolateral membranes (Brasitus et al., 1984a). The diet-induced increase in fluidity of proximal enterocyte microvillus membranes was confirmed by demonstrating that the lower critical temperature of the lipid thermotropic transition, as detected by fluorescence polarization studies, was decreased from 25°C (butter fat diet) to 20°C (corn oil diet). The major change in lipid composition which accounted for the fluidity difference was a reduction in acyl chain saturation index owing to the polyunsaturated diet. In the microvillus membranes this resulted from an increase in linoleic acid (18:2), whereas in the basolateral membranes both 18:2 and arachidonic acid (20:4) were increased.

The increase in fluidity owing to the polyunsaturated diet elicited compensatory mechanisms designed to minimize the fluidity change. Thus, after 6 weeks on this diet, enterocyte microvillus membranes contained more cholesterol and less total phospholipid, i.e., higher cholesterol/phospholipid molar ratios, than did the preparations from rats fed butter fat. Significantly, these changes were not observed in the basolateral membranes. Since this pattern of selective response of the luminal, as compared to the contraluminal, membrane is similar to that observed previously on altering enterocyte cholesterol biosynthesis (Brasitus and Schachter, 1982), we examined the possibility that the polyunsaturated diet stimulates cholesterol biosynthesis in the enterocyte as a homeostatic response. Slices of intestine from rats on each diet were incubated in vitro with [1-^{14}C]octanoate, and the incorporation into digitonin-precipitable sterols (essentially cholesterol) estimated (Brasitus et al., 1984a). The incorporation was approximately 76% greater ($P < 0.001$) in ileal slices from animals on the polyunsaturated as compared to the saturated diet. In summary, these observations provide another example of how modulation of the rate of cholesterol biosynthesis in the ileal enterocyte alters the cholesterol content and cholesterol/phospholipid molar ratio of the microvillus but not of the basolateral membrane. In this instance, the change in cholesterol biosynthesis (and possibly in total phospholipid synthesis) is apparently a compensatory mechanism to minimize membrane fluidity changes owing to the diet. That the diet-induced fluidity alterations can modulate membrane protein functions was shown by assay of microvillus *p*-nitrophenylphosphatase and basolateral membrane ($Na^+ + K^+$)ATPase and MgATPase. Maintenance on the polyunsaturated diet increased the *p*-nitrophenylphosphatase specific activity from 2- to 3-fold. (Arrhenius studies indicated no change in the break point temperature of the enzyme, 26 ± 1°C for both diet groups, in the face of a decrease of 5°C in the lipid transition temperature owing to the corn oil diet, as noted above. Thus, as in the delipidation-relipidation experiments (Brasitus et al., 1979), the break point temperature of the enzyme can remain constant in the face of a change in the transition temperature of the bulk lipid.) The polyunsaturated diet also increased the specific activity of the enterocyte basolateral membrane ($Na^+ + K^+$)ATPase by 45–126%, whereas the MgATPase was not changed significantly. Changes of this magnitude in the ($Na^+ + K^+$)ATPase could arise from both direct and indirect mechanisms. An increase in membrane fluidity might increase the maximal velocity, or turnover number, of each pump unit. Alternatively, increased fluidity of the enterocyte plasma membranes might enhance the passive flux of Na^+ into the cell and require increased pump activity to maintain the normal, low intracellular concentrations of the cation. Under these circumstances an increase in the number of pump units might well occur. Regardless of the precise mechanisms, the experiments demonstrate that dietary modulation of intestinal membrane lipid composition and fluidity can influence important protein-mediated activities.

3. IONIC REGULATION OF ENTEROCYTE MICROVILLUS MEMBRANE FLUIDITY

Although our information is still sparse, it seems clear that changes in ionic composition of the ambient media can alter the fluidity of enterocyte microvillus membranes. As discussed in section II.4, and illustrated in Table 5, pH strongly influences microvillus membrane fluidity, apparently by a mechanism which involves the membrane proteins. Preliminary observations in our laboratory indicate that the effects of pH on fluidity are relatively specific for microvillus as compared to basolateral membranes, and for duodenal (proximal) as compared to more distal microvillus membranes. Moreover, the effects of pH on fluidity may involve a displacement of membrane-bound calcium by ambient H^+. When the microvillus membranes are first treated with a calcium chelator, the effects of pH are largely eliminated. The experiments suggest the presence of a relatively specific control mechanism in which H^+ and Ca^{2+} interact in the regulation of the membrane fluidity of duodenal microvillus membranes.

VI. Summary and comments

Intestinal plasma membranes provide useful experimental preparations for the examination of protein-lipid interactions. Particular advantages include the high degree of functional specialization of the luminal and contraluminal membranes, and the feasibility of preparing relatively pure suspensions of each in quantities sufficient for examination by a variety of techniques, e.g., biophysical studies of lipid fluidity, biochemical analyses of composition, and assays of enzyme and transport activities. Studies described in this chapter illustrate how the integration of these experimental approaches can provide new understanding of the molecular aspects of membrane function and protein-lipid interactions, on the one hand, and of the systemic functions of the intestinal mucosa on the other. The enterocyte microvillus membrane is considerably less fluid than is the contraluminal basolateral membrane, and composition differences which account for this include a higher cholesterol/phospholipid molar ratio, a higher protein/lipid ratio and, in distal intestinal segments, a higher acyl chain saturation index in the microvillus membrane. Both plasma membranes exhibit a lipid thermotropic transition, and Arrhenius studies of enzyme and transport activities provide clear evidence of protein-lipid interactions and lipid modulation of the functions of intrinsic membrane proteins. Differential scanning calorimetry indicates that the enterocyte plasma membranes function in vivo within the transition. The enthalpy of the lipid transitions is decreased in the mature membranes owing to the presence of cholesterol and of protein-lipid interactions. Homeostatic mechanisms exist in the enterocyte to control the lipid fluidity and composition of the plasma membranes. The rate of cholesterol biosynthesis is one such mechanism, which affects selectively the cholesterol content and lipid fluidity of the microvillus but not of the basolateral membrane. Interactions with cations, including H^+ and Ca^{2+}, also modulate the lipid fluidity of the microvillus membrane by a relatively specific mechanism which involves one or more membrane proteins. Undoubtedly, additional regulatory mechanisms remain to be detected and characterized.

A striking aspect of intestinal function is the regional differentiation along the length of the organ, and it is reasonable to relate the lower fluidity of the distal ileal membranes to this functional specialization. It is known, for example, that the ileum is more efficient in net absorption of sodium and water than the more proximal intestine. The major driving force for this process is active transport of sodium out of the enterocyte across the basolateral membrane. Decreased lipid fluidity of the ileal enterocyte plasma membranes is expected to diminish passive back-fluxes of sodium and potassium and, thereby, to enhance the net unidirectional fluxes owing to the sodium-potassium pump. Since this ATPase mechanism is also known, however, to require a relatively fluid lipid microenvironment, it is not surprising that protein-lipid interactions assure such fluid microdomains, as indicated by Arrhenius plot break point temperatures significantly below that of the bulk membrane lipid (Table 10). That the ileal mechanism for sodium and water absorption is relatively efficient is further emphasized by recent studies of the maximal velocity, total enzyme sites and turnover number of the basolateral $(Na^+ + K^+)$ATPase (Brasitus and Schachter, 1984). The enzyme specific activity tested under maximal velocity conditions is greatest in the duodenal, and least in the ileal, preparations. The number of enzyme sites is similarly lower in the ileal as compared to the proximal segments. The turnover number, i.e., activity per enzyme site, is approximately 30% greater in the ileal as compared to the duodenal preparations.

Lastly, it is worth noting that further exploration of protein-lipid interactions in intestinal and other mammalian membranes should yield important information concerning basic molecular mechanisms, organ physiology and possible clinical applications. The demonstration that a diet rich in corn oil can markedly increase the specific activity of intestinal basolateral membrane ($Na^+ + K^+$)-ATPase (Brasitus et al., 1984a), or that a membrane-fluidizing regimen can partially relieve cholestasis owing to ethinyl estradiol in the rat (Storch and Schachter, 1984), suggests the utility of exploring dietary, drug and other modalities in the treatment of cell membrane-related disorders.

Acknowledgements

I am grateful for the support of National Institutes of Health grants AM21238, AM01483 and HL16851. It was my good fortune to collaborate with a group of outstanding co-workers in the studies described here. These individuals include Thomas A. Brasitus, Richard E. Abbott, Uri Cogan, Carol J. Livingstone, Meir Shinitzky, Judith Storch and Alan R. Tall. The expert technical assistance of Eugenia Dziopa and the unflagging secretarial profiency of Beulah Minkoff are gratefully acknowledged.

References

Andersen, J.M. and Dietschy, J.M. (1977) Regulation of sterol synthesis in 16 tissues of rat. 1. Effect of diurnal light cycling, fasting, stress manipulation of enterohepatic circulation, and administration of chylomicrons and Triton. J. Biol. Chem. 252, 3646–3651.

Anzai, K., Kirino, Y. and Shimizu, H. (1978) Temperature-induced change in the Ca^{2+}-dependent ATPase activity and in the state of the ATPase protein of sarcoplasmic reticulum membrane. J. Biochem. (Tokyo) 84, 815–821.

Bach, D., Bursuker, I. and Goldman, R. (1977) Differential scanning calorimetry and enzymic activity of rat liver microsomes in the presence and absence of Δ^1-tetrahydrocannabinol. Biochim. Biophys. Acta 469, 171–179.

Bloj, B. and Zilversmit, D.B. (1982) Heterogeneity of rabbit intestine brush border plasma membrane cholesterol. J. Biol. Chem. 257, 7608–7614.

Bloom, W. and Fawcett, D.W. (1968) *A Textbook of Histology*, pp. 560–569, W.B. Saunders, Philadelphia.

Boedeker, E.C., Higgins, Y.K., Donaldson, R.M., Jr. and Small, D.M. (1976) Selective solubilization of enzymic proteins from microvillus membranes (MVM) using non-ionic and anionic detergents. Gastroenterology 70, A-7/865.

Brasitus, T.A. and Keresztes, R.S. (1983) Isolation and partial characterization of basolateral membranes from rat proximal colonic epithelial cells. Biochim. Biophys. Acta 728, 11–19.

Brasitus, T.A. and Schachter, D. (1980a) Lipid dynamics and lipid-protein interactions in rat enterocyte basolateral and microvillus membranes. Biochemistry, 19, 2763–2769.

Brasitus, T.A. and Schachter, D. (1980b) Membrane lipids can modulate guanylate cyclase activity of rat intestinal microvillus membranes. Biochim. Biophys. Acta 630, 152–156.

Brasitus, T.A. and Schachter, D. (1982) Cholesterol biosynthesis and modulation of membrane cholesterol and lipid dynamics in rat intestinal microvillus membranes. Biochemistry 21, 2241–2246.

Brasitus, T.A. and Schachter, D. (1984) Lipid composition and fluidity of rat enterocyte basolateral membranes: regional differences. Biochim. Biophys. Acta 774, 138–146.

Brasitus, T.A., Schachter, D. and Mamouneas, T.G. (1979) Functional interactions of lipids and proteins in rat intestinal microvillus membranes. Biochemistry 18, 4136–4144.

Brasitus, T.A., Tall, A.R. and Schachter, D. (1980) Thermotropic transitions in rat intestinal plasma membranes studied by differential scanning calorimetry and fluorescence polarization. Biochemistry 19, 1256–1261.

Brasitus, T.A., Davidson, N.O. and Schachter, D. (1985) Variations in dietary triglycerol saturation alter the lipid composition and fluidity of rat intestinal plasma membranes. Biochim. Biophys. Acta 812, 460–472.

Brasitus, T.A., Yeh, K.-Y., Holt, P.R. and Schachter, D. (1984) Lipid fluidity and composition of intestinal microvillus membranes isolated from rats of different ages. Biochim. Biophys. Acta 778, 341–348.

Ceuterick, F., Peeters, J., Heremans, K., De Smedt, H. and Olbrechts, H. (1978) Effect of high pressure detergents and phospholipase on the break in the Arrhenius plot of *Azotobacter* nitrogenase. Eur. J. Biochem. 87, 401–407.

Colley, C.M. and Metcalfe, J.C. (1972) The localisation of small molecules in lipid bilayers. FEBS Lett. 24, 241–246.

De Kruijff, B., Demel, R.A. and Van Deenen, L.L.M. (1972) The effect of cholesterol and epicholesterol incorporation on the permeability and on the phase transition of intact *Acholeplasma laidlawii* cell membranes and derived liposomes. Biochim. Biophys. Acta 255, 331–347.

Dietschy, J.M. and Siperstein, M.D. (1965) Cholesterol synthesis by the gastrointestinal tract: localization and mechanisms of control. J. Clin. Invest. 44, 1311–1327.

Dipple, I. and Houslay, M.D. (1978) The activity of glucagon-stimulated adenylate cyclase from rat liver plasma membranes is modulated by the fluidity of its lipid environment. Biochim. J. 174, 179–190.

Douglas, A.P., Kerley, R. and Isselbacher, K.J. (1972) Preparation and characterization of the lateral and basal plasma membranes of the rat intestinal epithelial cell. Biochem. J. 128, 1329–1338.

Dowdle, E.B., Schachter, D. and Schenker, H. (1960) Active transport of Fe^{59} by everted segments of rat duodenum. Am. J. Physiol. 198, 609–613.

Estep, T.N., Mountcastle, D.B., Biltonen, R.L. and Thompson, T.E. (1978) Studies on the anomalous thermotropic behavior of aqueous dispersions of dipalmitoylphosphatidylcholine-cholesterol mixtures. Biochemistry 17, 1984–1989.

Farquhar, J.W. (1962) Human erythrocyte phosphoglycerides. I. Quantification of plasmalogens, fatty acids and fatty aldehydes. Biochem. Biophys. Acta 60, 80–89.

Finkelstein, J.D. and Schachter, D. (1962) Active transport of calcium by intestine: effects of hypophysectomy and growth hormone. Am. J. Physiol. 203, 873–880.

Forstner, C.G., Tanaka, K. and Isselbacher, K.J. (1968) Lipid composition of the isolated rat intestinal microvillus membrane. Biochem. J. 109, 51–59.

Fox, C.F. (1975) Phase transitions in model systems and membranes. In: *Biochemistry of Cell Walls and Membranes*, Vol. 2, pp. 279–306. Editor: C.F. Fox. University Park Press, Baltimore.

Fujita, M., Kawai, K., Asano, S. and Nakao, M. (1973) Protein components of two different regions of an intestinal epithelial cell membrane. Regional singularities. Biochim. Biophys. Acta 307, 141–151.

Glass, G.B.J. (1963) Gastric intrinsic factor and its function in the metabolism of vitamin B_{12}. Physiol. Rev. 43, 529–849.

Glynn, I.M. and Karlish, S.J.D. (1975) The sodium pump. Ann. Rev. Physiol. 37, 13–55.

Gómez-Fernández, J.C., Goñi, F.M., Bach, D., Restall, C. and Chapman, D. (1979) Protein-lipid interactions. A study of (Ca^{2+}-Mg^{2+})ATPase reconstituted with synthetic phospholipids. FEBS Lett. 98, 224–228.

Gray, J.P., Henderson, G.I., Dunn, G.D., Swift, L.L., Wilson, F.A. and Hoyumpa, A.M. (1981) Membrane fluidity and lipid composition of normal rat enterocytes: differentiation between brush border (BB) and basolateral membranes (BLM). Gastroenterology 80, 1162.

Grisham, C.M. and Barnett, R.E. (1972) The interrelationship of membrane and protein structure in the functioning of the (Na^+ + K^+)-activated ATPase. Biochim. Biophys. Acta 266, 613–624.

Haest, C.W.M., Verkleij, A.J., De Gier, J., Scheek, R., Ververgaert, P.H.J. and Van Deenen, L.L.M. (1974) The effect of lipid phase transitions on the architecture of bacterial membranes. Biochim. Biophys. Acta 356, 17–26.

Hauser, H., Gains, N., Semenza, G. and Spiess, M. (1982) Orientation and motion of spin-labels in rabbit small intestinal brush border vesicle membranes. Biochemistry 21, 5621–5628.

Hinze, H.-J. and Sturtevant, J.M. (1972a) Calorimetric investigation of the influence of cholesterol on the transition properties of bilayers formed from synthetic L-α-lecithins in aqueous suspension. J. Biol. Chem. 247, 3697–3700.

Hinz, H.-J. and Sturtevant, J.M. (1972b) Calorimetric studies of dilute aqueous suspension of bilayers formed from L-α-lecithins. J. Biol. Chem. 247, 6071–6075.

Hopfer, U., Nelson, K., Perotto, J. and Isselbacher, K.J. (1973) Glucose transport in isolated brush border from rat small intestine. J. Biol. Chem. 248, 25–32.

Johnson, C.F. (1966) Disaccharidase: localization in hamster intestine brush borders. Science 155, 1670–1672.

Kawai, K., Fujita, M. and Nakao, M. (1974) Lipid components of two different regions of an intestinal epithelial cell membrane of mouse. Biochim. Biophys. Acta 369, 222-233.

Kimberg, D.V., Schachter, D. and Schenker, H. (1961) Active transport of calcium by intestine: effects of dietary calcium. Am. J. Physiol. 200, 1256–1262.

Kimelberg, H.K. and Papahadjopoulos, D. (1974) Effects of phospholipid acyl chain fluidity, phase transitions and cholesterol on (Na^+ + K^+)-stimulated ATPase. J. Biol. Chem. 249,1071–1080.

Ladbrooke, B.D. and Chapman, D. (1969) Thermal analysis of lipids, proteins and biological membranes: a review and summary of some recent studies. Chem. Phys. Lipids 3, 304–356.

Ladbrooke, B.D., Williams, R.M. and Chapman, D. (1968) Studies on lecithin-cholesterol-water interactions by differential scanning calorimetry and X-ray diffraction. Biochim. Biophys. Acta 150, 333–340.

Lee, A.G. (1975) Functional properties of biological membranes: a physical-chemical approach. Prog. Biophys. Mol. Biol. 29, 3–56.

Lee, A.G. (1977a) Lipid phase transitions and phase diagrams. I. Lipid phase transitions. Biochim. Biophys. Acta 472, 237–281.

Lee, A.G. (1977b) Lipid phase transitions and phase diagrams. II. Mixture involving lipids. Biochim. Biophys. Acta 472, 285–344.

Le Grimellec, C., Giocondi, M.C., Carriere, B., Carriere, S. and Cardinal, J. (1982) Membrane fluidity and enzyme activities in brush border and basolateral membranes of the dog kidney. Am. J. Physiol. 242, 246–253.

Le Grimellec, C., Carriere, S., Cardinal, J. and Giocondi, M.C. (1983) Fluidity of brush border and basolateral membranes from human kidney cortex. Am. J. Physiol. 245, 227–231.

Lewis, B.A., Gray, G.M., Coleman, R. and Michell, R.H. (1975) Differences in the enzymic, polypeptide, glycopeptide, glycolipid and phospholipid compositions of plasma membranes from the two surfaces of intestinal epithelial cells. Biochem. Soc. Trans. 3, 752–753.

Linden, C.D., Wright, K.L., McConnell, H.M. and Fox, C.F. (1973) Lateral phase separations in membrane lipids and the mechanism of sugar transport in *Escherichia coli*. Proc. Natl. Acad. Sci. U.S.A. 70, 2271–2275.

Livingstone, C.J. and Schachter, D. (1980b) Lipid dynamics and lipid-protein interactions in rat hepatocyte plasma membranes. J. Biol. Chem. 225, 10902–10908.

Louvard, D., Maroux, S., Vannier, Ch. and Desnuelle, P. (1975) Topological studies on the hydrolases bound to the intestinal brush border membranes. 1. Solubilization by papain and Triton X-100. Biochim. Biophys. Acta 375, 236–248.

Lowe, P.J. and Coleman, R. (1982) Fluorescence anisotropy from diphenylhexatriene in rat liver plasma membranes. Biochim. Biophys. Acta 689, 403–409.

Maestracci, D. (1976) Enzymic solubilization of the human intestinal brush border membrane enzymes. Biochim. Biophys. Acta 433, 469–481.

Melchior, D.L. and Steim, J.M. (1976) Thermotropic transitions in biomembranes. Ann. Rev. Biophys. Bioeng. 5, 205–238.

Murer, H., Hopfer, U., Kinne-Safran, E. and Kinne, R. (1974) Glucose transport in isolated brush-border and lateral-basal plasma-membrane vesicles from intestinal epithelial cells. Biochim. Biophys. Acta 345, 170–179.

Murer, H., Ammann, E., Biber, J. and Hopfer, U. (1976) The surface membrane of the small intestinal epithelial cell. 1. Localization of adenyl cyclase. Biochim. Biophys. Acta 433, 509–519.

Nishi, Y., Yoshida, T.O. and Takesue, Y. (1968) Electron microscope studies on the structure of rabbit intestinal sucrase. J. Mol. Biol. 37, 441–444.

Oda, T. (1976) Molecular organization of cellular membranes. In: *Recent Progress in Electron Microscopy of Cells and Tissues*, pp. 1–23. Editors: E. Yamada, V. Mizuhira, K. Kurosumi and T. Nagano. University Park Press, Baltimore.

Okada, Y., Irimajiri, A. and Inouye, A. (1977) Electrical properties and active solute transport in rat small intestine. II. Conductive properties of transepithelial routes. J. Membr. Biol. 31, 221–232.

Oldfield, E. and Chapman, D. (1972) Dynamics of lipids in membranes: heterogeneity and the role of cholesterol. FEBS Lett. 23, 285–297.

Op den Kamp, J.A.F. (1979) Lipid asymmetry in membranes. Ann. Rev. Biochem. 48, 47–71.

Pang, K.-Y., Bresson, J.L. and Walker, W.A. (1983) Development of the gastrointestinal mucosal barrier. Evidence for structural differences in microvillus membranes from newborn and adult rabbits. Biochim. Biophys. Acta 727, 201–208.

Racker, E. (1972) Reconstitution of a calcium pump with phospholipids and a purified Ca^{2+}-adenosine triphosphatase from sarcoplasmic reticulum. J. Biol. Chem. 247, 8198–8200.

Razin, S. (1975) The mycoplasma membrane. Prog. Surf. Memb. Sci. 9, 257–342.

Reinert, J.C. and Steim, J.M. (1970) Calorimetric detection of a membrane-lipid phase transition in living cells. Science 168, 1580–1582.

Rose, R.C. and Schultz, S.G. (1971) Studies on the electrical potential profile across rabbit ileum. Effects of sugars, amino acids on transmural and transmucosal electrical potential differences. J. Gen. Physiol. 57, 639–663.

Rothman, J.E. and Lenard, J. (1977) Membrane asymmetry. Science 195, 743–753.

Rottem, S., Cirillo, V.P., De Kruijff, B., Shinitzky, M. and Razin, S. (1973) Cholesterol in Mycoplasma membranes. Correlation of enzymic and transport activities with physical state of lipids in membranes of *Mycoplasma mycoides* var. *capri* adapted to grow with low cholesterol concentrations. Biochim. Biophys. Acta 323, 509–519.

Saini, P.K. and Done, J. (1972) The diversity of alkaline phosphatase from rat intestine. Isolation and purification of the enzymes. Biochim. Biophys. Acta 258, 147–153.

Schachter, D., and Shinitzky, M. (1977) Fluorescence polarization studies of rat intestinal microvillus membranes. J. Clin. Invest. 59, 536–548.

Schachter, D., Dowdle, E.B. and Schenker, H. (1960) Active transport of calcium by the small intestine of the rat. Am. J. Physiol. 198, 263–268.

Schachter, D., Cogan, U. and Shinitzky, M. (1976) Interaction of retinol and intestinal microvillus membranes studied by fluorescence polarization. Biochim. Biophys. Acta 448, 620–624.

Schmitz, J., Preiser, H., Maestracci, D., Ghosh, B.K., Cerda, J.J. and Crane, R.K. (1973) Purification of the human intestinal brush border membrane. Biochim. Biophys. Acta 323, 98–112.

Schwarz, S.M., Hostetler, B., Ling, S., Lee, L. and Watkins, J.B. (1982) Fluorescence polarization studies of the small intestinal microvillus membrane during development. Gastroenterology 82, 1174a.

Seeman, P. (1972) The membrane actions of anesthetics and tranquilizers. Pharmacol. Rev. 24, 583–655.

Sinensky, M. (1977) Isolation of a mammalian cell mutant resistant to 25-hydroxy cholesterol. Biochem. Biophys. Res. Commun. 78, 863–867.

Sinensky, M. (1978) Defective regulation of cholesterol biosynthesis and plasma membrane fluidity in a Chinese hamster ovary cell mutant. Proc. Natl. Acad. Sci. U.S.A. 75, 1247–1249.

Steim, J.M., Tourtellotte, M.E., Reinert, J.C., McElhaney, R.N. and Rader, R.L. (1969) Calorimetric evidence for the liquid-crystalline state of lipids in a biomembrane. Proc. Natl. Acad. Sci. U.S.A. 63, 104–109.

Storch, J. and Schachter, D. (1984) A dietary regimen alters hepatocyte plasma membrane lipid fluidity and ameliorates ethinyl estradiol cholestasis in the rat. Biochim. Biophys. Acta 798, 137–140.

Storch, J., Schachter, D., Inoue, M. and Wolkoff, A.W. (1983) Lipid fluidity of hepatocyte plasma membrane subfractions and their differential regulation by calcium. Biochim. Biophys. Acta 727, 209–212.

Takesue, Y. and Sato, R. (1968) Biochemical and morphological characterization of microvilli isolated from intestinal mucosal cells. J. Biochem. (Tokyo) 64, 885–893.

Thorneley, R.N.F., Eady, R.R. and Yates, M.F. (1979) Nitrogenases of *Klebsiella pneumoniae* and *Azotobacter chroococcum*. Complex formation between the component proteins. Biochim. Biophys. Acta 403, 269–284.

Tourtelotte, M.E. (1972) Mycoplasma membranes: structure and function. In: *Membrane Molecular Biology*, pp. 439–470. Editors: C.F. Fox and A.D. Keith. Sinauer Associates, Stamford, CT.

Weiner, I.M. and Lack, L. (1968) Bile salt absorption; enterohepatic circulation. In: *Handbook of Physiology: Sect. 6. Alimentary Canal*, Vol. III, pp. 1439–1456. Editor: C.F. Code, American Physiological Society, Washington, D.C.

Watts/De Pont (Eds.)
Progress in Protein-Lipid Interactions

CHAPTER 7

Use of photoreactive phospholipids for the study of lipid-protein interactions

ROBERTO BISSON and CESARE MONTECUCCO

C.N.R. Center for the Study of the Physiology of Mitochondria and Laboratory of Molecular Biology and Pathology, Institute of General Pathology, University of Padova, Via Loredan 16, 35100 Padova, Italy

I. Introduction

The fundamental role of lipid-protein interactions in membrane structure and function has been demonstrated by such findings as the effect of different classes of lipids or perturbing agents on the activity of membrane-bound enzymes (Singer, 1974; Sanderman, 1978; Chapman et al., 1982).

Different techniques (most of them are discussed in the present book) have contributed to the present knowledge of the structural and dynamic aspects of membrane components. These approaches have mainly focused on the lipid part of membranes. For this reason and because of the great difficulty in obtaining crystals of membrane proteins amenable to high resolution X-ray diffraction, no information at the atomic level is at present available for any membrane protein. It should be pointed out that it is highly unlikely that this level of resolution will ever be attained for the lipid component because of its high mobility and of the necessary presence of detergents and other amphiphyles in integral protein crystals (Michel, 1983).

Chemical modification and cross-linking have provided highly valuable information on soluble proteins and can be used to identify the polypeptide segments involved in the interaction with lipids (De Pierre and Ernster, 1977; Ji, 1983). This approach has relied upon the presence of reactive functional groups on the protein surface. However, a common feature of integral protein primary sequences is the presence of unreactive aliphatic residues clustered in hydrophobic stretches of 18–24 amino acids. These segments are suspected to be embedded in the membrane to form the lipid-protein boundary region and are characterized by a very low chemical reactivity. Hence, new reagents have to be introduced in order to label specifically the hydrophobic domain of integral proteins. Such probes should fulfill two main requirements: (a) specific labelling of the peptide segments exposed to lipids; and (b) high reactivity and broad specificity versus the different lateral residues.

At the present time the reagents which more strictly meet these demands are photoreactive and hydrophobic. These probes are, in fact, stable in the dark, where they show little or no chemical reactivity. After dispersion in the membrane they can be converted by illumination to highly unstable intermediates, able to form covalent derivatives with the lipid-exposed protein segments. The photoactivatable group can be part of a small lipophilic molecule or anchored to a selected position of one fatty acid chain of a phospholipid.

At variance from spectroscopic methods this is a direct approach that transforms a short-lived non-covalent interaction between a phospholipid molecule and the protein in a stable permanent one. This chemical modification can be traced at the different levels of organization of the protein in the membrane.

Hydrophobic photolabelling has been recently reviewed (Brunner, 1981; Robson et al., 1982; Bayley, 1982, 1983; Bisson and Montecucco, 1984). The present paper will focus mainly on the use of photoreactive phospholipids to probe the hydrophobic domain of integral proteins and their interaction with lipids.

II. Photoreactive groups

Among the several photoreactive groups available only carbene and nitrene precursors have been used extensively. Some of them are shown in Table 1. Illumination at the proper wavelength leads to the release of nitrogen and the formation of a carbene or a nitrene. As it appears from Table 1 stability in the dark (absence of chemical reactivity) and the possibility of rearrangement of the photogenerated intermediate is largely influenced by the nature of the group directly linked to the photoactive moiety. This also determines the absorbance spectrum of the reagent and hence its wavelength of illumination. Due to the possible damage induced by irradiation on the biological system under study, groups absorbing in the visible/near U.V. are preferable.

Once formed, the unstable, electron-deficient carbene or nitrene can follow different pathways of reaction: (a) react with neighbouring molecules; (b) rearrange intramolecularly or (c) react with themselves as shown in Figs. 1 and 2.

a. Reaction with neighbouring molecules. It has been shown that both carbenes and nitrenes react preferentially with the electron-rich amino or thiol groups and to a lower extent with the rather inert primary C-H bonds (Knowles, 1972; Bayley

TABLE 1
Some photoreactive groups used in protein labelling.

Formula and name	Stability in the dark in neutral solution	Possibility of intra-molecular rearrangements after photolysis	Wavelength for photolysis (nm)
$R-N_3$ Alkylazides	Good	High	<300
$Ar-N_3$ Arylazides	Excellent	Low	$\leqslant 300$
$O_2N-Ar-N_3$ Nitroarylazides	Excellent	Low	>350
$-O-C(=O)-C(CF_3)=N=N$ Trifluorodiazopropionyloxy	Good	Moderate	<300
$Ar-CH\langle N{=}N$ (diazirine ring) Aryldiazirine	Good	Moderate	>350
$Ar-C(CF_3)\langle N{=}N$ (diazirine ring) Aryltrifluoromethyldiazirine	Good	Low	350

and Knowles, 1977; Chowdry and Westeimer, 1979; Staros, 1980). Reactive groups are present in the membrane only on protein, while hydrocarbon chains constitute the majority of the hydrophobic membrane phase. Hence, highly reactive intermediates are desirable since they allow one to label the aliphatic amino acid residues, largely present on the hydrophobic sector on intrinsic proteins. However, increasing reactivity leads to a parallel enhancement in the insertion of the probe into lipids instead of proteins. This is of great advantage in lipid labelling experiments, but results in a lower efficiency of protein labelling. The problem can be overcome by increasing the amount of probe or with reagents of very high specific radioactivity. The latter possibility is preferable since it minimizes the membrane perturbation introduced by the probe.

There is evidence indicating that, in general, carbenes are more reactive, and hence less selective, than nitrene. For example phenylcarbene reacts with secondary C-H bonds 6-times faster than with primary C-H bonds, while for phenylnitrene the difference is more than a hundred-fold (Gutsche et al., 1962; Hall et al., 1965). However, reactivity is strongly influenced by the group linked to the carbene or nitrene. For example it has been shown that the lifetime* of the

* The lifetime of an intermediate can be used as a measure of reactivity when the different compounds are compared in the same system.

intermediate is decreased orders of magnitude in going from antracene-nitrene to naphthyl-nitrene and to phenyl nitrene (Reiser et al., 1966, 1968). Moreover the presence of electron-withdrawing substituents in the phenyl ring such as a nitro group greatly decreases the lifetime of the nitrene intermediate (Reiser and Leyshon, 1970). It should be emphasized that the yield of cross-linking depends not only on the reactivity of the intermediate, but also on the nature of the

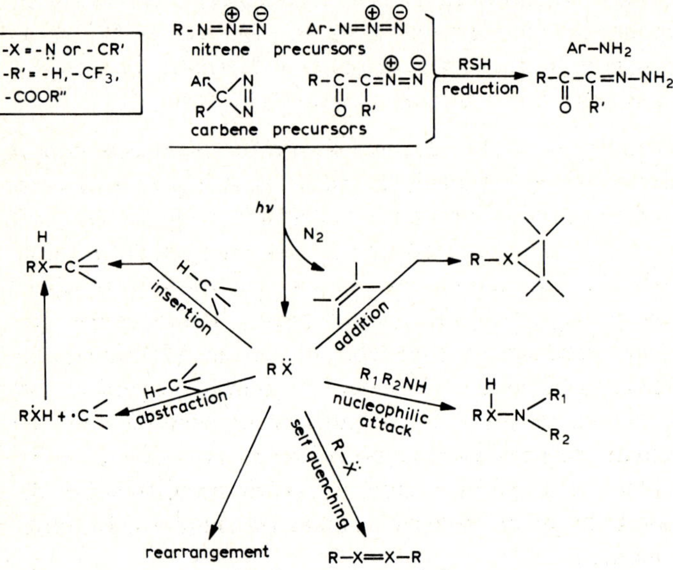

Fig. 1. Pathways of reaction for carbenes and nitrenes.

Fig. 2. Possible rearrangements of carbenes and nitrenes reported by: (a) Kyba and Abramovitch, 1980; (b) Chapman and LeRoux, 1978, Nielsen and Buchardt, 1982; (c) Chowdry and Westheimer, 1979; (d) Smith and Knowles, 1975.

reacting partners present in the membrane. Most studies indicate that this parameter is the governing factor in determining the yield of the different possible derivatives when more than one kind of bond is present in the reacting partner. For example even the most reactive trifluorophenylcarbene, when photolized in methanol, shows 95% insertion into the O-H bond with respect to the three-times more frequent primary C-H bonds (Richards and Brunner, 1980). Hence due to the presence of nucleophiles (S-H, S-CH_3, nitrogen derivatives, O-H and aromatic rings) on the membrane-embedded protein segments there may be little or no difference in the labelling patterns found with carbenes and nitrenes (see Sections V.2 and V.3 for a discussion of the available experimental data).

b. Intramolecular rearrangement. This reaction should be minimized since it may lead to a decreased efficiency of labelling and because it generates a less unstable intermediate which can react specifically as depicted in Fig. 2. This 'dark reaction' can be detected because it proceeds in the dark after photoactivation. An example of such a labelling has been reported for glycophorin and is discussed in Section V.3. The schemes reported in Fig. 2 have been described in the literature, but cannot be generalized since the tendency to rearrange is dramatically influenced by substituents and chemical environment. Moreover appropriate tests have been carried out only in few cases. For example it has been shown that a nitro group in the aromatic ring makes reaction b.I negligible and that a trifluoromethyl group in place of a hydrogen atom in aryl diazirine stabilizes the linear diazo intermediate reducing its chemical labelling (Brunner and Richards, 1980; Nielsen and Buchardt, 1982).

c. Self quenching. It has been demonstrated that a major route of decay of the nitrene intermediate (but this should be valid also for carbenes) in hydrocarbon solvents is the reaction with itself or with the still unconverted precursors (Reiser et al., 1968). The extent of this reaction increases with increasing frequency of collision, which is determined by the concentration of the probe in the membrane* and by its diffusion coefficient. This problem is largely overcome when probes of high specific radioactivity are used. In membrane studies self quenching acquires relevance in the case of poorly reactive components.

The carbene and nitrene precursors used in membrane studies are generally stable in the dark, but they have been shown to be easily reduced by thiols, as shown in Fig. 1 (Staros et al., 1978; Takagaki et al., 1980).

* Even though the probe concentration in the sample volume is low, its actual concentration in the membrane is usually higher than millimolar.

III. Small lipophilic photoreactive reagents

Historically, in membrane studies photoactivatable groups were first introduced in small molecules such as those reported in Fig. 3. Due to their hydrophobic character they partition readily in the membrane. Another advantage is their small size that causes only a minor modification on the protein facilitating the isolation and identification of the labelled fragments. On the other hand the determination of the integral nature of membrane proteins and the labelling of lipid-exposed sites may be affected by the absorption of the probe onto hydrophobic pockets present on the hydrophilic protein surface. This event has been shown to occur with hydrophobic molecules of different dimensions and chemical characteristics (Ueda et al., 1976; Halsey et al., 1978; Stefanini et al., 1979; Krebs et al., 1984). Moreover small reagents cannot be used in the study of lipid-protein interactions because they are unrelated to the structure of the lipids present in the membrane. For the same reason they could be excluded from areas of strong lipid-protein interaction. A further limitation is their undefined location in the lipid bilayer. To overcome these and other limitations discussed in detail elsewhere (Bisson and Montecucco, 1985) photoreactive lipid analogues have been introduced by Chakrabarti and Khorana (1975).

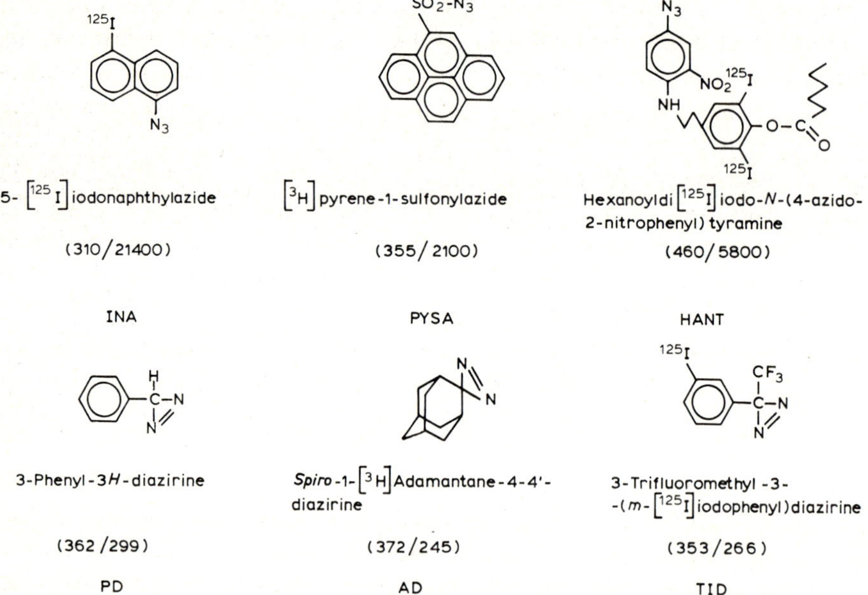

Fig. 3. Small lipophilic photoreactive reagents: the most used of these probes are shown together with their absorption maxima useful for irradiation and the corresponding εmM. INA, Bercovici and Gitler, 1978; PYSA, Sator et al., 1979; HANT, Hebdon et al., 1980; PD, Smith and Knowles, 1975; AD, Bayley and Knowles, 1980; TID, Brunner and Semenza, 1981.

IV. Photoreactive lipids

Table 2 lists the formulae and some properties of lipid analogues which have been used in membrane studies. The photoreactive group is located at different positions of one fatty acid chain. In principle this should allow one to label proteins at different depths in the membrane monitoring their penetration in the lipid bilayer (see Section V.2). As apparent from Table 2 all probes are made radioactive since this is the most sensitive method to determine the pattern of labelling, allowing the use of the probe in trace amounts.

1 SYNTHESIS OF PHOTOREACTIVE PHOSPHOLIPIDS

Among the various classes of phospholipids only photoactivatable lecithins and sphingomyelins have been prepared so far. Their synthesis involves the preparation of the photoreactive group in a form suitable to be inserted into a fatty acid chain. The photoreactive fatty acid is then coupled to lysolecithin. Radioactivity has been inserted at different steps in different protocols. Figures 4 and 5 show the preparitive schemes of aromatic and linear carbene precursors. Preparative yields, when available, are reported together with the final yield shown in square brackets. The nitrophenylazido group in the form of 4-fluoro-3-nitro azide, used for its coupling to amino or a hydroxyl fatty acid, is commercially available.

The insertion of a photoreactive group in a fatty acid chain does not present, in general, great difficulties as shown by the examples in Fig. 6. However, the complexity of the preparation of a photoreactive fatty acid increases if a radioactive derivative is needed. In general terms radioactivity can be inserted on the photoreactive group itself, on the fatty acid chain, or on the phospholipid head group. The first two possibilities are preferable because of the more stable linkage of radioactivity to the photocrosslinked derivative. In practical terms it is to be considered that the third way is much easier and that it has not precluded even the determination of the modified residues by amino acid sequencing (see below and Section V.3 for more details).

Figure 7 shows three possible ways of introducing radioactive isotopes in the fatty acid chain before coupling the photoreactive group. Different levels of specific radioactivities can be attained with the different methods as reported in Table 2.

To prepare a photoreactive phosphatidylcholine, only mild methods avoiding the presence of reducing agents can be used. In fact, most carbene and nitrene precursors are easily reduced. Figure 8 reports two synthetic routes, which have been used successfully in the preparation of several different phosphatidylcholine analogues. Figure 9 shows a method used to introduce radioactivity only at the last step of the synthesis. In this way, highly radioactive or deuterated derivatives can be prepared because all three methyl groups of the choline moiety are deut-

TABLE 2
Amphipathic photoactivatable phosphatidylcholines.

General structure:

$$\begin{array}{l} R_1\text{-O-CH}_2 \\ R_2\text{-O-CH} \\ \quad\ \ \text{CH}_2\text{-O-}\overset{O^{\ominus}}{\underset{O}{P}}\text{-O-CH}_2\text{-CH}_2\text{-}\overset{\oplus}{N}(CH_3)_3 \ \ [X] \end{array}$$

Code No.	R_1	R_2 (Structure)	X	Ci/mol	References[a]
I	palmitoyl	F_3C, N_2, O, * , O		18.4	
II	palmitoyl	N, N, O, *, O		48	1
III	myristoyl	*, O	⊕N, N, N	5.6	
IV	palmitoyl	N, N, CF_3, ^3H, O		3600	2
V	palmitoyl	N, N, CF_3, S-S, O		—	3
VI	palmitoyl	N_3, S-S, O		—	3
VII	palmitoyl	O_2N, N_3, O	$-N(C^3H_3)_3$	3900	4
VIII	myristoyl	N_3, NH, NO_2, O	$-N(^{14}CH_3)_3$	177	4
IX	palmitoyl	O, N_3, NO_2, *, O		0.74	5
X	palmitoyl or	^3H, ^3H, N_3, O		100	6
XI	stearoyl	N_3, ^3H, ^3H, O			6

* shows the location of the ^{14}C isotope.

[a] References: 1, Radhakrishnan et al. (1981); 2, Brunner et al. (1983); 3, Brunner and Richards (1980); 4, Bisson and Montecucco (1981); 5, Moonen et al. (1974); 6, Stoffel et al. (1982).

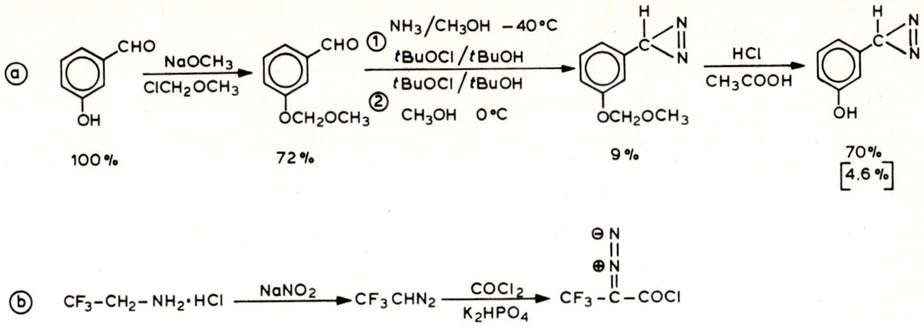

Fig. 4. Schemes of preparation of *m*-(methoxy-methylene oxy)-phenyl-3*H*-diazirine (a) (Radhakrishnan et al., 1981), and of 2-diazo-3,3,3-trifluoropropionyl chloride (b) (Chowdry et al., 1976).

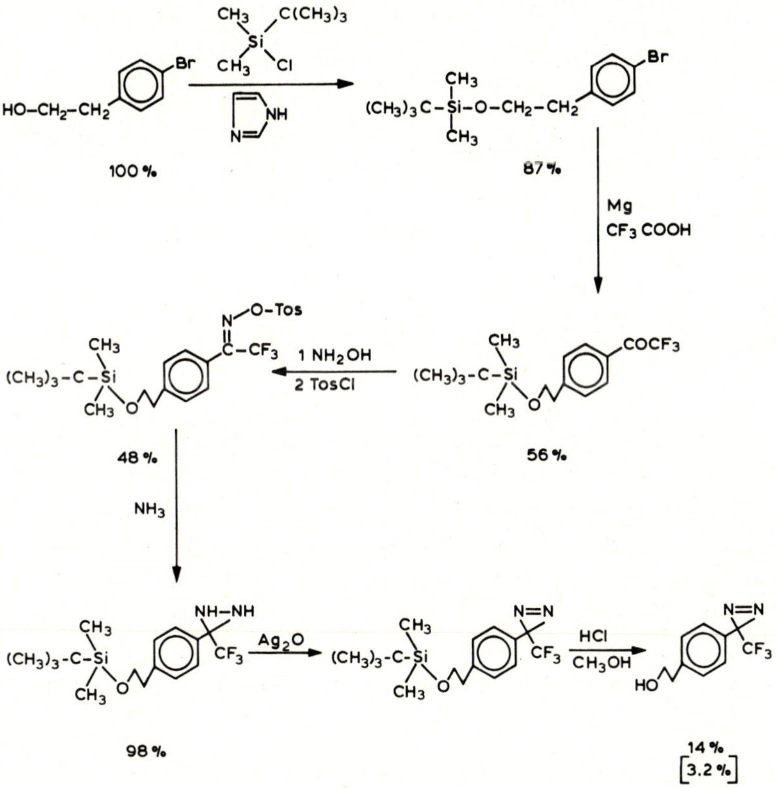

Fig. 5. Preparation scheme of 2-[4-[(trifluoromethyl)diazirinyl-phenyl]ethanol (Brunner et al., 1983).

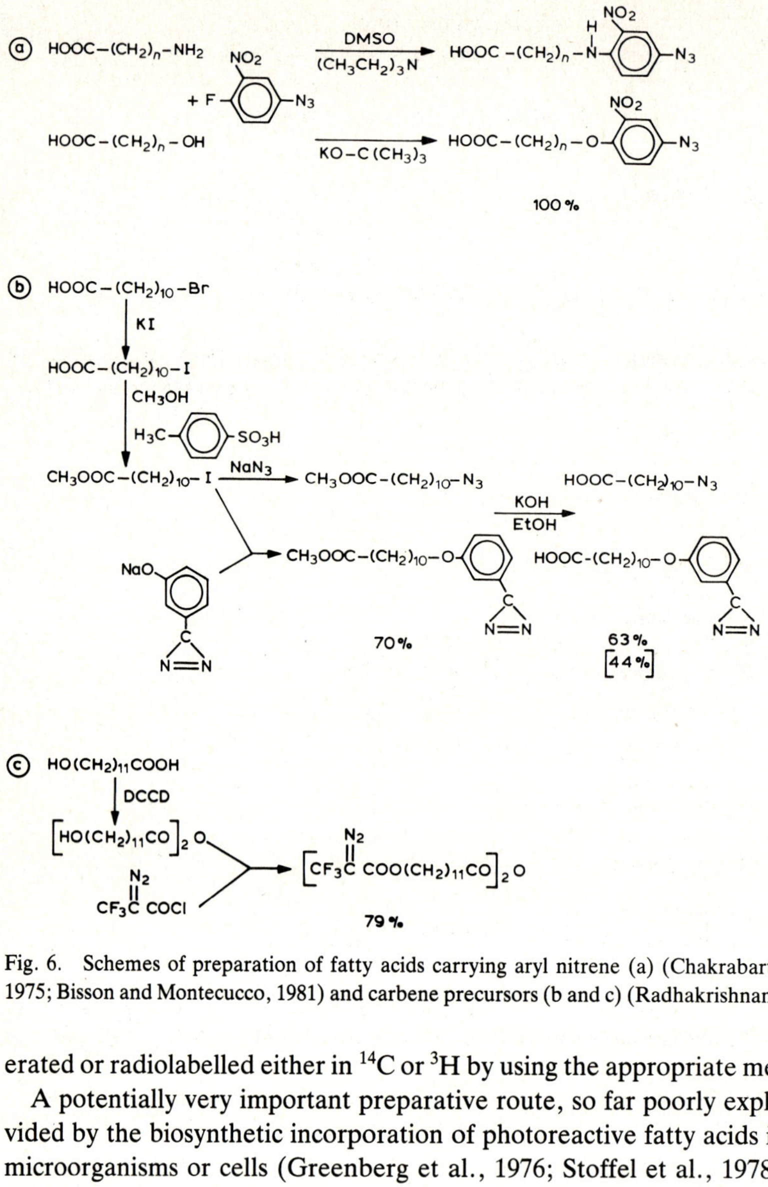

Fig. 6. Schemes of preparation of fatty acids carrying aryl nitrene (a) (Chakrabarti and Khorana, 1975; Bisson and Montecucco, 1981) and carbene precursors (b and c) (Radhakrishnan et al., 1981).

erated or radiolabelled either in ^{14}C or ^{3}H by using the appropriate methyl iodide.

A potentially very important preparative route, so far poorly explored, is provided by the biosynthetic incorporation of photoreactive fatty acids into cultured microorganisms or cells (Greenberg et al., 1976; Stoffel et al., 1978; Leblanc et al., 1982).

V. Use

Photoreactive phospholipids have been used to obtain information at the different levels of membrane organization. A possible limitation in their use is related to

(a) $Cl(CH_2)_6OH \xrightarrow{Na^{14}CN} N^{14}C(CH_2)_6OH \xrightarrow{NaOH} HOO^{14}C(CH_2)OH$

100 % — 55–65 % — 65 % [42 %]

(b) $X\text{-}CH_2\text{-}CH_2OH \xrightarrow{\text{dipyridine Cr(VI) oxide}} X\text{-}CH_2\text{-}CHO \xrightarrow{NaB^3H_4} X\text{-}CH_2\text{-}CH(^3H)\text{-}OH$

100 % — 12 %

$\xrightarrow{(F_3C\text{-}O\text{-}SO_2)_2O} X\text{-}CH_2\text{-}CH(^3H)\text{-}OSO_2CF_3 \xrightarrow{HO(CH_2)_6COOC(CH_3)_3,\ K_2CO_3} X\text{-}CH_2\text{-}CH(^3H)\text{-}O(CH_2)_6COOC(CH_3)_3$

$\xrightarrow{CF_3COOH} X\text{-}CH_2\text{-}CH(^3H)\text{-}O(CH_2)_6COOH$

56 % [6.7 %]

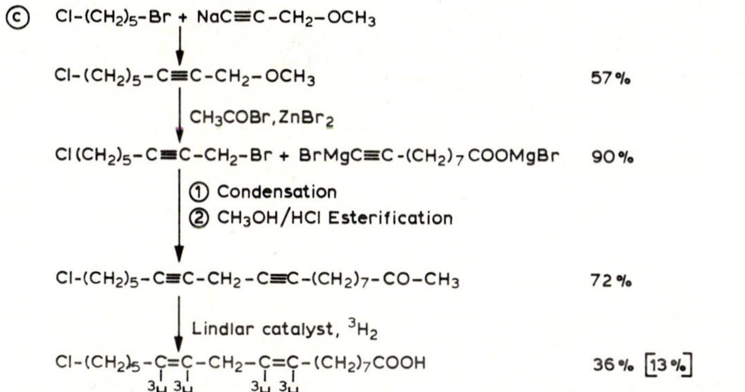

Fig. 7. Preparation of radioactive fatty acids with functional groups suitable to be coupled to photoreactive group (a) Chakrabarti and Khorana, 1975; (b) Brunner et al., 1983; (c) Stoffel et al., 1982.

the difficult insertion of exogenous phospholipids into preformed membranes.

This problem can be overcome with phospholipid exchange proteins or with the biosynthetic incorporation of the fatty acid analogue into the cell membrane phospholipids (Akeroyd and Wirtz, 1982; Greenberg et al., 1976; Leblanc et al., 1982). However, recently it has been found that photoreactive phosphatidylcholines can transfer with high efficiency from liposomes to red blood cell ghosts, sarcoplasmic reticulum vesicles and acetylcholine receptor-rich membranes (Brunner et al., 1983; Gutweniger and Montecucco, 1984; Giraudat et al., 1985). This is probably related to the higher free monomer concentration of these lecithin analogues with respect to the unmodified phosphatidylcholine.

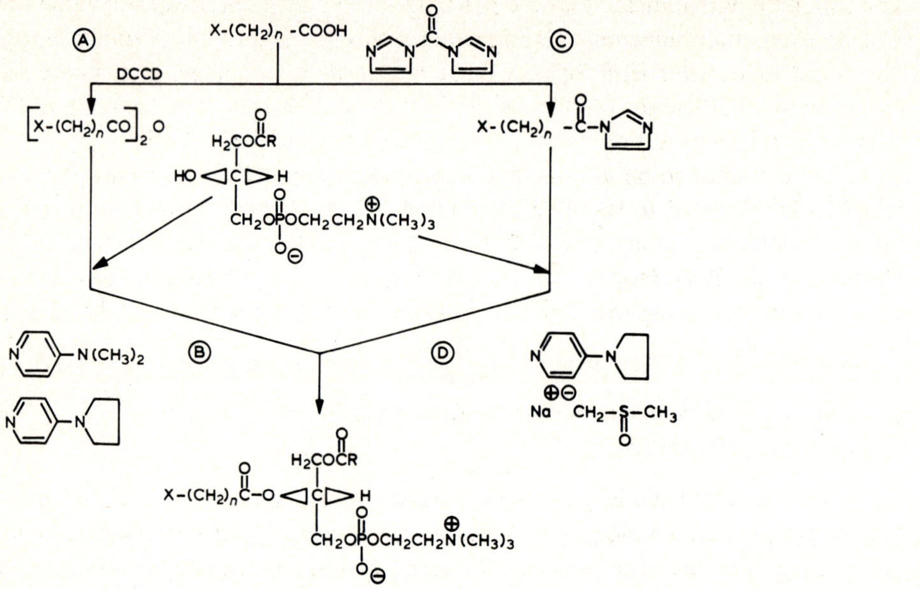

Fig. 8. Preparation of photoreactive lecithins. (A) Formation of fatty acid anhydrides with DCCD and (B) coupling to lysolecithin (Gupta et al., 1977; Patel et al., 1979); (C) formation of fatty acid imidazolides and (D) coupling to lysolecithin (Warner and Benson, 1977, and our unpublished observations).

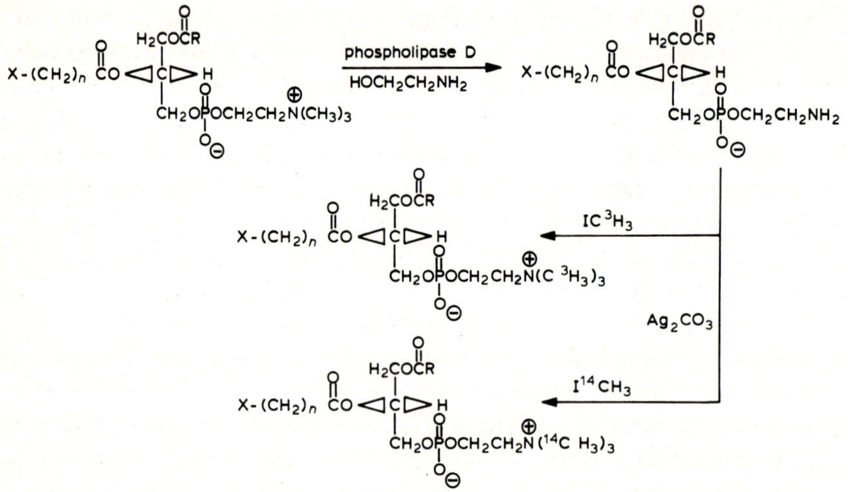

Fig. 9. Introduction of radioactivity at the methyl group of the choline moiety of lecithin (Smith et al., 1977).

The available data indicate that the position of the photoreactive group along the hydrocarbon chain appears to influence dramatically the ability of phospholipid analogues to transfer from liposomes to the erythrocyte membrane; probe IV inserts with an efficiency of 70%, while in the same conditions probes II and III show only a 20% insertion (Ross et al., 1982; Brunner et al., 1983).

All the evidence so far obtained demonstrates that photoreactive phospholipid labelling is restricted to the hydrophobic part of the lipid bilayer (Bisson et al., 1979b; Montecucco et al., 1980, 1983; Prochaska et al., 1980; Bisson et al., 1982; Brunner et al., 1983; Hoppe et al., 1983a; Takagaki et al., 1983a, b). Here, relevant examples showing the range of possibilities of this approach will be discussed.

1. LABELLING OF INTRINSIC PROTEINS AND DETERMINATION OF THEIR LIPID-EXPOSED SEGMENTS

The distribution of radioactivity of photoreacted phospholipids among the different polypeptides of a biological membrane allows one to determine which of them are of integral nature. The pattern of labelling is most conveniently determined by SDS-polyacrylamide gel electrophoresis (SDS-PAGE), the most widely used method to classify and characterize membrane proteins. With this procedure a most effective separation of non-covalently bound photoreacted lipids from proteins is obtained. The radioactivity associated with the gel can be determined either by fluorography (qualitatively) or by gel slicing and counting (quantitatively).

An application of these techniques is shown by the work of Stoffel et al. (1978) and Capone et al. (1983) who have used the biosynthetic approach to incorporate photoreactive fatty acids in the phospholipids of fibroblasts grown in media containing the fatty acid analogues. The cells were then infected with vesicular stomatitis virus and the virions recovered and illuminated. Both studies, which employed aliphatic nitrenes and aromatic carbenes, showed an almost exclusive labelling of the G protein, demonstrating that this polypeptide of the virus membrane is intrinsic and exposed to lipids. In this case the interpretation of the pattern of labelling was simple because of the reduced number of proteins present in the membrane. However, with most native membranes the interpretation of the pattern of labelling may be hampered by the limited resolving power of SDS-PAGE. The problem of protein bands overlapping can be overcome in several ways: (a) isolation of the protein under study (this procedure can be applied when a micromethod for the isolation of the protein is available, or when a large amount of membrane is available); (b) immunoprecipitation with specific antibodies (this method permits to determine if the protein under study is or is not labelled, even when present in a very small proportion with respect to the other proteins of the membrane); (c) when the protein is available in isolated form and when it can be reconstituted into an active lipid-protein complex the problem of polypeptide overlapping on the gel is largely avoided.

A membrane-bound enzyme, which has been labelled in an isolated and reconstituted form is the (Na^+-K^+)ATPase. This protein couples the transport of Na^+ and K^+ across the plasma membrane to the hydrolysis of ATP (Jørgensen, 1983). It can be isolated from different sources and it is composed of a large polypeptide termed α (M_r, 105 000) and of a sialoglycoprotein subunit termed β (M_r, 40 000). In some preparations a smaller component of M_r 10–12 000 is also present. This polypeptide is labelled with azido derivatives of ouabain and strophantidin, two specific inhibitors of the (Na^+-K^+)ATPase activity. Figure 10 shows the profile of labelling with probes VII and VIII of the ATPase extracted from two different unrelated sources, such as pig kidney and the electric organ of the eel *Electrophorous electricus*. Both the α and β subunits are labelled with both the 'shallow' probe VII and the 'deeper' probe VIII. This indicates that both subunits contribute to form the lipid-protein boundary of the (Na^+-K^+)ATPase. No radioactivity is present at the level of the 12 000 protein, whose presence is evident in the *Electrophorous* preparation. As in any other probe method when

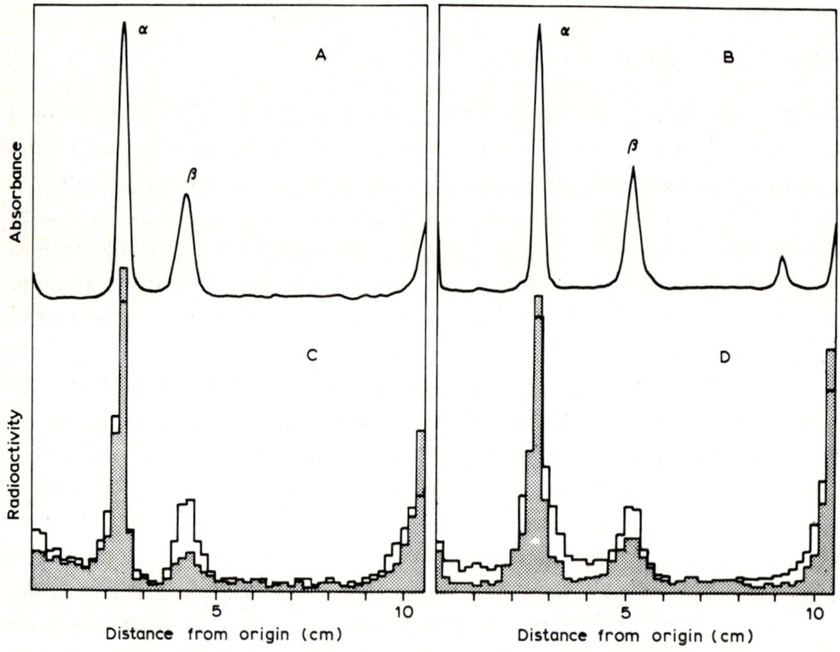

Fig. 10. Labelling of (Na^+,K^+)ATPase with arylazidophospholipids: (A) Coomassie Blue staining profile of the ATPase isolated from pig-kidney and (C) corresponding labelling patterns with probe VII (shaded area) and with probe VIII (open area); (B) Coomassie Blue staining profile of the enzyme isolated from *Electrophorus electricus* and (D) corresponding labelling patterns with probe VII (shaded area) and with probe VIII (open area) (after Montecucco et al., 1981).

a polypeptide is not labelled or modified nothing can be said about its location. In this case it could be either external to the membrane or shielded by the labelled subunits, and hence not exposed to lipids. A third possibility is that it does not contain reactive residues as those present in the membrane-embedded segments of the α and β subunits. This is less likely since this polypeptide is labelled by azido analogues of ouabain and strophantidin.

The two probes used in this study are expected to label different regions of the membrane, since in probe VII the photoreactive group is strictly localized at the polar head group level while in probe VIII the photoreactive group is placed at the fatty acid methyl terminus. This is supported by the different distribution of the radioactivity of the two probes among the α and β subunits. It is noteworthy that enzymes from such evolutionary distant sources as a fish and a mammal give rise to the same pattern of labelling. This kind of comparison may help to overcome the problem of selective labelling of certain residues (see Section II) since residues exposed to lipids are highly variable within the only constraint of being hydrophobic (Van Heyne, 1981). Moreover, the approach using enzymes isolated from unrelated sources such as different animals or organs may be very useful in establishing correlations among different subunits, particularly when highly complex multi-subunit enzymes are studied.

2 PHOTOREACTIVE PHOSPHOLIPIDS AS CHEMICAL RULERS

The possibility of anchoring the photoreactive group at a selected position of a hydrocarbon chain of a phospholipid molecule, in principle, allows one to label protein regions exposed to lipids at different depths of the membrane. However the mobilities of the various carbon atoms of the fatty acid chain increase with increasing distance from the carboxyl group. This lowers the mean residence time of the photoreactive group around the position occupied in the fully extended chain. In other words, the probability of the formation of kinks, which would bring the photoreactive group nearer to the membrane surface, increases along the chain. This problem is greatly emphasized when reactive residues are unevenly distributed across the membrane (see below).

An extensive analysis of the possibility of using photoreactive phospholipids as membrane chemical rulers has been carried out only on liposomal model systems (Gupta et al., 1979a, b; Curatolo et al., 1981). The analysis of the sites of intermolecular cross-linking of carbene-generating phospholipids in liposomes of pure saturated lecithin shows a broad distribution around the expected depth in the bilayer. The reactivity of both carbenes and nitrenes were tested toward saturated and unsaturated phospholipids. Arylnitrenes were found to be less efficient than the corresponding carbenes in forming phospholipid dimers with dipalmitoylphosphatidylcholine (0.6 versus 10–15%). In similar studies with probes V and VI Brunner and Richards found 2.8% for arylnitrene and 18.5% for the trifluoromethylarylcarbene. In the case of unsaturated phospholipids a high cross-linking efficiency was found also with aliphatic azides, which offer the advantage of a reduced size, and hence introduce a minimal perturbation compared to the relatively bulky substituted aromatic groups (Stoffel et al., 1982). This is particularly important in lipid model systems where the maximal attention has to be paid to the possible role of the probe in interfering with the phenomenon under investigation.

In mixed saturated and unsaturated phosphatidylcholine liposomes the carbene-phospholipid analogue forms phospholipid dimers three-times more efficiently than with the unsaturated lecithin, indicating that the photoreactive group can fold back to reach the more reactive olefinic bond. A reaction at a level different from that occupied in the fully extended conformation is emphasized by the presence in the membrane of reactive nucleophiles. This is the case of the sulfur and nitrogen derivatives and of the aromatic side chains, which are often present in the hydrophobic domain of integral membrane proteins. Brunner and Richards (1980) have tested this possibility with gramicidin A using probes V and VI carrying the photoactivatable group at the methyl terminus. These probes are expected to label predominantly those amino acid residues exposed to the central hydrophobic region of the lipid bilayer. On the contrary both the carbene- and nitrene-generating probes labelled exclusively the Trp residues present near the membrane surface. The much higher reactivity of the Trp residue with respect to the aliphatic sidechains present in the central membrane region of gramicidin A can explain this finding. It appears from this study that the large difference of reactivity between the trifluoromethylarylcarbene and the phenylnitrene, found in the labelling of primary C-H bonds, is abolished or minimized by the high reactivity of certain protein residues. The bulky tryptophanes may have themselves facilitated the folding back of the photoreactive group. An irregular and rough surface exposed to lipids appears from the available primary sequences to be a general occurrence in membrane proteins. These two facts suggest that photoreactive phospholipids may be unreliable as depth probes. Within these limits useful information can still be obtained with lipid derivatives bearing the photoreactive group at very different positions such as probes II and III or probes VII and VIII. If the two phospholipid analogues, labelled with two different isotopes, are used simultaneously, different patterns of labelling can be obtained in the same experiment. In fact, if the protein hydrophobic domain is made up of α-helices running across the membrane, similar profiles are expected from the two probes. On the other hand, when the protein is peripheral, a labelling with a superficial probe, but not with a deeper probe, should be obtained. Here, two examples of this approach will be discussed.

Complex II of the inner mitochondrial membrane transfers electron from succinate to ubiquinol. It is formed by four subunits: SDH_1 (M_r, 70 000), SDH_2 (M_r, 27 000), $C_{II\text{-}3}$ (M_r, 13 000) and $C_{II\text{-}4}$ (M_r, 7000). Succinate dehydrogenase (SDH_1

+ SDH_2) can be reversibly dissociated from C_{II-3} and C_{II-4} and maintained in solution in a lipid-free form without detergents (Trumpower and Katki, 1979). After incubation of the soluble enzyme with liposomes tagged with probes VII and VIII and illumination, SDH_2 was labelled to a considerable extent, whereas significant labelling of SDH_1 only occurred with the shallow probe VII (Girdlestone et al., 1981). This result, although not related to the physiological situation, indicates that the two subunits penetrate the lipid bilayer to a different extent (Fig. 11A), even though the presence of SDH_1 in the hydrophobic core of the membrane with unreactive residues exposed to lipids cannot be excluded (Fig. 11B). When bound to C_{II-3} and C_{II-4}, as in the inner mitochondrial membrane, SDH_1 was not labelled by any of the probes. The smaller subunits C_{II-3} and C_{II-4} appear to act as an anchor for the dehydrogenase subunits. The possible arrangements of polypeptides, that can fit the experimental data are shown and discussed in Fig. 11.

Another example is given by cholera toxin. This protein is composed of three different subunits: α (M_r 21 000), β (M_r 11 590) and γ (M_r 5395) with the stoichiometry $\alpha\beta\gamma_5$. The β subunits bind specifically to ganglioside G_{M1}, the

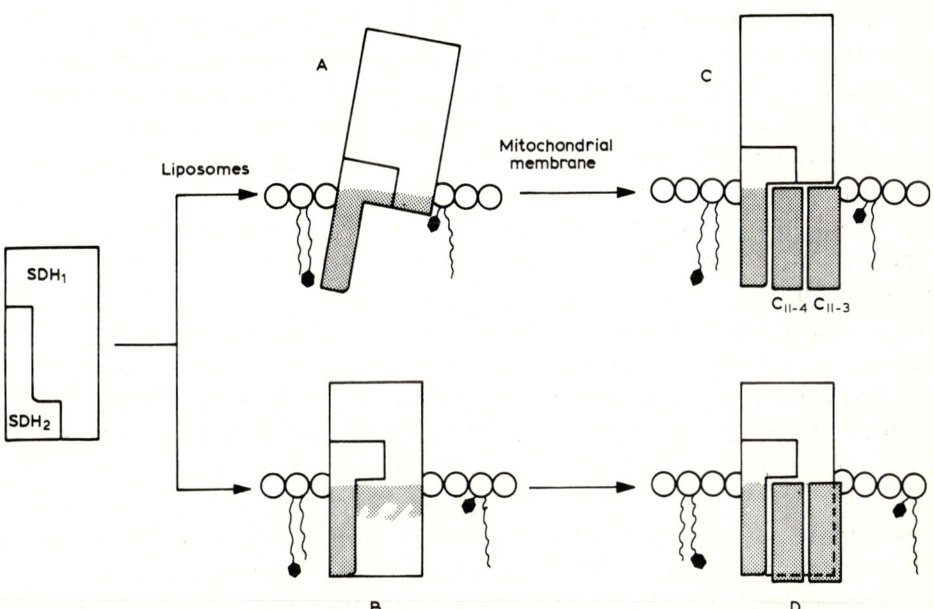

Fig. 11. Labelling of succinate dehydrogenase. The position of the photoreactive group is shown by ⬥, while shaded areas represent protein labelling. Following a conformational change, soluble succinate dehydrogenase interacts with liposomes: (A) shows the most likely situation with SDH_1 interacting only superficially with the lipid bilayer; (B) depicts the alternative possibility of SDH_1, penetrating deeply into the lipid bilayer, but with poorly reactive residues; the smaller C_{II-3} and C_{II-4} subunits, present in complex II, could either raise (C) or shield (D) SDH_1, from contact with lipids (after Girdlestone et al., 1981).

specific cellular receptor of cholera toxin. The enzymic α subunit is linked to the cell binding region $\gamma\beta_5$ via a single disulfide bridge (van Heyningen, 1977). After binding to G_{M1}-containing liposomes, tagged with photoreactive phospholipids VII and VIII and illumination, the β and γ polypeptides were labelled only with the shallow probe VII (Tomasi and Montecucco, 1981; Tomasi et al., 1982). This indicates that the membrane-toxin interaction is mediated by subunits β and γ and that this interaction is restricted to the membrane surface. The enzymatic activity of the α subunit is freed after reduction of the $\beta\gamma$ disulfide bridge, and is exerted on the cytoplasmic side of the plasma membrane. Hence, somehow, the α subunit has to cross the membrane. The pattern of labelling under reducing conditions is very different. While again the β and γ subunits were labelled only by the shallow probe, the α subunit was then labelled by both the superficial and the deep probes. This indicates that, after reduction, the α subunit becomes a hydrophobic entity and penetrates deeply into the lipid bilayer. As in the case of the dehydrogenase complex, the lack of labelling of the $\beta\gamma$ subunits with the deep probe could be interpreted both as due to their superficial location on the membrane, or to a lack of reactive residues in the protein surface present in the membrane hydrophobic core. The latter possibility is less likely because, in the sequence of these polypeptides, there are no hydrophobic segments. However, conclusive information can be obtained only after determination of the modified residues. So far this has been accomplished only for the few examples discussed in the next section.

3. PEPTIDE FOLDING IN THE MEMBRANE

Experimental data on the conformation of the peptide segments embedded in the lipid bilayer are available only for few proteins (Henderson and Unwin, 1975; Schulte and Marchesi, 1979). Most integral membrane proteins show the presence in their sequences of hydrophobic stretches 18–22 residues long. Assuming a membrane thickness of 45 Å this is the number of residues needed to form a *trans*-membrane α-helix. However, on theoretical grounds other arrangements are possible for the hydrophobic domain (Kennedy, 1978; Henderson, 1981). Moreover one cannot determine on the base of the sequence which of these segments are at the lipid-protein boundary, and which of them form the core of the hydrophobic sector of the protein. For the exposed segments hydrophobic photolabelling with photoreactive lipids and determination of the precise site(s) of cross-linking can give valuable information on their identity and arrangement in the membrane.

Cytochrome *c* oxidase is the terminal component of the electron transfer chain of the inner mitochondrial membrane. It is composed of eigth to twelve different polypeptides of M_r ranging from 5000 to 57 000 (Azzi, 1980). Figure 12 reports the pattern of labelling with probes VII and VIII of bovine heart cytochrome *c* oxidase reconstituted into an active lipid-protein complex. Subunits I, II, III, IV, VII and component b are labelled; noteworthy, only those subunits containing hydrophobic stretches are derivatized (Bisson et al., 1979a). The distribution of labelling on subunits II and IV was further analyzed by protein fragmentation (Bisson et al., 1982; Malatesta et al., 1983). Radioactivity was confined exclusively to their hydrophobic peptide segments indicating that they are exposed to lipids, and providing further evidence for the labelling being selectively restricted to the hydrophobic domain. The determination of the precise site(s) of cross-linking with the superficial phospholipid probe VII was attempted on subunit II, whose primary structure is outlined in Figure 13. The 'shallow' probe was chosen because its labelling is restricted to those fragments intercalated in the probe head group region on the membrane. A cluster of residues at the NH_2 terminus of the first hydrophobic stretch was labelled while no radioactivity was found on the intermediate part of the segment. These data can be most easily accommodated assuming an α-helical conformation, which is supported by other topological investigations. This conclusion could have been based on more solid ground had the sequence analysis been carried out further to determine the radioactivity

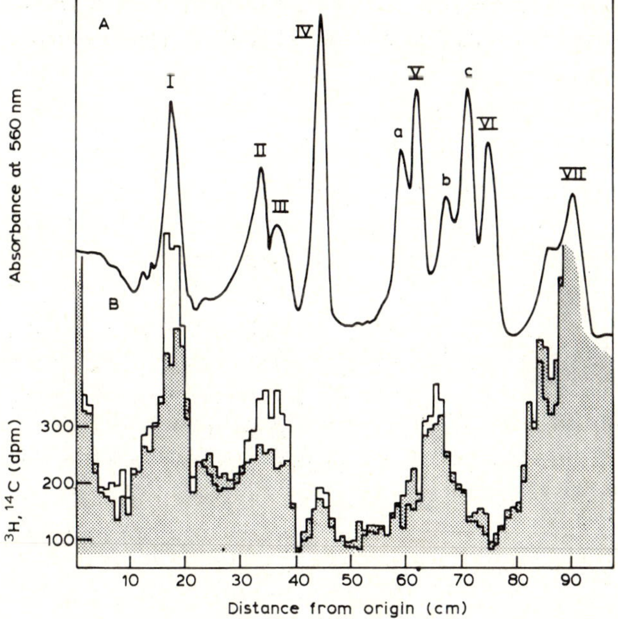

Fig. 12. Pattern of labelling of cytochrome *c* oxidase. (A) Coomassie Blue staining profile of bovine heart cytochrome *c* oxidase; subunits are indicated with roman numbers. Panel B shows the radioactivity distribution along the gel of probe VII (shaded area) and of probe VIII (open areas). A shift of the radioactive peak with respect to the corresponding unlabelled protein band due to the covalent addition of the probe is noticeable on the low M_r polypeptides.

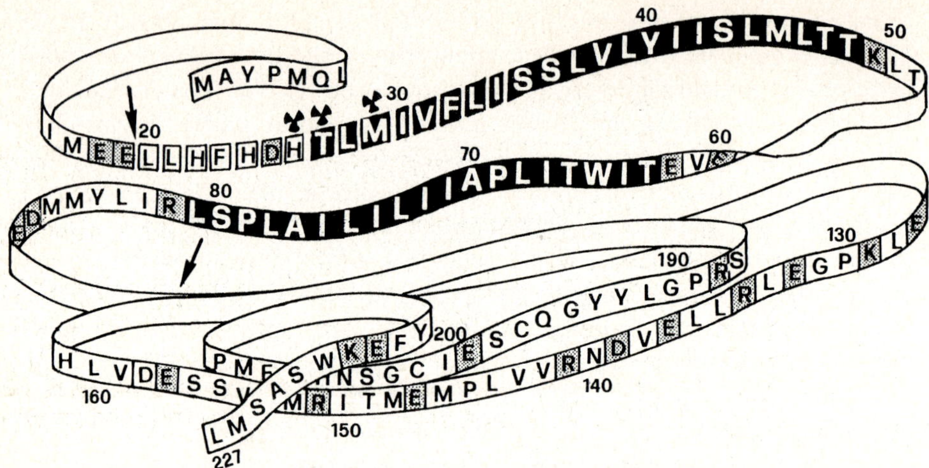

Fig. 13. Sequence of subunit II of cytochrome *c* oxidase. Arrows indicate fragment 20-98, containing the two hydrophobic stretches (shown in black), which has been partially sequenced to determine the sites of cross-linking with probe VII. ☢ indicate the heavily labelled amino acids. Sequenced residues are boxed (after Bisson et al., 1982).

associated with the COOH terminus. This was not done due to the use of a spinning cup sequenator where hydrophobic peptides leak out during washes, and because of other limitations discussed at the end of this section.

Another example of the kind of information on the arrangement of peptides in the membrane, obtained with photoreactive phospholipids is provided by the study on subunit b of the *E. coli* ATP-synthase of Hoppe et al. (1983a). This enzyme uses the energy of a proton gradient to form ATP. It is composed of a water-soluble part (five different subunits) bound to a hydrophobic domain (three different subunits), which forms a *trans*-membrane proton channel. The enzyme was labelled with probe VII to derivatize those segments intercalated to the superficial part of the membrane. The intact ATP synthase complex was labelled in a reconstituted form, and the highly radioactive subunit b of the hydrophobic domain was purified and fragmented. Radioactivity was confined exclusively to an NH_2 terminal stretch of hydrophobic residues. Sequence analysis showed that the probe was bound to two short regions at the extremities of a segment of 15 unlabelled residues. This label distribution supports the model shown in Fig. 14A where the NH_2 terminal part of subunit b is proposed to be arranged as an α-helix spanning the membrane. The labelling profile is also consistent with a folding back of the peptide chain which would bring the two labelled sequences on the same side of the membrane. However, this possibility is less likely on the basis of some considerations that can be made on the energetics of protein folding in a hydrophobic phase (Kennedy, 1978; Henderson, 1981). Since the protein has also been labelled with a small carbene precursor, which can diffuse freely in the

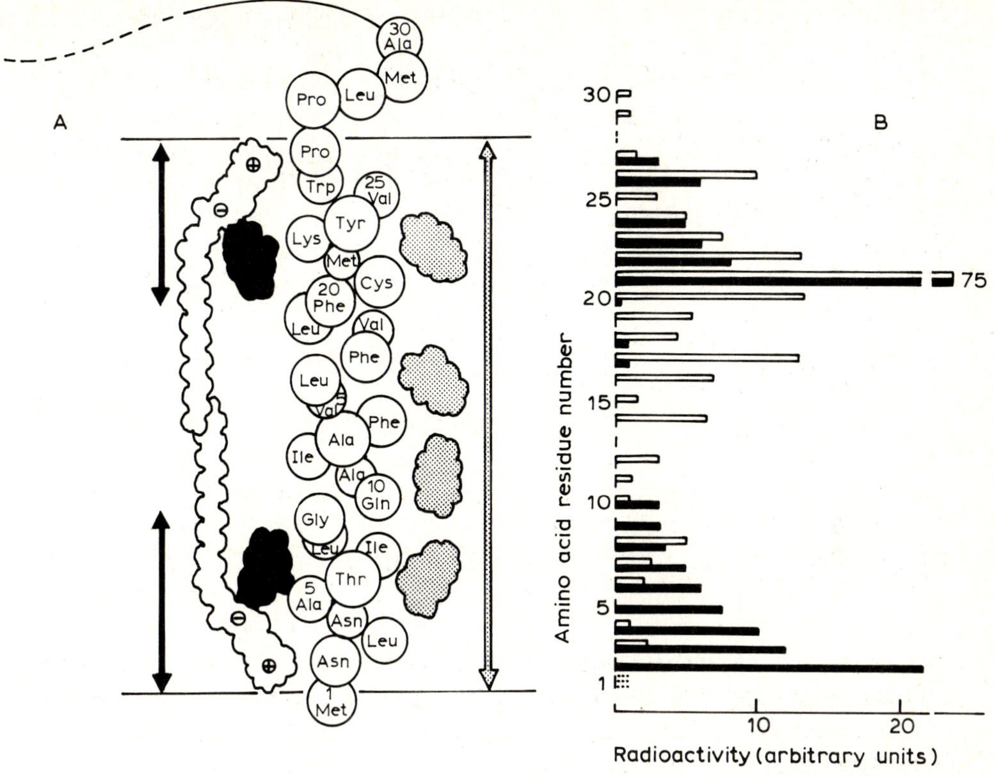

Fig. 14. Hydrophobic photolabelling of subunit b of ATP synthase and suggested arrangement in the membrane. Panel A shows schematically the location of the photoreactive group of phospholipid probe VII (black) and of TID (shaded): arrows encompass the different regions of the membrane accessible to the two probes. Panel B compares the amount of radioactivity of probe VII (full bars) and of TID (shaded bars) associated to the single residues. The figure also shows in panel A the α-helical arrangement proposed for the NH_2 terminal segment of subunit b in the membrane. Data were taken from Hoppe et al. (1983 a, b) and normalized on the amount of radioactivity bound to Cys 21 after background subtraction.

lipid bilayer (Hoppe et al., 1983b), an interesting comparison between the two probes, limited to the two commonly accessible flanking regions of the NH_2-terminal segment can be made. The comparison of the amount of radioactivity associated with the single residues in the two experiments, shown in Fig. 14B, is very significant because the sequence analysis was performed with identical procedures. It clearly appears that the large difference of protein labelling expected on the base of the differing reactivities of carbenes and nitrenes towards C-H bonds is not found. This appears to be related to the common presence on the hydrophobic surface of integral proteins of electron-rich groups that can react with photogenerated intermediates much faster than aliphatic amino acid residues

(Richards and Brunner, 1980). This minimizes the difference in reactivity of carbenes and nitrenes as protein hydrophobic labelling reagents.

The extreme possibility that a single site of a protein is so much more reactive that it takes up virtually all the radioactivity of the photogenerated intermediate is illustrated by the labelling of glycophorin with probe II (Ross et al., 1982). It was found that a single glutamyl residue present in one flanking region of the hydrophobic sector was derivatized. The labelling was not related to the carbene intermediate but to the rearranged linear diazo-intermediate (see Fig. 2d), which is long-lived and gives rise to a dark reaction. This is a further example of the unreliability of 'deep' probes as membrane chemical rulers.

As discussed above superficial probes such as the phospholipid analogue III, which was also used in this study are more reliable because of their restricted location in the membrane. Two protein regions at the two sides of the hydrophobic segments were labelled. Unfortunately the analysis of the modified residues could not be carried out further because of the dramatic loss of label during the first Edman degradation cycle.

Loss of radioactivity is not an uncommon occurrence in the analysis of cross-linked sites between proteins and carbenes or nitrenes. It should be emphasized that a whole range of different ill-characterized bonds are formed upon photoactivation. Many of them are unstable to CNBr degradation particularly in the case of nitrene derivatives (Bisson et al., 1982; Hoppe et al., 1983), and this poses a severe limit to the strategy applicable to photolabelled proteins. It has been shown that 50% of carbene derivatized bonds are hydrolized under strongly basic conditions, while they appear to be more stable to acid conditions (but these measurements were made after a preliminary protein isolation in an ethanol-formic acid mixture) (Takagaki et al., 1983a). The danger of loss of radioactivity is expected to be greater when additional phosphoester bonds are present between the photoreactive group and the radioactive tag, as in probes VII and VIII. However, it was found that in most acid conditions used in gel fixation, protein isolation and sequencing these carboxylic and phosphate ester bonds are not appreciably hydrolized.

4. HALF-BILAYER LABELLING

A further possibility offered by photoreactive phospholipids is the selective labelling of the lipids and proteins present in one of the two lipid bilayer halves. Only membranes with a very low flip-flop rate, where probe equilibration during the experiment is minimal, can be studied. It should be noticed that the rate of trans-bilayer movement of a phospholipid analogue may be very different from that of the parent compound. Hence, this possibility should be carefully evaluated in each experimental system by appropriate controls. Brunner et al. (1983) have shown that probe IV, after transfer from donor liposomes to sealed erythrocyte ghosts, was localized on the outer membrane leaflet. In fact, a different segment

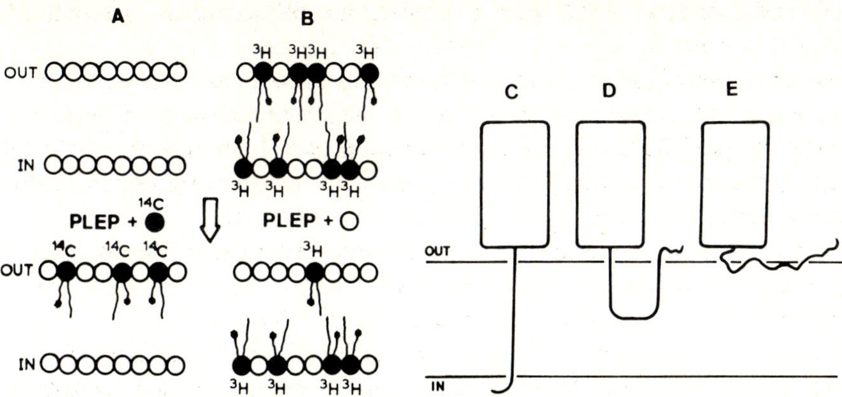

Fig. 15. Half-bilayer labelling used to study cyt b_5 folding in the membrane. Panel (A) shows the selective enrichment of the outer layer of membranes with PLEP and donor liposomes, containing radioactive photoreactive lecithin analogues. Panel (B) shows the selective depletion of the outer leaflet of liposomes tagged with lecithin probes by action of PLEP in the presence of cold donor lecithin vesicles. This approach has been used to discriminate between the different arrangements of cyt b_5 in the membrane, shown in C, D and E, by Takagaki et al. (1983b).

of the hydrophobic sector of glycophorin was labelled in this case with respect to the even labelling found when the probe was distributed among the two membrane halves. An alternative elegant approach of more general applicability uses PLEP to enrich or deplete the outer layer of biological or model membranes with a phospholipid probe as shown in Fig. 15A and B. The second possibility (panel B) is probably limited only to reconstituted systems where the probe can be easily and uniformly distributed among the two layers. The power of this approach is well illustrated in the work of Takagaki et al. (1983) on the different forms of insertion of cyt b_5 in membranes. Liver microsomal cyt b_5 is formed by a large heme-containing hydrophilic domain (over 70% of the protein mass) bound to the membrane via a COOH terminal segment of 30 residues. Depending on the reconstitution procedure adopted it inserts in the membrane in a loosely bound intermembrane transferable form or in a stable non-transferable form (Strittmatter and Dailey, 1982). The various possibilities corresponding to the different situations are outlined in the right panel of Fig. 15. With the rabbit liver protein it was found that probes I and II cross-linked to several residues comprised in the region between Ser 104 and Met 130. These data were interpreted as indicating that the membranous segment loops back as depicted in Fig. 15D. In another experiment, performed with probes VII and VIII on the loosely bound form of calf liver cyt b_5, labelling was restricted to the 18 residue COOH terminal segment supporting model E (Bisson et al., 1979b). However all these experiments do not definitively prove any of the models of Fig. 15. Moreover when the non-transferable form of cyt b_5 was tested with probe II the distribution of modified residues was found to be similar to that of the transferable form. The resolution of the approach was improved with the aid of the phosphatidylcholine exchange protein. In one case the outer layer was enriched in a ^{14}C-labelled probe (Fig. 15A), while in the other the inner layer was enriched with a ^{2}H-labelled probe as shown in Fig. 15B. After photolysis and purification the ^{14}C- and ^{3}H-labelled cyt b_5 samples were mixed and fragmented. The variation of the $^{3}H/^{14}C$ ratio on the different peptide fragments indicates their position in the membrane. The result clearly shows that the COOH terminal hydrophobic segment spans the lipid bilayer when cyt b_5 is present in the non-transferable form.

These kind of experiments have so far been carried out with phosphatidylcholine analogues carrying the photoreactive group at the fatty acid methyl terminus. Hence, no matter in which leaflet the probe is localized, the central region of the protein is the most likely to be labelled. For this reason phospholipid probes with the photoreactive moiety nearer to the polar lipid head group are expected to provide more valuable results.

5. PHOTOACTIVE PHOSPHOLIPIDS AS PHOTOAFFINITY REAGENTS

A particular case of lipid-protein interaction is that of phospholipids with the enzymes involved in their synthesis, degradation and transfer. Nitrene and carbene-generating phosphatidylcholines were used to map the binding site of this phospholipid on PLEP (Moonen et al., 1979; Westerman et al., 1983) extracted from bovine liver. The different probes were found to label the protein with similar high efficiency (more than 30%). This is probably related to the tight interaction between the phospholipid and the protein and hence to the long residence time of the probe on the protein. Both nitrenes and carbenes label a short sequence of seven residues close to the COOH terminus, indicating once more that carbenes and nitrenes, despite their different reactivity towards saturated hydrocarbons, may give very similar results in protein labelling. In their study Westerman et al. (1983) used probes with the photoreactive diazirinophenoxy group placed at different distances from the carboxyl ester bond, in the hope of labelling different protein regions. Nonetheless, the same amino acids were labelled indicating that the photoactivated group can reorient during its lifetime and reach the same reactive protein residues.

Another system devoted to phospholipid transport is constituted by the various classes of blood lipoproteins (Scanu, 1978). They have been labelled with aliphatic nitrene probes X and XI with high efficiency, demonstrating that fatty acid chains are in contact with the protein in the lipoprotein complex (Stoffel and Metz, 1979, 1982).

VI. Conclusions

Phospholipids are now available with a variety of photoactivatable groups and labelled with different isotopes of differing specific activity. Most of them are phosphatidylcholines and their syntheses appear to vary greatly in difficulty. Other phospholipid classes can be prepared biosynthetically or by suitable synthetic schemes. They have been demonstrated to be very useful in determining the integral nature of several membrane proteins, and to identify those subunits forming the lipid-protein boundary in multi-subunit membrane proteins. In fact, all available data show that the labelling with photoreactive lipids is selectively restricted to the membrane. Hence they can also be safely used to trace the insertion of soluble proteins in the membrane. However, the main use they were designed for, that of membrane chemical rulers, had been hampered by the possibility of reorientation of the photoreactive group in the membrane and by its non-unlimited reactivity. This problem could be overcome by preparing more reactive groups linked to the phospholipid polar head through a rigid chain. In general terms, highly reactive photogenerated intermediates will give more reliable information on all the above problems.

If the lifetime of the intermediate in the membrane can be evaluated this will open the possibility of using this technique to study dynamic events. Specific lipid binding sites have been suggested to exist on several membrane-bound enzymes. Photoreactive phospholipids of the various classes should allow onc to label and map those protein regions forming these specific sites.

Acknowledgements

We thank Drs. J. Brunner, H. Monaco and M. Zoratti for criticism and suggestions, and Prof. G.F. Azzone for support and encouragement. The skilful secretarial assistance of Mrs. F. Mazzari and the careful drawing of Mr. G.P. Schiavo are gratefully acknowledged. We also thank Drs. H. Bayley and J. Brunner for making available to us manuscripts prior to publication.

References

Akeroyd, R. and Wirtz, K.W.A. (1982) Properties of phospholipid transfer proteins. In: *Membranes and Transport*, Vol. 1, pp. 93–97. Plenum Press, New York.

Azzi, A. (1980) Cytochrome *c* oxidase: towards a clarification of its structure, interactions and mechanisms. Biochim. Biophys. Acta 594, 231–252.

Bayley, H. (1982) Photoactivated hydrophobic reagents for integral membrane proteins: a critical discussion. In: *Membranes and Transport*, Vol. 1, pp. 185–194, Plenum Press, New York.

Bayley, H. (1983). *Photogenerated Reagents in Biochemistry and Molecular Biology*, pp. 1–187, Elsevier Biomedical Press, Amsterdam.

Bayley, H. and Knowles, R. (1977) Photoaffinity labeling. Methods Enzymol. 46, 69–114.

Bayley, H. and Knowles, J.R. (1980) Photogenerated reagents for membranes. Selective labeling of intrinsic membrane proteins in the human erythrocyte membrane. Biochemistry 19, 3883–3892.

Bercovici, T. and Gitler, G. (1978) 5- [^{125}I] -Iodonapthyl azide, a reagent to determine the penetration of proteins into the lipid bilayer of biological membranes. Biochemistry 17, 1484–1489.

Bisson, R., Montecucco, C., Gutweniger, H. and Azzi, A. (1979a) Cytochrome *c* oxidase subunits in contact with phospholipids: hydrophobic photolabelling with azidophospholipids. J. Biol. Chem. 254, 9962–9965.

Bisson, R., Montecucco, C. and Capaldi, R.A. (1979b) Interaction of membranous cytochrome b_5 with arylazidophospholipids. FEBS Lett. 106, 317–320.

Bisson, R. and Montecucco, C. (1981) Photolabelling of membrane proteins with photoactive phospholipids. Biochem. J. 193, 757–763.

Bisson, R., Steffens, G.C.M. and Buse, G. (1982) Localization of lipid binding domain(s) on subunit II of beef heart cytochrome *c* oxidase. J. Biol. Chem. 257, 6716–6720.

Bisson, R. and Montecucco, C. (1985) Topology of membrane proteins: determination of regions exposed to the lipid bilayer. In: *Techniques for the Analysis of Membrane Proteins*. Editors: R. Cherry and C.I. Ragan, Chapman & Hall, London, in press.

Brunner, J. (1981) Labelling the hydrophobic core of membranes. Trends Biochem. Sci. 6, 44–46.

Brunner, J. and Richards, F.M. (1980) Analysis of membranes photolabelled with lipid analogues. J. Biol. Chem. 255, 3319–3329.

Brunner, J. and Semenza, G. (1981) Selective labeling of the hydrophobic core of membranes with 3-(trifluoromethyl)-3-(*m*- [^{125}I]-iodophenyl)diazirine, a carbene-generating reagent. Biochemistry 20, 7174–7182.

Brunner, J., Spiess, M., Aggeler, R., Huber, P. and Semenza, G. (1983) Hydrophobic labeling of a single leaflet of the human erythrocyte membrane. Biochemistry 22, 3812–3820.

Capone, J., Leblanc, P., Gerber, G.E. and Gosh, H.P. (1983) Localization of membrane proteins by the use as a photoreactive fatty acid incorporated in vivo into vesicular stomatitis virus. J. Biol. Chem. 258, 1395–1398.

Chapman, D., Goméz-Fernández, J.C. and Goni, F.M. (1982) The interaction of intrinsic proteins and lipids in biomembranes. Trends Biochem. Sci. 7, 67–69.

Chapman, O.L. and Leroux, J.P. (1978) 1-aza-1,2,4,6-cycloheptatetraene. J. Am. Chem. Soc. 100, 282–285.

Chakrabarti, P. and Khorana, H.G. (1975) A new approach to the study of phospholipid-protein interactions in biological membranes. Synthesis of fatty acids and phospholipids containing photosensitive group. Biochemistry 14, 5021–5033.

Chowdhry, V. and Westheimer, F.H. (1979) Photoaffinity labeling of biological systems. Ann. Rev. Biochem. 48, 293–325.

Chowdhry, V., Vaughan, R. and Westheimer, F.H. (1976) 2-Diazo-3,3,3-trifluoropropionyl chloride: reagent for photoaffinity labeling. Proc. Natl. Acad. Sci. U.S.A. 73, 1406–1408.

Curatolo, W., Radhakrishnan, R., Gupta, C.M. and Khorana, H.G. (1981) Photoactivatable carbene-generating phospholipids: physical properties and use in detection of phase separations in lipid mixtures. Biochemistry 20, 1374–1378.

De Pierre, J.W. and Ernster, L. (1977) Enzyme topology of intracellular membranes. Ann. Rev. Biochem. 46, 201–262.

Giraudat, G., Montecucco, C., Bisson, R. and Changeux J.P. (1985) Transmembrane topology of acetylcholine receptor subunits probed with photoreactive phospholipids. Biochemistry, in press.

Girdlestone, J., Bisson, R. and Capaldi, R.A. (1981) Interaction of succinate-ubiquinone reductase (complex II) with (arylazido) phospholipids. Biochemistry 20, 152–156.

Greenberg, G.R., Chakrabarti, P. and Khorana, H.G. (1976) Incorporation of fatty acids containing photosensitive group into phospholipids of *Escherichia coli*. Proc. Natl. Acad. Sci. U.S.A. 73, 86–90.

Gupta, C.M., Radhakrishnan, R. and Khorana, H.G. (1977) Glycerophospholipid synthesis: improved general method and new analogs containing photoactivatable groups. Proc. Natl. Acad. Sci. U.S.A. 74, 4315–4319.

Gupta, C.M., Radhakrishnan, R., Gerber, G.F., Olsen, W.L., Quay, S.C. and Khorana, H.G. (1979a) Intermolecular crosslinking of fatty acyl chains in phospholipids: use of photoactivatable carbene precursors. Proc. Natl. Acad. Sci. U.S.A. 76, 2595–2599.

Gupta, C.M., Costello, C.E. and Khorana, H.G. (1979b) Sites of intermolecular crosslinking of fatty acyl chains in phospholipids carrying a photoactivatable carbene precursor. Proc. Natl. Acad. Sci. U.S.A. 76, 3139–3143.

Gutsche, C.D., Bachman, G.L. and Coffey, R.S. (1962) Chemistry of bivalent carbon intermediates: comparative intermolecular and intramolecular reactivities of phenylcarbene to various bond types. Tetrahedron 18, 617–627.

Gutweniger, H.E. and Montecucco, C. (1984) Labelling of the integral proteins of sarcoplasmic reticulum membranes. Biochem. J., 220, 613–616.

Hall, J.H., Hill, J.W. and Tsai, H.C. (1965) Insertion of reactions of aryl nitrenes. Tetrahedron Lett. 26, 2211–2216.

Halsey, M.J., Brown, F.F. and Richards, R.E. (1978) Perturbations of model protein systems as a basis for the central and peripheral mechanisms of general anaesthesia. In: *Molecular Interaction and Activity in Proteins*. CIBA Found. Symp. 60, pp. 123–136.

Hebdon, G.M., Knott, J.C.A. and Green, N.M. (1980) Preparation and properties of hexanoyl-diiodo-*N*-(4-azido-2-nitrophenyl) tyramine. Biochemistry 19, 6216.

Henderson, R. (1981) Membrane protein structure. In: *Biophysics of Membranes and Intercellular Communication*, pp. 232–249. Editors: Balian et al., Elsevier/North-Holland, Amsterdam.

Henderson, R. and Unwin, P.N.T. (1975) Three-dimensional model of purple membrane obtained by electron microscopy. Nature 257, 28–32.

Hoppe, J., Montecucco, C. and Friedl, P. (1983a) Labeling of subunit b of the ATP synthase from *Escherichia coli* with a photoreactive phospholipid analogue. J. Biol. Chem. 258, 2882–2885.

Hoppe, J., Brunner, J. and Jorgensen, B.B. (1983b) Structure determination of membrane integrated polypeptides of the F_0 part of ATP synthase from *E. coli* by surface labeling with 3-trifluoromethyl-3-(*m* [^{125}I]iodophenyl) diazirine. In: *Structure and Function of Membrane Proteins*, Editors: F. Palmieri and E. Quagliariello, Elsevier/North-Holland, Amsterdam.

Ji, T.H. (1983) Bifunctional reagents. Methods Enzymol. 91, 580–609.

Jørgensen, P.L. (1982) Mechanism of the Na^+,K^+ pump: protein structure and conformation of the pure (Na^+ + K^+) ATPase. Biochim. Biophys. Acta 694, 27–68.

Kennedy, S.J. (1978) Structures of membrane proteins. J. Membr. Biol. 42, 265–279.

Krebs, J., Buerkler, J., Guerini, D., Brunner, J. and Carafoli, E. (1984) 3-(trifluoromethyl)-3-(*m*-[^{125}I]iodophenyl)diazirine, a hydrophobic photoreactive probe, labels calmodulin and calmodulin fragments in a Ca^{2+}-dependent way. Biochemistry 23, 400–403.

Knowles, J.R. (1972) Photogenerated reagents for biological receptor-site labeling. Accounts Chem. Res. 5, 155–160.

Kyba, P.E. and Abramovitch, R.A. (1980) Photolysis of aryl azides. Evidence for a non nitrene mechanism. J. Am. Chem. Soc. 102, 735–740.

Leblanc, P., Capone, J. and Gerber, E.G. (1982) Synthesis and biosynthetic utilization of radioactive photoreactive fatty acids. J. Biol. Chem. 257, 14586–14589.

Malatesta, F., Darley-Usmar, V., De Jong, C., Prochaska, L.J., Bisson, R., Capaldi, R.A., Steffens, G.C.M. and Buse, G. (1983) Arrangement of subunit IV in beef-heart cytochrome *c* oxidase probed by chemical labelling and protease digestion experiments. Biochemistry 22, 4405–4411.

Michel, H. (1983) Crystallization of membrane proteins. Trends Biochem. Sci. 7, 56–58.

Montecucco, C., Bisson, R., Dabbeni-Sala, F., Pitotti, A. and Gutweniger, H. (1980) Interaction of the mitochondrial ATPase complex with phospholipids. J. Biol. Chem. 255, 10040–10043.

Montecucco, C., Bisson, R., Gache, C. and Johansson, A. (1981) Labelling of the hydrophobic domain of Na^+,K^+-ATPase. FEBS Lett. 128, 17–20.

Montecucco, C., Dabbeni-Sala, F., Friedl, P. and Galante, Y.M. (1983) Membrane topology of ATP synthase from bovine heart mitochondria and *Escherichia coli*. Eur. J. Biochem. 132, 189–194.

Moonen, P., Haagman, H.P., van Deenen, L.L.M. and Wirtz, K.W.A. (1979) Determination of the hydrophobic binding site of phosphatidylcholine exchange protein with photosensitive phosphatidylcholine. Eur. J. Biochem. 99, 439–445.

Nielsen, P.E. and Buchardt, O. (1982) Aryl azides as photoaffinity labels. A photochemical study of some 4-substituted aryl azides. Photochem. Photobiol. 35, 317–323.

Patel, K.M., Morrisett, J.D. and Sparrow, J.T. (1979) A convenient synthesis of phosphatidylcholines: acylation of *sn*-glycero-3-phosphocholine with fatty acid anhydride and 4-pyrrolidinopyridine. J. Lipid Res. 20, 674–677.

Prochaska, L.J., Bisson, R. and Capaldi, R.A. (1980) Structure of the cytochrome *c* oxidase complex: labeling by hydrophilic and hydrophobic protein modifying reagents. Biochemistry 19, 3174–3179.

Radhakrishnan, R., Robson, R.J., Takagaki, Y. and Khorana, H.G. (1981) Synthesis of modified fatty acids and glycerophospholipid analogs. Methods Enzymol. 72, 408–433.

Reiser, A., Terry, G.C. and Willets, F.W. (1966) Observation of the nitrene in the flash-photolysis of 1-azidoanthracene. Nature 211, 410.

Reiser, A., Willets, F.W., Terry, G.C., Williams, V. and Marley, R. (1968) Photolysis of aromatic azides. Lifetimes of aromatic nitrenes and absolute rates of some of their reactions. Trans. Faraday Soc. 64, 3265–3275.

Reiser, A., Leyshon, L. (1970) A correlation between negative charge on nitrogen and the reactivity of aromatic nitrenes. J. Am. Chem. Soc. 92, 7487.

Richards, F.M. and Brunner, J. (1980) General labeling of membrane proteins. Ann. N.Y. Acad. Sci. U.S.A. 346, 144–164.

Robson, R.J., Radhakrishnan, R., Ross, A.H., Takagaki, Y. and Khorana, H.G. (1982) Photochemical cross-linking in studies of lipid-protein interactions. In: *Lipid-Protein Interactions*, Vol. 2, pp. 149–192. Wiley and Sons, New York.

Ross, A.H., Radhakrishnan, R., Robson, R.J. and Khorana, H.G. (1982) The transmembrane domain of glycophorin A as studied by cross-linking using photoactivatable phospholipids. J. Biol. Chem. 257, 4152–4161.

Sandermann, H. (1978) Regulation of membrane enzymes by lipids. Biochim. Biophys. Acta 515, 209–237.

Sator, V., Gonzalez-Ros, J.M., Calvo-Fernandez, P. and Martinez-Carrion, M. (1979) Pyrenesulfonyl azide: a marker of acetylcholine receptor subunits in contact with membrane hydrophobic environment. Biochemistry 18, 1200–1206.

Scanu, A.M. (1978) Plasma lipoproteins: structure, function and regulation. Trends Biochem. Sci. 3, 202–205.

Schulte, T.H. and Marchesi, V.T. (1979) Conformation of human erythrocyte glycophorin A and its constituent peptides. Biochemistry 18, 275–280.

Singer, S.J. (1974) The molecular organization of membranes. Ann. Rev. Biochem. 43, 805–833.

Smith, G.A., Montecucco, C. and Bennett, J.P. (1977) Isotopic labeling of phosphatidylcholine in the choline moiety. Lipids 13, 92–94.

Smith, R.A.G. and Knowles, J.R. (1975) The preparation and photolysis of 3-Aryl-3H-Diazirines. J. Chem. Soc. Perkin Trans. 2, 686–694.

Staros, J.V. (1980) Aryl azide photolabels in biochemistry. Trends Biochem. Sci. 5, 320–322.

Staros, J.V., Bayley, H., Standring, D.N. and Knowles, J.R. (1978) Reduction of aryl azides by thiols: implications for the use of photoaffinity reagents. Biochem. Biophys. Res. Commun. 80, 568–572.

Stefanini, S., Finazzi-Agrò, A., Chiancone, E. and Antonini, E. (1979) Binding of hydrophobic compounds to apoferritin subunits. FEBS Lett. 100, 296–300.

Stoffel, W., Schreiber, C. and Scheefers, H. (1978) Lipids with photosensitive group as chemical probes for the structural analysis of biological membranes. On the localization of the G and M-protein of vesicular stomatitis Virus. Hoppe-Seyler's Z. Physiol. Chem. 359, 923–931.

Stoffel, W. and Metz, P. (1979) Covalent cross-linking of photosensitive phospholipids to human serum high density apolipoproteins. Hoppe-Seyler's Z. Physiol. Chem. 360, 197–206.

Stoffel, W. and Metz, P. (1982) Chemical studies on the structure of human serum high-density lipoprotein (HDL): photochemical crosslinking of azido-labelled lipids in HDL. Hoppe Seyler's Z. Physiol. Chem. 363, 19–31.

Stoffel, W., Saem, K.P. and Muller, M. (1982) Syntheses of phosphatidylcholines, sphingomyelins and cholesterol substituted with azido fatty acids. Hoppe Seyler's Z. Physiol. Chem. 363, 1–18.

Strittmatter, P. and Dailey, H.A. (1982) Essential structural features and orientation of cytochrome b_5 in membranes. In: *Membranes and Transport*, Vol. 1, pp. 71–82. Plenum, New York.

Takagaki, Y., Gupta, C.M. and Khorana, H.G. (1980) Thiols and the diazo group in photoaffinity labels. Biochem. Biophys. Res. Commun. 95, 589–595.

Takagaki, Y., Radhakrishnan, R., Gupta, G.M. and Khorana, H.G. (1983a) The membrane-embedded segment of cytochrome b_5 as studied by cross-linking with photoactivatable phospholipids. J. Biol. Chem. 258, 9128–9135.

Takagaki, Y., Radhakrishnan, R., Wirtz, K.W.A. and Khorana, H.G. (1983b) The membrane-embedded segment of cytochrome b_5 as studied by cross-linking with photoactivatable phospholipids. J. Biol. Chem. 258, 9136–9142.

Tomasi, M. and Montecucco, C. (1981) Lipid insertion of cholera toxin after binding to G_{M1}-containing liposomes. J. Biol. Chem. 256, 11177–11181.

Tomasi, M., D'Agnolo, G. and Montecucco, C. (1982) Micellar gangliosides mediate the lipid insertion of cholera toxin protomer A. Biochim. Biophys. Acta 692, 339–344.

Trumpower, B.L. and Katki, A.G. (1979) Succinate-cytochrome *c* reductase complex of the mitochondria electron transport chain. In: *Membrane Proteins in Energy Transduction*, pp. 89–206. Dekker, New York.

Ueda, I., Kamaya, H. and Eyring, H. (1976) Molecular mechanism of inhibition of firefly luminescence by local anesthetics. Proc. Natl. Acad. Sci. U.S.A. 73, 481–485.

Van Heyne, G. (1981) Membrane proteins: the amino acid composition of membrane-penetrating segments. Eur. J. Biochem. 120, 275–278.

Van Heyningen, S. (1977) Cholera toxin. Biol. Rev. 52, 509–549.

Warner, T.G. and Benson, A.A. (1977) An improved method for the preparation of unsaturated phosphatidylcholines: acylation of *sn*-glycero-3-phosphorylcholine in the presence of sodium methylsulfinylmethide. J. Lipid. Res. 18, 548–552.

Westerman, J., Wirtz, K.W., Berkhout, T., van Deenen, L.L.M., Radhakrishnan, R. and Khorana, H.G. (1983) Identification of the lipid-binding site of phosphatidylcholine-transfer protein with phosphatidylcholine analogs containing photoactivatable carbene precursors. Eur. J. Biochem. 132, 441–449.

Note added in proof

A further example that the combined use of shallow, and a deeper, probes can provide reliable information on the depth of penetration and arrangement of a peptide segment in the membrane is provided by a recent study of Giraudat et al. (1985) on the acetylcholine receptor of *Torpedo marmorata* and *californica*.

Subject Index

A

B

C